10th Anniversary of *Catalysts*: Biocatalysis in Analysis and Synthesis—Past, Present and Future

10th Anniversary of *Catalysts*: Biocatalysis in Analysis and Synthesis—Past, Present and Future

Editors

Evangelos Topakas
Roland Wohlgemuth
David D. Boehr

MDPI

Basel • Beijing • Wuhan • Barcelona • Belgrade • Novi Sad • Cluj • Manchester

Editors

Evangelos Topakas
School of Chemical Engineering
National Technical University of Athens
Athens
Greece

Roland Wohlgemuth
Faculty of Chemistry
Lodz University of Technology
Lodz
Poland

David D. Boehr
Department of Chemistry
The Pennsylvania State University
University Park
United States

Editorial Office
MDPI
St. Alban-Anlage 66
4052 Basel, Switzerland

This is a reprint of articles from the Special Issue published online in the open access journal *Catalysts* (ISSN 2073-4344) (available at: www.mdpi.com/journal/catalysts/special_issues/10th_biocataly).

For citation purposes, cite each article independently as indicated on the article page online and as indicated below:

Lastname, A.A.; Lastname, B.B. Article Title. *Journal Name* **Year**, *Volume Number*, Page Range.

ISBN 978-3-7258-0644-7 (Hbk)
ISBN 978-3-7258-0643-0 (PDF)
doi.org/10.3390/books978-3-7258-0643-0

Contents

About the Editors

Evangelos Topakas

Evangelos Topakas is a Professor in the School of Chemical Engineering at the National Technical University of Athens and a visiting Associate Professor in the Department of Civil, Environmental, and Natural Resources Engineering at Luleå University of Technology in the field of industrial biotechnology. He leads the Industrial Biotechnology and Biocatalysis Group (IndBioCat). Dr. Topakas is at the forefront of industrial enzyme discovery, focusing on utilizing residual biomass for second-generation liquid biofuels and high-value compounds within the realm of biorefineries. His innovative approach involves employing conventional and modern omics technologies, along with genetic and protein engineering tools, to investigate and modify biocatalytic systems in bacteria and fungi for the production of energy, chemicals, and materials from waste biomass. Notably, his recent work involves leveraging knowledge from plant cell wall degrading enzymes to combat plastic pollution by converting and upcycling synthetic polymers. His interdisciplinary research activities transcend conventional boundaries, requiring collaboration among engineers, biochemists, molecular biologists, and physicists. Dr. Topakas excels at orchestrating diverse talents to achieve common goals, a skill honed through his extensive experience in academia, where he has supervised researchers at the BSc, MSc, and Ph.D. levels. At the helm of the Industrial Biotechnology and Biocatalysis Group (IndBioCat), Dr. Topakas leads a pioneering research group inspiring young chemical engineers and biologists in the field of bioeconomy.

Roland Wohlgemuth

Roland Wohlgemuth studied chemistry and biology at the University of Basel in Switzerland and obtained his Ph.D. with Prof. Joachim Seelig in 1979 at the Biocenter of the University of Basel. He did postdoc work at the Lawrence Berkeley Laboratory and UC Berkeley with a Swiss National Science Foundation Award in 1980 and US Department of Energy employment with Prof. Melvin Calvin from 1981–1983. Since 1983, Roland Wohlgemuth has led and built up the bioanalytical and biochemistry/biotechnology laboratories at Fluka in Switzerland. In 1989, Fluka became part of the Sigma-Aldrich Corporation, which, in 2015, was itself acquired by Merck KGaA. His research interests are at the molecular and engineering interface of medicine, biology, and chemistry, in the areas of biocatalysis, glycobiology, metabolomics, protein analysis, bioanalytical technologies, molecular medicine, technical biochemistry, sustainable chemistry, industrial biotechnology, and bioeconomy. In 2017, Roland Wohlgemuth moved from industry to academia, working as a Professor at Lodz University of Technology. Roland Wohlgemuth is cofounder of the Swiss Industrial Biocatalysis Consortium (SIBC), member of the STRENDA Commission, President of the European Society of Applied Biocatalysis (ESAB), and President of the Swiss Coordination Committee on Biotechnology (SKB).

David D. Boehr

Dr. David Boehr is the principal investigator at the Boehr Laboratory, which is renowned for its research on the role of protein dynamics in enzyme function, coordination, and regulation. Currently holding a position at Penn State University, Dr. Boehr's general interests encompass a multi-disciplinary approach to biochemistry, integrating in vivo assays with biochemical and biophysical methods to explore enzyme activity. Under Dr. Boehr's guidance, the lab is pioneering the use of nuclear magnetic resonance (NMR) to analyze enzyme dynamics, providing site-specific structural and dynamic details across a vast timescale from 10^{12} to 10^{5} seconds.

This innovative technique, combined with traditional enzyme methodologies, has significantly advanced the understanding of the interplay between enzyme function, structure, and dynamics. The Boehr lab's current projects focus on enzymes involved in viral and bacterial pathogenesis, with the aim of informing rational drug and vaccine design. These efforts have not only contributed to the field of protein engineering and structure-based drug design but have also shed light on the molecular evolution of protein function. Dr. Boehr's scientific interests are reflected in his numerous awards and accolades, which recognize his contributions to resolving the ongoing controversy regarding the importance of protein motion to enzyme function and its practical applications.

Preface

Welcome to this Special Issue commemorating the 10th anniversary of *Catalysts*. Over the past decade, biocatalysis has evolved from a niche field to a cornerstone of the field of catalysis, driving advancements in organic synthesis and analysis. In this Special Issue, leading researchers showcase groundbreaking work that highlights the versatility and potential of biocatalysts. From enzyme discovery to innovative applications, each contribution underscores the transformative impact of biocatalysis on diverse fields. We delve into enzymatic transformations, exploring substrate specificity, reaction mechanisms, and protein engineering. Interdisciplinary collaborations across microbiology, bioinformatics, and materials science further expand the horizons of biocatalysis research. As we reflect on a decade of innovation in biocatalysis, we extend our gratitude to the authors, reviewers, and editorial team for their invaluable contributions. Together, we continue to push the boundaries of biocatalysis, shaping the future of this exciting field. We invite you to explore this Special Issue and join us in celebrating a decade of excellence in *Biocatalysis*.

Evangelos Topakas, Roland Wohlgemuth, and David D. Boehr
Editors

catalysts

MDPI

Editorial

Special Issue "10th Anniversary of Catalysts: Biocatalysis in Analysis and Synthesis—Past, Present and Future"

Evangelos Topakas [1], David Boehr [2] and Roland Wohlgemuth [3,*]

[1] Biotechnology Laboratory, School of Chemical Engineering, National Technical University of Athens, 9 Iroon Polytech Str., Zografou Campus, 15780 Athens, Greece
[2] Department of Chemistry, The Pennsylvania State University, University Park, PA 16802, USA
[3] Institute of Molecular and Industrial Biotechnology, Lodz University of Technology, 90-537 Lodz, Poland
* Correspondence: roland.wohlgemuth.1@p.lodz.pl

The milestone of the 10th anniversary of *Catalysts* is a great time to reflect on past accomplishments, present progress and challenges, as well as to identify future challenges and opportunities. Biocatalysis has moved from a niche area of catalysis to the forefront as a key enabling technology for successfully addressing challenges in the fields of organic synthesis and analysis. This is also illustrated by the continuous growth of the "Biocatalysis" section of *Catalysts*, with a total number of 130 Special Issues, from which 32 are active online and 98 have been completed, and a total number of 761 articles published to date. As a way of celebrating the 10th anniversary of *Catalysts*, and in view of the key importance of biocatalysis, the "Biocatalysis" section has therefore launched a Special Issue entitled "10th Anniversary of *Catalysts*: Biocatalysis in Analysis and Synthesis—Past, Present and Future".

The enzymatic monoacetylation of diols catalyzed by *Candida antarctica* lipase B is a valuable desymmetrization methodology and has been applied by Madalińska et al. [1] to prochiral phosphines and phosphine *P*-sulfides as a route towards P-chiral catalysts. An enantiomeric excess of 98% and 10% yield could be achieved in the case of bis(2-hydroxymethylphenyl)phenylphosphine when using *C. antarctica* lipase B as a catalyst and *t*-butyl methyl ether/pyridine as a solvent, while 77% enantiomeric excess and 60% yield was the best result achieved in the case of bis(2-hydroxymethylphenyl)phenylphosphine-*P*-sulfide when using lipase from *Pseudomonas fluorescens* as a catalyst and *t*-butyl methyl ether as a solvent [1].

The low-cost liquid lipase Eversa Transform, a variant lipase from *Thermomyces lanuginosus*, was applied by Vieira et al. [2] in the hydrolysis of acylglycerols from soybean oil deodorizer distillate to free fatty acids in high yields, and for the simultaneous esterification/transesterification of soybean oil deodorizer distillate to fatty acid ethylesters in high yields using ethanol as an acyl acceptor.

A simple mathematical tool has been developed by Rodrigues de Sousa et al. [3] for optimizing the syntheses of short, medium or long-chain esters from acids and alcohols using immobilized lipase and solvent-free systems.

The substrate scope, crystal structure, kinetic properties and thermostability of the recombinantly expressed L-amino acid oxidase from *Pseudoalteromonas luteoviolacea* have been determined by Savino et al. [4]. The high expression level, ease of purification, high thermostability and activity on many different L-amino acids make this enzyme not only attractive for the synthesis of enantiopure amino acids or related compounds but also for detection, due to its high catalytic efficiency on a subset of amino acids [4]. The determined crystal structure provides a solid basis for engineering tailor-made variants of L-amino acid oxidase for activity on specific amino acids [4].

The hydrogen-dependent carbon dioxide reductase from *Thermoanaerobacter kivui* was immobilized in a redox polymer on a cathode, and its activity was investigated by Ruth et al. [5] regarding H_2 formation from electricity. A 340-fold increase in the current density

check for updates

Citation: Topakas, E.; Boehr, D.; Wohlgemuth, R. Special Issue "10th Anniversary of Catalysts: Biocatalysis in Analysis and Synthesis—Past, Present and Future". *Catalysts* **2022**, *12*, 1626. https://doi.org/10.3390/catal12121626

Received: 29 November 2022
Accepted: 8 December 2022
Published: 12 December 2022

has been demonstrated in a rotating disk electrode system using cobaltocene-functionalized polyallylamine as a redox polymer paired with the hydrogen-dependent carbondioxide reductase from *T. kivui*, which resulted in significantly higher maximum current densities than for previously reported systems [5].

Cell-free protein synthesis has been used by Rolf et al. [6] as a valuable tool for simplifying and accelerating the identification of novel non-heme Fe^{2+}/α-ketoglutarate-dependent dioxygenases, which can be applied for the selective hydroxylation of L-lysine in the 3- and 4-position in whole-cell biotransformations. Six novel and yet uncharacterized non-heme Fe^{2+}/α-ketoglutarate-dependent dioxygenases from *Kineococcus rhizosphaerae*, *Mycobacterium interjectum*, *Photorhabdus luminescens*, *Burkholderia* sp. MSMB617WGS, *Burkholderia pseudomallei* and *Burkholderia plantarii* with suitable activities have been found and extend the range of enzymes for catalyzing the hydroxylation of L-lysine, whereby further investigations will be of interest for providing the absolute configuration of the resulting 3-hydroxy-L-lysine and 4-hydroxy-L-lysine [6].

The whole genome sequencing of three *Streptomyces* sp. strains, different identification approaches for transaminases and laccases and the functional expression of the corresponding genes have been combined by Ferrandi et al. [7]. They enabled the characterization of a novel transaminase and a novel laccase, which were shown to be exceptionally thermostable. The novel transaminase Sbv333-TA was demonstrated to have a broad substrate scope, including β-ketoesters such as methyl acetoacetate and ethyl benzoylacetate, while improved activity in the presence of the organic solvent acetonitrile was found for the novel laccase Sbv286-LAC [7].

The substrate scope of silicatein-α, a hydrolytic enzyme from siliceous marine sponges of interest for biocatalytic silylation, has been investigated by Sparkes et al. [8] in a series of condensation reactions of triethylsilanol with various aromatic and aliphatic alcohols. The preference of silicatein-α for the silylation of the *S*-enantiomers of aliphatic alcohols and the high degree of conversion in the nonpolar solvents *n*-octane and toluene are good starting points for further evolution as valuable biocatalysts for the synthesis of organosiloxanes [8].

Laccases from *Trametes versicolor*, *Myceliophthora thermophila*, *Bacillus subtilis* and laccase-like multicopper oxidase from *T. thermophila* have been investigated by Milovanovic et al. in the oxidation of 1,4-dihydropyridine-based hexahydroquinolines to the corresponding pyridine-containing tetrahydroquinolines and in the oxidation of 1,4-dihydropyridine-based decahydroacridines to the corresponding pyridine-based octahydroacridines [9].

Phosphotransferases, phosphohydrolases, phosphorylases and phosphomutases are powerful biocatalysts for highly selective and efficient phosphorylation reactions, and their applications have been highlighted by Wohlgemuth [10], including useful phosphoryl donors and systems for their regeneration, reaction engineering, product recovery and purification. Examples of valuable analytical and synthetic applications of phosphorylation biocatalysts are provided, which illustrate the resource efficiency of highly selective phosphorylation reactions proceeding with complete conversion [10].

Biocatalysts, including both protein-based and nucleic acid-based enzymes, can also be utilized for constructing catalyst-based biomolecular logic gates that can read various molecular inputs and provide chemical, optical, and electrical outputs. Progress in constructing logic gates that take advantage of biological catalysts is discussed by Winston and Boehr [11].

The biological synthesis of biodegradable short-chain-length, medium-chain-length and short-medium-chain-length polyhydroxyalkanoates and their applications and recycling have been discussed by Dalton et al. [12].

We would like to thank all of the authors for their contributions and the editorial staff at *Catalysts*, particularly Pamela Li, M.Sc., Section Managing Editor, and Ellia Zhang, Assistance Editor, for their efforts. We hope that you enjoy this Special Issue to commemorate the 10-Year Anniversary of Catalysts.

Funding: This research received no external funding.

Conflicts of Interest: The authors declare no conflict of interest.

References

1. Madalińska, L.; Kiełbasiński, P.; Kwiatkowska, M. Enzymatic Desymmetrisation of Prochiral Phosphines and Phosphine P-Sulfides as a Route to P-Chiral Catalysts. *Catalysts* **2022**, *12*, 171. [CrossRef]
2. Vieira, A.-C.; Moulin Cansian, A.B.; Guimarães, J.R.; Marquettotti Salcedo Vieira, A.; Fernandez-Lafuente, R.; Waldir Tardioli, P. Performance of Liquid Eversa on Fatty Acid Ethyl Esters Production by Simultaneous Esterification/Transesterification of Low-to-High Acidity Feedstocks. *Catalysts* **2021**, *11*, 1486. [CrossRef]
3. Rodrigues de Sousa, R.; Sant'Ana da Silva, A.; Fernandez-Lafuente, R.; Santana Ferreira-Leitão, V. Simplified Method to Optimize Enzymatic Esters Syntheses in Solvent-Free Systems: Validation Using Literature and Experimental Data. *Catalysts* **2021**, *11*, 1357. [CrossRef]
4. Savino, S.; Daniël-Moráh Meijer, J.; Rozeboom, H.J.; van Beek, H.L.; Fraaije, M.W. Kinetic and Structural Properties of a Robust Bacterial L-Amino Acid Oxidase. *Catalysts* **2021**, *11*, 1309. [CrossRef]
5. Ruth, J.C.; Schwarz, F.M.; Müller, V.; Spormann, A.M. Enzymatic Hydrogen Electrosynthesis at Enhanced Current Density Using a Redox Polymer. *Catalysts* **2021**, *11*, 1197. [CrossRef]
6. Rolf, J.; Nerke, P.; Britner, A.; Krick, S.; Lütz, S.; Rosenthal, K. From Cell-Free Protein Synthesis to Whole-Cell Biotransformation: Screening and Identification of Novel α-Ketoglutarate-Dependent Dioxygenases for Preparative-Scale Synthesis of Hydroxy-l-Lysine. *Catalysts* **2021**, *11*, 1038. [CrossRef]
7. Ferrandi, E.E.; Spasic, J.; Djokic, L.; Vainshtein, Y.; Senthamaraikannan, R.; Vojnovic, S.; Grumaz, C.; Monti, D.; Nikodinovic-Runic, J. Novel Transaminase and Laccase from *Streptomyces* spp. Using Combined Identification Approaches. *Catalysts* **2021**, *11*, 919. [CrossRef]
8. Sparkes, E.I.; Egedeuzu, C.S.; Lias, B.; Sung, R.; Caslin, S.A.; Tabatabaei Dakhili, S.Y.; Taylor, P.G.; Quayle, P.; Wong, L.S. Biocatalytic Silylation: The Condensation of Phenols and Alcohols with Triethylsilanol. *Catalysts* **2021**, *11*, 879. [CrossRef]
9. Milovanovic, J.; Gözde Gündüz, M.; Zerva, A.; Petkovic, M.; Beskoski, V.; Thomaidis, N.S.; Topakas, E.; Nikodinovic-Runic, J. Synthesis and Laccase-Mediated Oxidation of New Condensed 1,4-Dihydropyridine Derivatives. *Catalysts* **2021**, *11*, 727. [CrossRef]
10. Wohlgemuth, R. The Power of Biocatalysts for Highly Selective and Efficient Phosphorylation Reactions. *Catalysts* **2022**, *12*, 1436. [CrossRef]
11. Winston, D.S.; Boehr, D.D. Catalyst-Based Biomolecular Logic Gates. *Catalysts* **2022**, *12*, 712. [CrossRef]
12. Dalton, B.; Bhagabati, P.; De Micco, J.; Padamati, R.B.; O'Connor, K. A Review on Biological Synthesis of the Biodegradable Polymers Polyhydroxyalkanoates and the Development of Multiple Applications. *Catalysts* **2022**, *12*, 319. [CrossRef]

catalysts

Article

Enzymatic Desymmetrisation of Prochiral Phosphines and Phosphine *P*-Sulfides as a Route to *P*-Chiral Catalysts

Lidia Madalińska, Piotr Kiełbasiński and Małgorzata Kwiatkowska *

Division of Organic Chemistry, Centre of Molecular and Macromolecular Studies, Polish Academy of Sciences, Sienkiewicza 112, 90-363 Łódź, Poland; lmad@cbmm.lodz.pl (L.M.); piokiel@cbmm.lodz.pl (P.K.)
* Correspondence: gosia@cbmm.lodz.pl

Abstract: The enzyme-catalyzed monoacetylation of prochiral bis (2-hydroxymethylphenyl)methyl phosphine and bis (2-hydroxymethylphenyl)phenylphosphine and their *P*-sulfides gave, in one single step, as a result of desymmetrisation, the corresponding monoacetates in moderate yields and with an enantiomeric excess of 16 to 98%, depending on the substrate structure and enzyme applied. The absolute configurations of the selected products were determined by a chemical correlation. This led to the conclusion that, in the case of phosphines, phosphine oxides and phosphine sulfides enzymes preferentially produce compounds of the same spatial arrangement. The new compounds obtained will be transformed into chiral catalysts/ligands.

Keywords: biotransformations; configuration determination; desymmetrisation; enzyme catalysis; phosphorus compounds

Citation: Madalińska, L.; Kiełbasiński, P.; Kwiatkowska, M. Enzymatic Desymmetrisation of Prochiral Phosphines and Phosphine *P*-Sulfides as a Route to *P*-Chiral Catalysts. *Catalysts* **2022**, *12*, 171. https://doi.org/10.3390/catal12020171

Academic Editors: Evangelos Topakas, Roland Wohlgemuth and David D. Boehr

Received: 20 December 2021
Accepted: 26 January 2022
Published: 28 January 2022

1. Introduction

Since the pioneering times of the mid-1970s, when the first practical and generally applicable methods in asymmetric synthesis [1] were developed, there has been a tremendous growth in this research field.

One of the fundamental research goals in modern chemistry is the development of efficient and selective procedures to access organic compounds. Among all the methodologies developed to date, catalysis offers an efficient and economical approach to enantiomerically pure substances. In particular, organocatalysis, transition metal catalysis and enzymes, and biotechnology methods are in high demand due to their application in stereoselective carbon–hydrogen, carbon–carbon or carbon–heteroatom bond formations.

Among the commonly used catalysts, there is a substantial number of hetero-organic compounds, especially organic phosphorus and sulfur compounds. Concerning the former ones, there are mainly phosphines and phosphine oxides [2], whose synthesis requires the application of various methodologies. Among them, the desymmetrisation of prochiral phosphorus substrates seems particularly interesting. Such transformations were performed mainly in a chemical manner, using chiral catalysts (for some recent results see [3,4]). There are only a few examples of the use of enzymes as catalysts, such as desymmetrisation [5–8], three of which came from our laboratory [5,7,8] (vide infra).

The chiral sulfur derivatives that were applied in asymmetric catalysis are variously functionalized sulfinyl compounds, sulfoximines and others (for a recent overview on the use of *S*-chiral sulfur ligands/catalysts in asymmetric synthesis, see [9]).

Searching for new chiral catalysts, we synthesized some time ago a series of organosulfur chiral compounds **3**, containing a stereogenic sulfinyl moiety, an enantiomeric amine fragment and the hydroxyl group (Scheme 1) [10,11]. They proved to be excellent catalysts in a variety of reactions of asymmetric synthesis [9,12].

Following these positive results, we decided to synthesize catalysts in which the stereogenic sulfinyl moiety was replaced by a stereogenic phosphorus-containing group.

In the first approach, attempts were made at the synthesis of the phosphinyl (P = O) analogues. We synthesized the enantiomerically pure (2-acetoxymethylphenyl)(2′-hydroxymethylphenyl)phosphine oxide **5**, which was planned to be a precursor of the desired ligands **6** (Scheme 2).

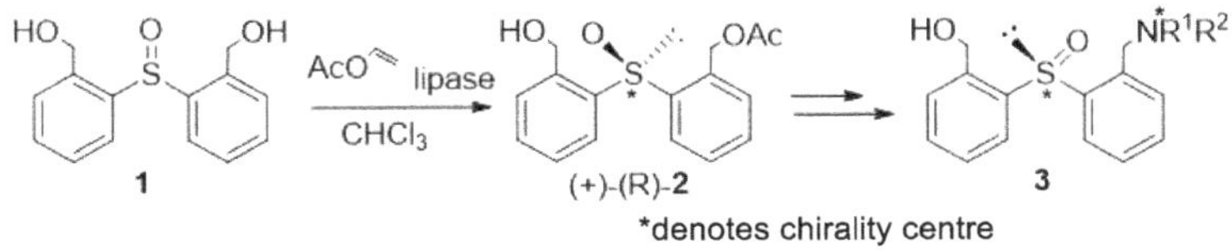

Scheme 1. Synthesis of chiral tridentate sulfinyl catalysts.

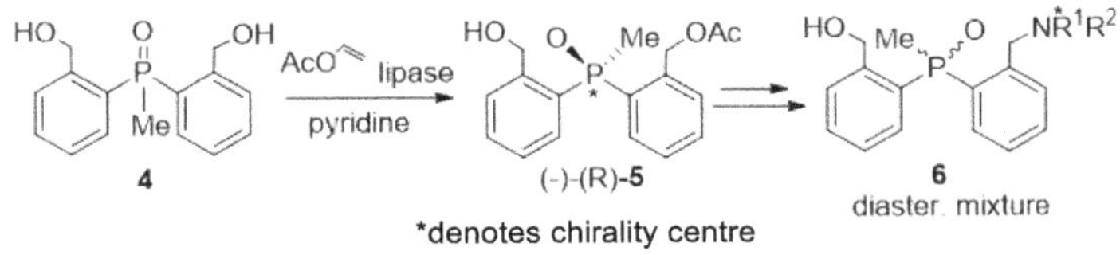

Scheme 2. Synthesis of chiral phosphine oxides as substrates for potential tridentate ligands.

The initial steps involving the enzymatic desymmetrisation of **4** (Scheme 2) were fully successful [8]. In all cases, the catalyst precursor **5** was obtained in high yields and with *ee*s up to 99.5%. Unexpectedly, attempts to replace the hydroxyl group with an enantiomerically pure chiral amine moiety via the mesylation of enantiomerically pure **5** always led to mixtures of diastereomers (which was not the case for the corresponding sulfinyl analogs **2**), which means that racemization must have taken place at one of the stages of the synthesis. The partial racemization of enantiomeric precursor **5** ultimately led to the formation of inseparable diastereomeric mixtures of the desired ligands **6** [13]. Moreover, the latter proved to be totally ineffective as catalysts for the several reactions of asymmetric synthesis [14].

Later, we found that the phosphinyl group (P = O) is responsible for both negative features of ligands **6** **[13,14]**. Therefore, we decided to use, instead of the phosphinyl substrates, the corresponding phosphines **12**, **13** and sulfides **22**, **23**, in the hope of avoiding the above problems.

2. Results and Discussion

2.1. Synthesis of Bis (2-hydroxymethylphenyl)phosphines 12 *and* 13

The syntheses of starting phosphines **12** and **13** were based on the reactions of *o*-tetrahydropyranyloxymethylphenylmagnesium bromide (Mg-THPBB) **7** with the appropriate dichlorophosphines **8** and **9**. Thus, the synthesis of bis (2-hydroxymethylphenyl)methyl phosphine **12** was achieved by the reaction of dichloro (methyl) phosphine **8** with Mg-THPBB **7** to give **10**, which was then treated with pyridinium *p*-toluenesulfonate (PPTS) to remove the THP protecting group. In turn, the synthesis of bis (2-hydroxymethylphenyl) phenylphosphine **13** was accomplished in a similar way, starting from dichloro (phenyl) phosphine and treatment of the resulting **11** with PPTS [8] (Scheme 3).

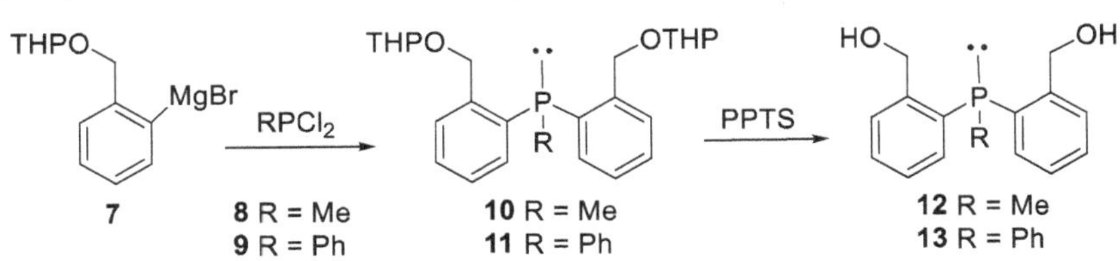

Scheme 3. Synthesis of bis (2-hydroxymethylphenyl)phosphines **12** and **13**.

2.2. *Desymmetrisation of Bis (2-hydroxymethylphenyl)phosphines* **12** *and* **13**

Due to the fact that in our laboratory we have successfully used enzymes for the synthesis of non-racemic hetero-organic compounds, particularly those containing stereogenic phosphorus and sulfur atoms, we decided to use also in this case such a procedure for desymmetrisation of the appropriate substrates. Thus, prochiral phosphines **12** and **13** were subjected to asymmetric acetylation with an excess of vinyl acetate in various solvents at 30 °C, using selected enzymes as biocatalysts (Scheme 4).

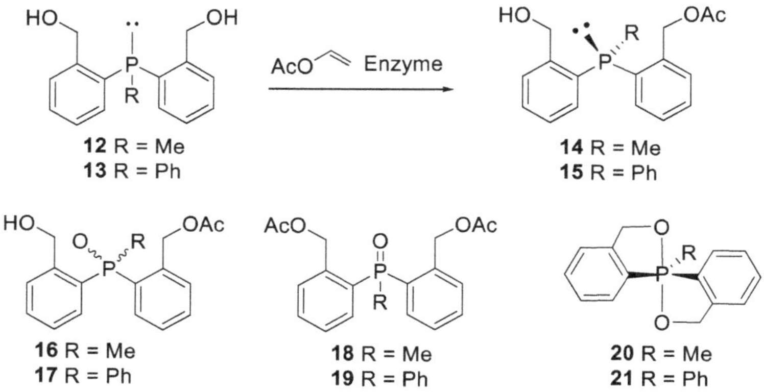

Scheme 4. Desymmetrisation of bis (2-hydroxiymethylphenyl)phosphines **12** and **13**.

The reactions were monitored by ^{31}P NMR. After completion, the enzymes were filtered off, the solvent and excess of vinyl acetate were evaporated to leave the residue, which, in addition to the desired enantiomerically enriched monoacetates **14** and **15**, contained some byproducts. These were the corresponding oxidized derivatives **16** and **17**, diacetates **18** and **19** and phosphoranes **20** and **21**, whose formation proved inevitable in spite of all the undertaken protection from air (as shown in Scheme 4)

The residue was separated via column chromatography and the results are collected in Table 1. The inspection of Table 1 clearly shows that the products exhibiting the highest enantiomeric excess were obtained using CAL-B as the biocatalyst and toluene as the solvent in acetylation of diol **12** and CAL-B as the biocatalyst and *t*-butyl methyl ether as the solvent in acetylation of diol **13**. The relatively low yield of the desired monoacetates may be due to the formation of byproducts (vide supra). Some attempts to improve the results, using increased amounts of enzymes, did not lead to the expected better outcome.

The comparison of the above results with those obtained in our previous paper devoted to the desymmetrisation of the P = O analogs (vide supra and Scheme 2) [8] clearly shows that the replacement of the phosphinyl (P = O) moiety by the phosphino (P-:) one results in a lower stereoselectivity of the reaction and lower yields of the products.

2.3. *Desymmetrisation of Bis (2-hydroxymethylphenyl)phosphine Sulfides*

Because the results obtained during desymmetrisation of bis (2-hydroxymethyl-phenyl) phosphines **12** and **13** were acceptable, but not very satisfactory and, moreover, some byproducts arising from oxidation phosphines were formed, we decided to synthesize appropriate phosphine sulfides **22** and **23**. Prochiral bis (2-hydroxymethylphenyl)phosphines **12** and **13** were treated with elemental sulfur to give the appropriate phosphine sulfides **22** and **23** (Scheme 5).

Table 1. Monoacetates **14** and **15** via Desymmetrisation of bis (2-hydroxymethylphenyl)phosphines **12** and **13**.

Entry	Diol	R	Enzyme [(a)]	Solvent	Time [Days]	Monoacetate Symbol	Yield [%]	$[\alpha]_D$ [(b)]	Ee [%] [(c)]
1	**12**	Me	CAL-B	CH_2Cl_2	8	**14**	53	−5.40	40
2	**12**	Me	CAL-B	*i*-Pr_2O	1	**14**	52	−7.10	53
3	**12**	Me	CAL-B	TBME	1	**14**	49	−8.70	64
4	**12**	Me	CAL-B	toluene	11	**14**	25	−9.85	73
5	**13**	Ph	CAL-B	TBME + pyridine	21	**15**	10	4.49	98
6	**13**	Ph	CAL-B	*i*-Pr_2O + toluene + pyridine	23	**15**	10	4.42	95
7	**13**	Ph	CAL-B	acetone	14	**15**	52	2.74	60
8	**13**	Ph	CAL-B	toluene + pyridine	15	**15**	38	2.74	60
9	**13**	Ph	CAL-B	toluene	6	**15**	27	2.28	50
10	**13**	Ph	CAL-B	acetonitrile	22	**15**	26	1.26	28
11	**13**	Ph	CAL-B	$CHCl_3$	12	**15**	16	−1.50	33
12	**13**	Ph	CAL-B	CH_2Cl_2	29	**15**	5	−1.40	31
13	**13**	Ph	PFL	$CHCl_3$	12	**15**	6	−2.26	49
14	**13**	Ph	LPL	toluene	28	**15**	<5	-	-
15	**13**	Ph	CR	toluene	33	**15**	11	2.54	56
16	**13**	Ph	PS	$CHCl_3$	29	**15**	10	-	-
17	**13**	Ph	-	Toluene	24	-	-	-	-

[(a)] Enzyme: CAL-B: *Candida antarctica* lipase B (Novozym 435); PFL: lipase from *Pseudomonas fluorescens*; CR: lipase from *Candida rugosa*; PS: Lipase PS (AMANO); LPL: Lipoprotein lipase; [(b)] in chloroform (c = 1); [(c)] the *ee* values were determined by chiral HPLC: OD, *n*-Hexane: (*i*-PrOH: EtOH 4:1) 98%: 2%, Fl. 0.5 mL/min.

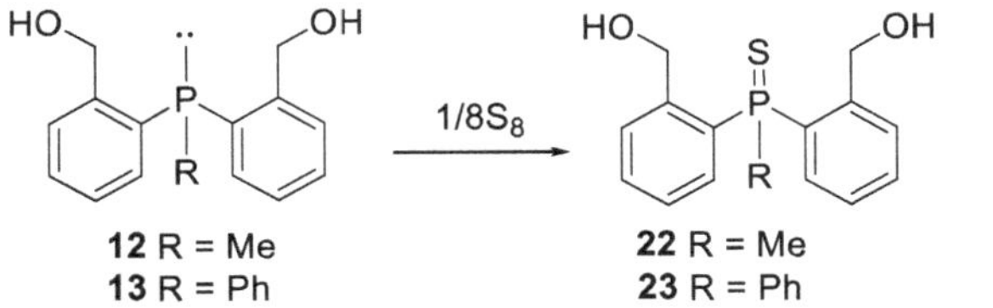

Scheme 5. Synthesis of bis (2-hydroxymethylphenyl)phosphine sulfides **22** and **23**.

The latter were subjected to asymmetric acetylation with an excess of vinyl acetate in various solvents at 30 °C, using a number of lipases (Scheme 6).

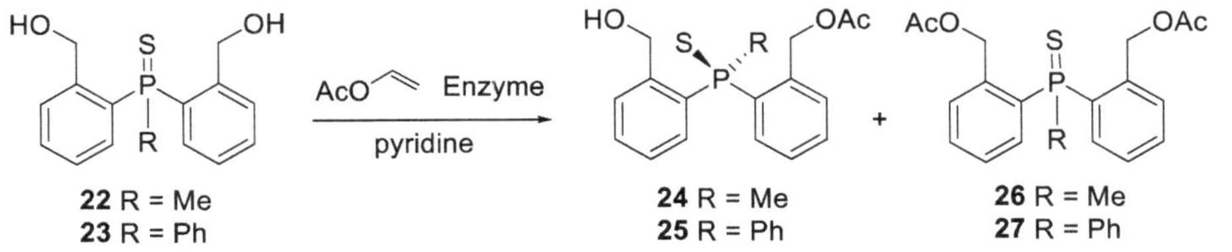

Scheme 6. Desymmetrisation of bis (2-hydroxymethylphenyl)phosphine sulfides **22** and **23**.

The reactions were monitored by ^{31}P NMR. After completion, the enzymes were filtered off, the solvent and excess of vinyl acetate were evaporated, and the residue was separated via column chromatography. Interestingly, also in this case, the formation of achiral diacetates **26** and **27** was observed, which, in some cases, may be responsible for a decreased yield of monoacetates. The results are collected in Table 2.

Table 2. Monoacetates **24** and **25** via desymmetrisation of bis (2-hydroxymethylphenyl)phosphine sulfides **22** and **23**.

Entry	Diol	R	Enzyme [(a)]	Solvent + Pyridine	Time [Days]	Monoacetate Symbol	Yield [%]	$[\alpha]_D$ [(c)]	Ee [%] [(d)]
1	**22**	Me	CAL-B	TBME [(b)]	4	**24**	60	4.7	72 [(e)]
2	**22**	Me	CAL-B	toluene	4	**24**	80	3.25	50 [(e)]
3	**22**	Me	CAL-B	acetone	4	**24**	70	1.65	25 [(e)]
4	**22**	Me	CAL-B	CH_2Cl_2	8	**24**	49	0	0
5	**22**	Me	CR	TBME	11	**24**	55	0	0
6	**22**	Me	PFL	TBME	11	**24**	60	0	0
7	**23**	Ph	PFL	TBME	38	**25**	60	−7.57	77
8	**23**	Ph	PFL	Et_2O	38	**25**	55	−4.44	45
9	**23**	Ph	PFL	toluene	38	**25**	66	−1.56	16
10	**23**	Ph	PFL	CH_2Cl_2	38	**25**	44	−1.62	16
11	**23**	Ph	CAL-B	Et_2O	38	**25**	44	3.25	54
12	**23**	Ph	CAL-B	i-Pr_2O	38	**25**	37	1.1	0
13	**23**	Ph	CAL-B	acetone	38	**25**	14	2.25	23
14	**23**	Ph	CAL-B	toluene	26		only substrate		
15	**23**	Ph	CAL-B	TBME	16	**25**	10	2.2	23
16	**23**	Ph	CR	TBME	38	**25**	23	1.8	18

[(a)] Enzyme: CAL-B: *Candida antarctica* lipase B (Novozym 435); PFL: lipase from *Pseudomonas fluorescens*; CR: lipase from *Candida rugose*; [(b)] TBME: *t*-butyl methyl ether; [(c)] in chloroform (c = 1); [(d)] the *ee* values were determined by chiral HPLC: AS, *n*-Hexane: (*i*-PrOH: EtOH 4:1) 96%: 4%, Fl. 0.4 mL/min; [(e)] the ee values were calculated on the basis of the optical rotation, compared to the one shown in Scheme 7 (vide infra).

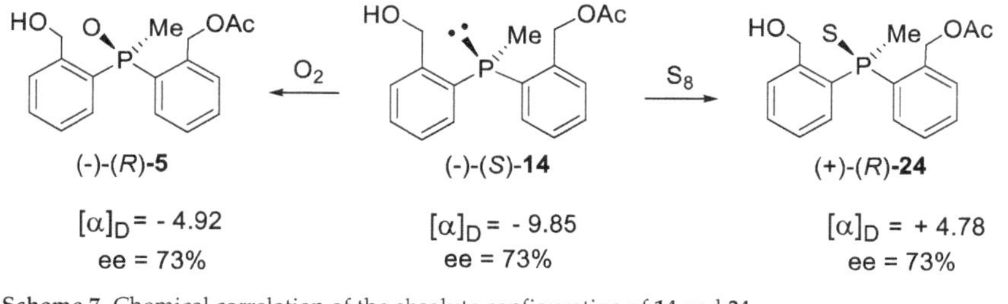

Scheme 7. Chemical correlation of the absolute configuration of **14** and **24**.

The inspection of Table 2 clearly shows that the reaction time, yields and enantioselectivity strongly depended on the lipase and the solvent used. Monoacetate **24** was obtained in good yield, but with moderate enantioselectivity. In turn, monoacetate **25** was formed in good yield and enantioselectivity. The best results were obtained using PFL as the biocatalyst and *t*-butyl methyl ether as the solvent. Again, attempts to use increased amounts of enzymes did not lead to higher outcomes. Nevertheless, it should be stressed that in both cases presented above, the stereoselective enzymatic transformation was achieved using the substrates in which the prostereogenic phosphorus atom and the reacting hydroxy oxygen were distant from each other by four bonds.

2.4. Determination of the Absolute Configuration of Monoacetates **14** and **24**

The absolute configuration of the newly obtained enantiomerically enriched monoacetates was determined as follows: phosphine **14** (Table 1, entry 4) was oxidized with air to give phosphine oxide (-)-(*R*)-**5**, whose absolute configuration and optical rotation are known from our previous work [8]. Since oxidation with air proceeds with the retention of the configuration at phosphorus [15], (*S*) configuration was ascribed to (-)-phosphine **14** obtained. In turn, the reaction of phosphine **14** with elemental sulfur, proceeding also with retention of configuration at phosphorus [15], gave phosphine sulfide **24**. Hence, the absolute configuration of the latter is (+)-(*R*) (Scheme 7), that is, the same as the one obtained in the enzymatic desymmetrisation.

Unfortunately, no such correlation could be made for phosphine monoacetate **15** and phosphine sulfide acetate **25** due to the lack of the corresponding phosphine oxides of known absolute configuration. Moreover, all the chiral products were oils, which excluded X-ray analysis.

3. Materials and Methods

3.1. General Information

The synthesized products were purified by column chromatography on Merck 60 silica gel (0.063–0.200 mm) or preparative plate chromatography using Merck 60 F_{254} silica gel plate (2.5 mm). TLC was performed on Merck 60 F_{254} silica gel plate (0.25 mm). Solvents were dried using general procedures and distilled prior to use. The NMR spectra were recorded in $CDCl_3$ with a Bruker AV 200 spectrometer. The chemical shifts (δ) are expressed in ppm, the coupling constants (*J*) are given in Hz. Mass spectra were recorded with a Finnigan MAT95 Voyager Elite spectrometer or Synapt G2-S*i* mass spectrometer (Waters) equipped with an ESI source and quadrupole-time-of-Flight mass analyser. Optical rotations were measured on a Perkin-Elmer 241 MC polarimeter. HPLC analysis was made using Varian Pro Star 210 instrument using column with chiral filling Chiralcel OD or Chiralpak AS. The enzymes were purchased from AMANO or SIGMA. Enzymes: CAL-B (Novozym 435)–lipase acrylic resin from *Candida antarctica* (E.C. 3.1.1.3), SIGMA-ALDRICH; PFL–lipase from *Pseudomonas fluorescens* (E.C. 3.1.1.3), SIGMA-ALDRICH; PS–lipase from *Pseudomonas species* (E.C. 3.1.1.3), AMANO, CR–lipase from *Candida rugosa* (E.C. 3.1.1.3), SIGMA-ALDRICH; LPL–Lipoprotein lipase (E.C. 3.1.1.34), SIGMA-ALDRICH. All the NMR spectra are collected in the "Supplementary Materials".

3.2. Synthesis of Bis (2-hydroxymethylphenyl)phosphines **12** and **13**

3.2.1. Synthesis of Bis [2-(2′-tetrahydropyranyloxy)methylphenyl]phosphine **10** and **11**

To magnesium (0.264 g, 0.01 mol) under nitrogen was added a solution of 2′-(2-tetrahydropyranyloxymethyl)bromobenzene **7**, obtained according to the known procedure [8] (3 g, 0.01 mol) in THF (8 mL) followed by a small crystal of iodine. The mixture was gently heated to initiate the Grignard reagent formation. After the magnesium completely dissolved, appropriate dichlorophosphine **8** or **9** (0.649 g for **8** or 0.985 g for **9**, 0.0055 mol) dissolved in THF (5 mL) was added and the solution was stirred for 3 h. THF was evaporated, saturated aqueous solution of NH_4Cl (10 mL) was added to the residue and the mixture was extracted with CH_2Cl_2 (3 × 10 mL). The combined organic layers were dried over anhydrous $MgSO_4$ and the solvent was removed to give crude phosphine **10** or **11**.

Bis [2-(2′-tetrahydropyranyloxy)methylphenyl]methylphosphine 10
Crude yield: 1.970 g, 90%
^{31}P NMR ($CDCl_3$): δ = −48.3

Bis [2-(2′-tetrahydropyranyloxy)methylphenyl]phenylphosphine 11
Crude yield: 2.535 g, 94%
^{31}P NMR ($CDCl_3$): δ = −25.7

3.2.2. Synthesis of Bis (2-hydroxymethylphenyl)phosphines **12** and **13**

To a solution of crude phosphine **10** or **11** (0.0054 mol) in EtOH (40 mL) pyridinium *p*-toluenesulfonate (PPTS) (0.2 eq., 0.27 g, 0.0011 mol) was added and the mixture was stirred at 55 °C for 7 h. EtOH was evaporated, to the residue saturated aqueous solution of $NaHCO_3$ (10 mL) was added and the mixture was extracted with CH_2Cl_2 (3 × 10 mL). The combined organic layers were dried over anhydrous $MgSO_4$. After the evaporation of the solvent, the crude products were purified by column chromatography using dichloromethane-acetone in a gradient from 5:1 to 1:1 as eluent to give products **12** and **13**, respectively.

Bis (2-hydroxymethylphenyl)methylphosphine 12
Oil, isolated yield: 0.439 g, 25%
^{31}P NMR ($CDCl_3$): δ = −50.8;
^{1}H NMR ($CDCl_3$): δ = 1.59 (d, *J* = 3.8 Hz, 3H), 4.67–4.90 (m, 4H), 7.27–7.39 (m, 8H);
MS (+ESI): m/z = 261 (M + H);
HRMS (+ESI): m/z = 261.1049, calcd for $C_{15}H_{18}PO_2$ (M + H), 261.1044.
Bis (2-hydroxymethylphenyl)phenylphosphine 13
Oil, isolated yield: 0.098 g, 56%
^{31}P NMR ($CDCl_3$): δ = −27.5;
^{1}H NMR ($CDCl_3$): δ = 4.75–5.04 (m, 4H), 6.94–7.52 (m, 13H);
MS (+ESI): m/z = 323 (M + H);
HRMS (+ESI): m/z = 323.1208, calcd for $C_{20}H_{20}PO_2$ (M + H), 323.1201.

3.3. *Synthesis of Bis (2-hydroxymethylphenyl)phosphine Sulfides* **22** *and* **23**

To obtain bis(2-hydroxymethylphenyl)phosphine sulfides **22** and **23**, to the solution of phosphine **12** and **13** (0.180 g, 0.692 mmol for **12** and 0.134 g, 0.416 mmol for **13**) in dichloromethane (20 mL) under nitrogen elemental sulfur (1 eq., 0.022 g, 0.692 mmol for **12** and 0.014 g, 0.416 mmol for **13**) was added. The mixture was refluxed until the substrate disappeared, which was found by ^{31}P NMR. Then, the reaction mixture was filtered through celite and the solvent was evaporated. The crude reaction mixture was purified by column chromatography using dichloromethane-acetone 6:1 with the addition of triethylamine (0.03% vol.) to give pure **22** and **23**.

Bis (2-hydroxymethylphenyl)methylphosphine sulfide 22
Yellowish solid, m. p. 92–94 °C, isolated yield: 0.124 g, 61%
^{31}P NMR ($CDCl_3$): δ = 34.2;
^{1}H NMR ($CDCl_3$): δ = 2.34 (d, J = 13.2 Hz, 3H), 3.71 (br. s, 1H), 4.39–4.75 (m, 4H), 7.11–7.72 (m, 8H);
^{13}C NMR ($CDCl_3$): δ = 23.68 (d, $J_{P\text{-}Me}$ = 60.5 Hz), 62.93 (d, $J_{P\text{-}CH2OH}$ = 6.0 Hz), 128.15, 131.03, 131.12, 131.73, 132.22, 132.30, 132.36, 132.74, 143.97, 144.04 (Aryl);
MS (+ESI): m/z = 293 (M + H);
HRMS (+ESI): m/z = 293.0771, calcd for $C_{15}H_{18}PO_2S$ (M + H), 293.0765.
Bis (2-hydroxymethylphenyl)phenylphosphine sulfide 23
Yellowish solid, m. p. 132–134 °C, isolated yield: 0.111 g, 76%
^{31}P NMR ($CDCl_3$): δ = 40.7;
^{1}H NMR ($CDCl_3$): δ = 3.85 (br. s, 1H), 4.67–4.82 (m, 4H), 5.29–5.49 (m, 2H), 6.78–7.72 (m, 13H);
MS (CI): m/z = 355 (M + H);
HRMS (CI): m/z = 354.0845, calcd for $C_{20}H_{19}PO_2S$ (M + H), 354.0843.

3.4. *General Procedure for the Enzymatic Desymmetrization of Bis (2-hydroxymethylphenyl)phosphines* **12** *and* **13** *and Bis (2-hydroxymethylphenyl)phosphine Sulfides* **22** *and* **23**

To a solution of the prochiral phosphine diol (**12** or **13**, 0.1 mmol) or prochiral *P*-sulfide diol (**22** or **23**, 0.1 mmol) in a solvent (5 mL) pyridine (3 eq., 0.024 mL, 0.3 mmol), vinyl acetate (0.5 mL) and an enzyme (20 mg) were added. In the case of phosphines **12** and **13**, the reaction was carried out under a nitrogen atmosphere. The whole mixture was stirred at room temperature. The conversion degree was determined by ^{31}P NMR. Then, the enzyme was filtered off and the solvents were evaporated. The crude reaction mixture was separated by column chromatography using dichloromethane-acetone in gradient from 100:1 to 1:1 with the addition of triethylamine (0.03% vol.) as eluent, to give pure enantiomerically enriched phosphine monoacetates **14** and **15** and *P*-sulfide monoacetates **24** and **25**. The results are collected in Table 1 for phosphines **12** and **13** and in Table 2 for sulfides **22** and **23**.

(2-acetoxymethylphenyl)(2′-hydroxymethylphenyl)methylphosphine 14
^{31}P NMR (CDCl$_3$): δ = −48.4;
^{1}H NMR (CDCl$_3$): δ = 1.59 (d, J = 3.6 Hz, 3H), 1.86 (s, 3H), 4.69–4.89 (m, 4H), 7.19–7.48 (m, 8H);
MS (CI): m/z = 303 (M + H);
HRMS (+ESI): m/z = 303.1158, calcd for $C_{17}H_{20}PO_3$ (M + H), 303.1150.

(2-acetoxymethylphenyl)(2′-hydroxymethylphenyl)phenylphosphine 15
^{31}P NMR (CDCl$_3$): δ = −26.3;
^{1}H NMR (CDCl$_3$): δ = 1.62 (s, 3H), 4.71–4.82 (m, 2H), 5.29–5.49 (m, 2H), 6.81–7.44 (m, 13H);
MS (FAB): m/z = 365 (M + H);
HRMS (FAB): m/z = 365.1317, calcd for $C_{22}H_{22}PO_3$ (M + H), 365.1306.

(2-acetoxymethylphenyl)(2′-hydroxymethylphenyl)methylphosphine sulfide 24
^{31}P NMR (CDCl$_3$): δ = 33.7;
^{1}H NMR (CDCl$_3$): δ = 1.89 (s, 3H), 2.38 (d, J = 13.2 Hz, 3H), 4.31–4.75 (m, 2H), 4.94–5.09 (m, 2H), 7.36–8.15 (m, 8H);
MS (+ESI): m/z = 335 (M + H);
HRMS (+ESI): m/z = 335.0876, calcd for $C_{17}H_{20}PO_3S$ (M + H), 335.0871.

(2-acetoxymethylphenyl)(2′-hydroxymethylphenyl)phenylphosphine sulfide 25
^{31}P NMR (CDCl$_3$): δ = 41.2;
^{1}H NMR (CDCl$_3$): δ = 1.84 (s, 3H), 4.01 (br. s, 1H), 4.60–4.82 (m, 2H), 5.38–5.55 (m, 2H), 6.89–7.73 (m, 13H);
MS (CI): m/z = 397 (M + H);
HRMS (+ESI): m/z = 397.1023, calcd for $C_{22}H_{22}PO_3S$ (M + H), 397.1027.

4. Conclusions

Prochiral bis (2-hydroxymethylphenyl) phosphines and their sulfides could be successfully transformed into enantiomerically enriched monoacetyl derivatives by a desymmetrisation procedure using an enzymatic acetylation reaction. The use of enzymes proved to be useful in obtaining the desired products with high stereoselectivity in one step. The determination of their absolute configuration proved that, in the case of phosphines, phosphine oxides and phosphine sulfides enzymes preferentially produce compounds of the same spatial arrangement. The new compounds obtained will be transformed into chiral catalysts/ligands. The appropriate investigations are underway.

Supplementary Materials: The following supporting information can be downloaded at: https://www.mdpi.com/article/10.3390/catal12020171/s1: NMR spectra of all the new compounds.

Author Contributions: Conceptualization, M.K. and P.K.; methodology, M.K.; investigation, L.M.; writing—original draft preparation, L.M.; writing—review and editing, M.K.; supervision, P.K.; funding acquisition, L.M. All authors have read and agreed to the published version of the manuscript.

Funding: Financial support by the National Research Centre (NCN), Poland, grant PRELUDIUM: UMO-2014/13/N/ST5/03481 for L.M., is gratefully acknowledged.

Data Availability Statement: Some or all data, models, or code that support the findings of this study are available from the corresponding author upon reasonable request.

Conflicts of Interest: The authors declare no conflict of interest.

References

1. Enders, D.; Hoffmann, R.W. Asymmetrische Synthese. *Chem. Unserer Zeit* **1985**, *19*, 177–199. [CrossRef]
2. Börner, A. (Ed.) *Phosphorus Ligands in Asymmetric Catalysis*; Wiley-VCH: Karlsruhe, Germany, 2008.
3. Gammon, J.J.; Viktoria, H.; Gessner, V.H.; Barker, G.R.; Granander, J.; Whitwood, A.C.; Strohmann, C.; O'Brien, P.; Brian, K.B. Synthesis of P-Stereogenic Compounds via Kinetic Deprotonation and Dynamic Thermodynamic Resolution of Phosphine Sulfides: Opposite Sense of Induction Using (-)-Sparteine. *J. Am. Chem. Soc.* **2010**, *132*, 13922–13927. [CrossRef] [PubMed]
4. Huang, Q.-H.; Zhou, Q.-Y.; Yang, C.; Chen, L.; Cheng, J.-P.; Li, X. Access to P-stereogenic compounds via desymmetrizing enantioselective bromination. *Chem. Sci.* **2021**, *12*, 4582. [CrossRef] [PubMed]
5. Kiełbasiński, P.; Żurawiński, R.; Albrycht, M.; Mikołajczyk, M. The first enzymatic desymmetrizations of prochiral phosphine oxides. *Tetrahedron Asymmetry* **2003**, *14*, 3379–3384. [CrossRef]
6. Wiktelius, D.; Johansson, M.J.; Luthmann, K.; Kann, N. A Biocatalytic Route to P-Chirogenic Compounds by Lipase-Catalyzed Desymmetrization of a Prochiral Phosphine−Borane. *Org. Lett.* **2005**, *7*, 4991–4994. [CrossRef] [PubMed]
7. Kiełbasiński, P.; Rachwalski, M.; Kwiatkowska, M.; Mikołajczyk, M.; Wieczorek, W.M.; Szyrej, M.; Sieroń, L.; Rutjes, F.P.J.T. Enzyme-promoted desymmetrisation of prochiral bis(cyanomethyl)phenylphosphine oxide. *Tetrahedron Asymmetry* **2007**, *18*, 2108–2112. [CrossRef]
8. Kaczmarczyk, S.; Kwiatkowska, M.; Madalińska, L.; Barbachowska, A.; Rachwalski, M.; Błaszczyk, J.; Sieroń, L.; Kiełbasiński, P. Enzymatic Synthesis of Enantiopure Precursors of Chiral Bidentate and Tridentate Phosphorus Catalysts. *Adv. Synth. Catal.* **2011**, *353*, 2446–2454. [CrossRef]
9. Otocka, S.; Kwiatkowska, M.; Madalińska, L.; Kiełbasiński, P. Chiral Organosulfur Ligands/Catalysts with a Stereogenic Sulfur Atom: Applications in Asymmetric Synthesis. *Chem. Rev.* **2017**, *117*, 4147–4181. [CrossRef] [PubMed]
10. Rachwalski, M.; Kwiatkowska, M.; Drabowicz, J.; Kłos, M.; Wieczorek, W.M.; Szyrej, M.; Sieroń, L.; Kiełbasiński, P. Enzyme-promoted Desymmetrization of Bis(2-hydroxymethylphenyl)sulfoxide as a Route to Tridentate Chiral Catalysts. *Tetrahedron Asymmetry* **2008**, *19*, 2096–2101. [CrossRef]
11. Leśniak, S.; Rachwalski, M.; Sznajder, E.; Kiełbasiński, P. New Highly Efficient Aziridine-functionalized Tridentate Sulfinyl Catalysts for Enantioselective Diethylzinc Addition to Carbonyl Compounds. *Tetrahedron Asymmetry* **2009**, *20*, 2311–2314. [CrossRef]
12. Kiełbasiński, P.; Kwiatkowska, M.; Cierpiał, T.; Rachwalski, M.; Leśniak, S. The Sulfinyl Group: Its Importance for Asymmetric Synthesis and Biological Activity. *Phosphorus Sulfur Silicon Relat. Elem.* **2019**, *194*, 649–653. [CrossRef]
13. Kaczmarczyk, S.; Madalińska, L.; Kiełbasiński, P. Unexpected Racemization of 2-Hydroxymethylphenylphosphine Oxides. *Phosphorus Sulfur Silicon Relat. Elem.* **2013**, *188*, 249–253. [CrossRef]
14. Madalińska, L.; Kwiatkowska, M.; Kaczmarczyk, S.; Kiełbasiński, P.; Division of Organic Chemistry, Centre of Molecular and Macromolecular Studies, Polish Academy of Sciences, Sienkiewicza, Poland, 2022. (*unpublished results*).
15. Kolodiazhna, A.O.; Kolodiazhnyi, O.I. Asymmetric Electrophilic Reactions in Phosphorus Chemistry. *Symmetry* **2020**, *12*, 108. [CrossRef]

catalysts

Article

Performance of Liquid Eversa on Fatty Acid Ethyl Esters Production by Simultaneous Esterification/Transesterification of Low-to-High Acidity Feedstocks

Ana Carolina Vieira [1], Ana Bárbara Moulin Cansian [1], José Renato Guimarães [1], Angelica Marquettotti Salcedo Vieira [2], Roberto Fernandez-Lafuente [3,4,*] and Paulo Waldir Tardioli [1,*]

1 Graduate Program in Chemical Engineering (PPGEQ), Laboratory of Enzyme Technologies (LabEnz), Department of Chemical Engineering, Federal University of São Carlos (DEQ/UFSCar), km 235 Washington Luís Road, São Carlos 13565-905, SP, Brazil; ana.acv@live.com (A.C.V.); ana_barbara_moulin@hotmail.com (A.B.M.C.); renatoge74@gmail.com (J.R.G.)
2 Department of Food Engineering, State University of Maringá (DAL/UEM), 5790 Colombo Ave, Maringá 87020-900, PR, Brazil; amsvieira@uem.br
3 Departamento de Biocatálisis, ICP-CSIC, Universidad Autónoma de Madrid (UAM), 28049 Madrid, Spain
4 Center of Excellence in Bionanoscience Research, External Scientific Advisory Academics, King Abdulaziz University, Jeddah 21589, Saudi Arabia
* Correspondence: rfl@icp.csic.es (R.F.-L.); pwtardioli@ufscar.br (P.W.T.); Tel.: +34-91-594-804 (R.F.-L.); +55-16-3351-9362 (P.W.T.)

Citation: Vieira, A.C.; Cansian, A.B.M.; Guimarães, J.R.; Vieira, A.M.S.; Fernandez-Lafuente, R.; Tardioli, P.W. Performance of Liquid Eversa on Fatty Acid Ethyl Esters Production by Simultaneous Esterification/Transesterification of Low-to-High Acidity Feedstocks. *Catalysts* **2021**, *11*, 1486. https://doi.org/10.3390/catal11121486

Academic Editors: Evangelos Topakas, Roland Wohlgemuth and David D. Boehr

Received: 15 November 2021
Accepted: 2 December 2021
Published: 3 December 2021

Abstract: Liquid Eversa was evaluated in hydrolysis of acylglycerols from soybean oil deodorizer distillate (SODD), as well as simultaneous esterification/transesterification of SODD with low-to-high free fatty acids (FFAs) content using ethanol as acyl acceptor. Hydrolysis of SODD at mild temperature (37 °C) and without pH control (water:SODD mass ratio of 4:1) increased its FFAs content from 17.2 wt.% to 72.5 wt.% after 48 h reaction. A cold saponification of SODD allowed a saponification phase (SODD-SP) to be recovered with 93 wt.% saponification index and 2.25 wt.% FFAs content, which was used to find the experimental conditions for simultaneous esterification/transesterification reactions by experimental design. Temperature of 35 °C, enzyme concentration of 8.36 wt.%, and molar ratio of 3.64:1 (ethanol:SODD-SP) were found as the best conditions for fatty acid ethyl esters (FAEEs) production from SODD-SP (86.56 wt.% ester yield after 23 h reaction). Under the same reaction conditions, crude SODD (17.2 wt.% FFAs) and hydrolyzed SODD (72.5 wt.% FFAs) yielded products containing around 80 wt.% FAEEs. Caustic treatment could increase the ester content to around 90 wt.% and reduce the FFAs content to less than 1 wt.%. Our results show the good performance of liquid Eversa in aqueous (hydrolysis reactions) and organic (esterification/transesterification reactions) media.

Keywords: liquid Eversa; SODD; hydrolysis; simultaneous esterification and transesterification; substrate low to high acidity

1. Introduction

There is in the market a low-cost liquid lipase (Eversa Transform, a variant lipase from *Thermomyces lanuginosus*) specially formulated for the biodiesel industry [1–4]. It has been reported that this enzyme exhibits excellent performance in both liquid and immobilized forms, with both methanol and ethanol as acyl acceptors, and using refined and acid feedstocks (alkyl esters yields up to 99%) (Table A1 in Appendix A) [1,5–19].

Among the several fatty-rich industrial co-products, the soybean oil deodorizer distillate (SODD) is an interesting low-cost raw material as source of fatty acids. The SODD is a by-product of the soy oil refining, generated in the oil deodorization step [20] that is carried out to remove volatile compounds responsible for unacceptable odor, color and taste in the quality standard of oils for commercialization [20,21]. This by-product is mostly composed of FFAs, acylglycerols (monoglycerides (MAGs), diglycerides (DAGs) and triglycerides (TAGs)) and smaller amounts of tocopherols, free sterols and scalene [22–24].

Due to the high content of saponifiable materials (up to 90 wt.%), deodorizer distillates (DDs) of vegetable oils (soy, palm, rapeseed, etc.) have been exploited as raw material for biodiesel production (Table A2 in Appendix A) [25–30], reaching ester yields from 88% to about 98%, mainly using commercial immobilized lipases (Lipozyme IM, Novozym 435 and Lipozyme RM-IM) [25–30]. According to our research in the scientific literature, there are still no works on the production of fatty acids ethyl esters (FAEEs) from SODD using liquid Eversa.

This work aimed to evaluate the performance of liquid Eversa in the simultaneous esterification of FFAs and transesterification of acylglycerols in feedstocks containing low-to-high FFAs content, employing SODD as model substrate. For this purpose, crude SODD (containing around 17 wt.% FFAs, as described below) was pretreated by two processes, aiming to reduce (cold saponification) or increase the FFAs (Eversa hydrolysis) content in the SODD. The experimental conditions for obtaining high ester yields by simultaneous esterification/transesterification using liquid Eversa as biocatalyst were evaluated by statistical design. At the defined conditions, FAEEs were produced from SODD containing low, medium, and high FFAs content and using ethanol as acyl acceptor. Finally, as an attempt to increase the FAEEs content in our product, a caustic treatment was adopt under the previously reported conditions [1,5].

2. Results and Discussion

2.1. Soybean Oil Deodorizer Distillate (SODD) Characterization

As the SODD is a by-product of oil refining, its composition depends on the oil source and the processing steps. In this way, its physical-chemical characterization is important to verify the particularities of the raw material under study. The physical-chemical properties of the SODD (Table 1) show a saponification index (181.62 ± 0.96 mg KOH/g) close to that reported by Yin et al. [31,32] (154.87 mg KOH/g ± 2.62), who used SODD for biodiesel production. The saponification index of refined soybean oil is in the range 180–200 mg KOH/g [33], therefore, our results were very close to these values, indicating that the SODD had similar characteristics to refined oil in terms of that index.

Table 1. Physicochemical properties of the soybean oil deodorizer distillate (SODD).

Parameter	Results	Method
Acidity index (mgKOH/g) (25 °C)	34.19 ± 0.52 (17.18 wt.%)	[34]
Iodine index by the Wijs method (gI_2/100 g) (25 °C)	112.98 ± 0.32	[35]
Saponification index (mgKOH/g) (25 °C)	181.62 ± 0.96 (91.27 wt.%)	[36]
Kinematic viscosity (mPa.s) (40 °C)	32.74 ± 0.01	Note 1
Kinematic viscosity (mPa.s) (25 °C)	57.10 ± 0.01	Note 1
Moisture (%) (130 °C)	1.33 ± 0.04	[37]
Saponifiable matter as fatty acid methyl esters (FAME) (wt.%)	85.15 ± 0.53	[38,39]
α-Tocopherol (g/100 g) (%)	0.37 ± 0.001	[40]
β-Tocopherol (g/100 g) (%)	0.09 ± 0.005	[40]
γ-Tocopherol (g/100 g) (%)	1.01 ± 0.003	[40]
δ-Tocopherol (g/100 g) (%)	0.36 ± 0.003	[40]
Total of tocopherols (g/100 g) (%)	1.83	[40]
Palmitic acid (C16:0) (g/100 g) (%)	3.19 ± 0.01	[41]
Stearic acid (C18:0) (g/100 g) (%)	0.99 ± 0.03	[41]
Oleic acid (C18:1) (g/100 g) (%)	5.43 ± 0.04	[41]

Table 1. *Cont.*

Parameter	Results	Method
Linoleic acid (C18:2) (g/100 g) (%)	7.91 ± 0.15	[41]
Linolenic acid (C18:3) (g/100 g) (%)	0.95 ± 0.06	[41]
Total of free fatty acids (g/100 g) (%)	18.47	[41]

Note 1: Rheometer (Brookfield DV-III Ultra with bath TC-650. Brookfield Brazil. Middleboro. MA. EUA) and spindle SC4–27. program Rheocalc V3.3 Build 49–1.

Regarding the FFAs content, the SODD acidity index (34.19 ± 0.52 mg KOH/g) was lower than those previously reported for deodorizer distillates of soybean oil (107.64 ± 228 mg KOH/g) [31,32], palm oil (191.69 mg KOH/g) [27], rice oil (163.66 ± 0.57 mg KOH/g) [42], and rapeseed oil (97.61 ± 1.87 mg KOH/g) [43]. In general, the content of FFAs of vegetable oil by-products depends on the composition of the original oil, as well as the deodorization conditions; thus, the physical-chemical parameters of deodorizer distillates can vary according to the oilseed [20].

The iodine index is a parameter related to the degree of the oil unsaturation [35]. The value obtained for the SODD (112.98 ± 0.32 gI_2/100 g) was about twice higher than that obtained for palm oil deodorizer distillate (63.8 gI_2/100 g) [27], since soybean oil is rich in polyunsaturated fatty acids, while palm oil is rich in saturated fatty acids. However, for biodiesel production it is recommended an iodine index less than 115 gI_2/100 g [44]. The moisture of SODD (1.33 ± 0.04%) was higher than those of refined oils (less than 0.5%) [45], but this parameter can be corrected for biodiesel production, since the presence of water may favor the hydrolysis of the esters produced, thus reducing the reaction yield. The kinematic viscosity of the SODD at 25 °C (57.10 ± 0.01 mPa.s) was very close to that of refined soybean oil [46] that is consistent with the high content of acylglycerols (MAGs, DAGs and TAGs) in the SODD (~70 wt.%).

The total content of FFAs and acylglycerides was also quantified by gas chromatography in terms of fatty acid methyl esters (FAMEs). For that, all glyceridic matter was previously submitted to an alkaline transesterification with methyl alcohol, converting it totally into FAMEs. Table 1 shows that the SODD is mainly composed of saponifiable matter (85.15 ± 0.53 wt.%), which makes it an excellent raw material to produce biodiesel.

Soybean oil is a good source of tocopherols (α-, β-, γ- and δ-tocopherols) [24], but part of these compounds are lost in the oil deodorization, making the SODD a good source of tocopherols too. The total of tocopherols in the SODD under study (around 1.8 wt.%) was around 10-fold lower than those reported in the literature (16.3–18.2%) [20]. This difference can be attributed to the fact that the degradation of tocopherols occurs quickly [23]. Tocopherols are prone to degradation at alkaline conditions and at high temperature (up to 61% degradation at 300 °C during the oil distillation process) [47].

The FFAs in the SODD were majority composed of linoleic, oleic and palmitic acids, since soybean oil is composed of linoleic acid (51%), oleic acid (23%), palmitic acid (10%), linolenic acid (7–10%) and stearic acid (4%) [48]. However, the total content of FFAs (18.47 wt.%) found in this work was lower than those reported by Kasim et al. [49] and Gunawan et al. [22], 41.15 ± 0.39% and 45.38 ± 2.13%, respectively. However, this parameter depends on the operational conditions of the deodorization process (temperature and pressure).

2.2. SODD Saponification

A cold saponification method was adopted in this study aiming to recovery the saponifiable matter of the SODD. The saponifiable phase (93.03 wt.% saponification index, 2.25 wt.% FFA and 1.33 wt.% tocopherols) was selected to carry out the statistical design to define better reaction conditions to produce FAEEs (discussed below). The saponification method adopted here degraded a large amount of tocopherols, mainly α-tocopherol, as will be discussed below. Tocopherol's degradation during saponification processes was also reported by Maniet et al. [50].

2.3. SODD Hydrolysis Reaction

SODD was hydrolyzed using free Eversa as biocatalyst. As the *Thermomyces lanuginosus* lipase, Eversa (a genetically-modified variant of *Thermomyces lanuginosus* lipase) is a 1,3-specific lipase [1] that mainly releases acyl moieties linked to the sn-1 and sn-3 positions at the glycerol backbone, being the hydrolysis of TAGs to DAGs faster than the hydrolysis of MAGs [51,52].

Figure 1 shows that after 48 h hydrolysis, a FFA yield of 72.48 ± 2.91 wt.% was reached, remaining approximately constant after 72 h reaction (FFA yield of 72.64 ± 3.76 wt.%). This FFA yield represents a reaction conversion of around 80% (based on the initial saponification index, 91.27%). Other works using *Thermomyces lanuginosus* lipase also report hydrolysis conversions in the same magnitude order, viz. 89% [53] and 94% [54], using soybean oil and waste cooking oil, respectively. However, the hydrolysis conversions are not directly comparable because of different reaction conditions. Besides FFAs, the hydrolysis product also contained non-converted acylglycerols, esters and tocopherols, as follows (in wt.%): 8.02 ± 0.05 MAGs, 7.12 ± 0.01 DAGs, 8.62 ± 0.03 TAGs, 3.76 ± 0.01 esters, and 0.28 total tocopherols (0.09 ± 0.004 α-tocopherol, 0.12 ± 0.11 β-tocopherol, 0.04 ± 0.001 δ-tocopherol, and 0.03 ± 0.001 γ-tocopherol).

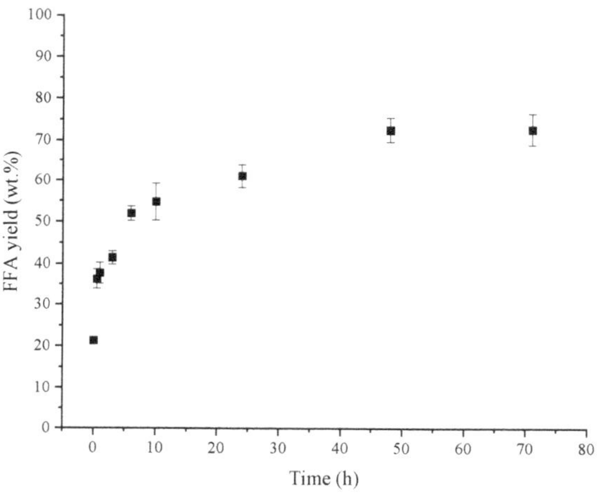

Figure 1. Free fatty acid (FFA) yield profile (wt.%) vs. time for the hydrolysis of SODD by the enzyme Eversa Transform 2.0. Reaction conditions: up to 72 h reaction; 37 °C; 4:1 mass ratio (H_2O:SODD) and 5% enzyme (m/mSODD).

2.4. Simultaneous Transesterification and Esterification Reactions

As commented above, the SODD saponifiable phase (SODD-SP, 93.03% saponification index) was used as substrate in the statistical design assays. The levels of the variables of the statistical design were based on reaction conditions previously reported for the synthesis of biodiesel using Eversa Transform (Table A2 in Appendix A) to establish a range of parameters that encompassed a large part of those works. The response variable was analyzed in terms of the reaction ester yield (FAEEs, mass basis).

The experimental runs resulted ester yields (in wt.%) ranging from 59.30 to 82.06% (Table 2 showing experimental and predict ester yields). Analysis of variance (ANOVA) (Table A3 in Appendix A) allowed determination of the significant parameters. The F-values of model and lack of fit were 24.448 (p-value < 0.0002, 95%-confidence level) and 14.402 (p-value > 0.08) (Table A3 in Appendix A), respectively, indicating that the fitted model (Equation (1)) represents well the behavior of our system (R-squared value of 0.9692). The difference between adjusted and predicted R-squared is recommended to be less than 0.20. The adjusted R-squared and predicted R-squared values were 0.9295 and 0.8890, indicating good representativity of experimental dataset and capability of extrapolating the model, respectively.

Table 2. Experimental design for the influences of three independent variables (real and coded) on the ester yield of the esterification/transesterification reaction with SODD saponifiable phase (SODD-SP) (experimental and predicted values).

Assays	Molar Ratio (Ethanol:SODD-SP)	Enzyme Concentration (wt. %)	Temperature (°C)	Ester Yield Experimental (wt.%)	Ester Yield Predicted (wt.%)
1	2.3:1 (−1)	3 (−1)	30 (−1)	73.00 ± 2.55	72.38
2	3.3:1 (+1)	3 (−1)	30 (−1)	70.77 ± 0.79	70.46
3	2.3:1 (−1)	7 (+1)	30 (−1)	81.91 ± 0.80	82.40
4	3.3:1 (+1)	7 (+1)	30 (−1)	80.10 ± 1.37	80.48
5	2.3:1 (−1)	3 (−1)	40 (+1)	59.30 ± 1.60	61.14
6	3.3:1 (+1)	3 (−1)	40 (+1)	69.36 ± 1.03	69.70
7	2.3:1 (−1)	7 (+1)	40 (+1)	67.57 ± 1.68	67.28
8	3.3:1 (+1)	7 (+1)	40 (+1)	74.40 ± 1.34	75.84
9	1.96:1 (−1.68)	5 (0)	35 (0)	67.98 ± 0.72	69.70
10	3.64:1 (+1.68)	5 (0)	35 (0)	73.90 ± 2.06	75.28
11	2.8:1 (0)	1.64 (−1.68)	35(0)	67.99 ± 4.44	68.36
12	2.8:1 (0)	8.36 (+1.68)	35 (0)	82.06 ± 4.87	81.93
13	2.8:1 (0)	5 (0)	26.6 (−1.68)	72.94 ± 1.61	76.42
14	2.8:1 (0)	5 (0)	43.4 (+1.68)	65.14 ± 4.32	63.08
15	2.8:1 (0)	5 (0)	35 (0)	72.03 ± 0.45	72.49
16	2.8:1 (0)	5 (0)	35 (0)	72.74 ± 1.01	72.49
17	2.8:1 (0)	5 (0)	35 (0)	72.76 ± 0.25	72.49

A second order (Equation (1)) model was fitted to the experimental data of ester yield vs. the coded independent variables X_1 (ethanol:SODD-SP molar ratio), X_2 (enzyme concentration, in wt. %) and X_3 (temperature, in °C):

$$Yield = 72.50 + 1.67X_1 - 0.50X_1^2 + 4.04X_2 + 0.94X_2^2 - 3.67X_3 - 0.98X_3^2 - 0.35X_1X_2 + 2.62X_1X_3 - 0.62X_2X_3 \quad (1)$$

A good agreement was observed between experimental and predicted (from Equation (1)) responses (Figure A1 in Appendix A).

Figure 2 shows the response surfaces constructed from the model (Equation (1)). As expected, both enzyme concentration (more expressive) and molar ratio (ethanol:SODD-SP) influenced positively the ester yield, i.e., an increase in these parameters leads to higher ester yields. In particular, as SODD has a high content of FFAs and its esterification generates water as a by-product, the reaction was favored using an excess of ethanol, which shifts the reaction equilibrium towards the products [17,55]. On the other hand, temperature negatively affected the ester yield, i.e., the reaction was favored at lower temperatures because high temperatures can inactivate the enzyme. These effects and their magnitudes are clearly shown in the Pareto chart (Figure A2 in Appendix A).

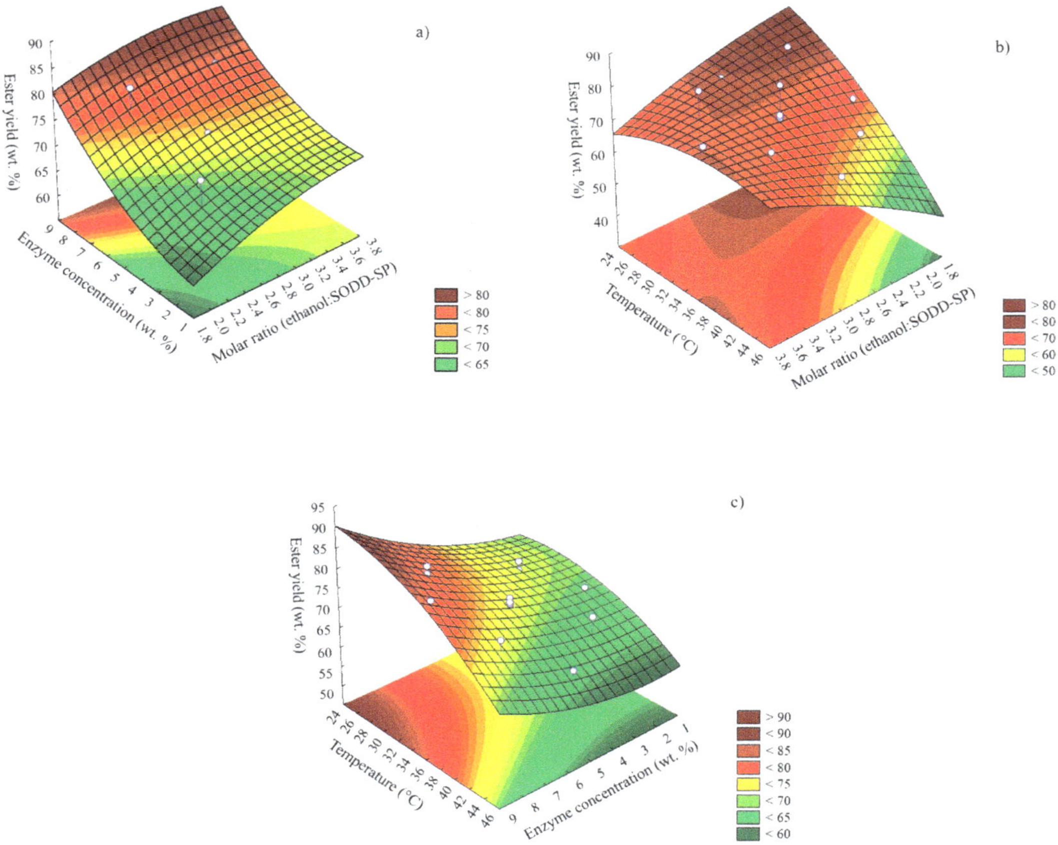

Figure 2. Response surfaces of biodiesel reaction with SODD saponifiable phase (SODD-SP). (**a**) Enzyme concentration vs. ethanol:SODD-SP molar ratio. (**b**) Temperature vs. ethanol:SODD-SP molar ratio. (**c**) Temperature vs. enzyme concentration. Z-axis indicates the ester yield (wt. %).

The highest FAEE yield was reached at 35 °C, 8.36 wt.% enzyme concentration and 3.64:1 molar ratio (ethanol:SODD-SP) (Figure A3 in Appendix A). Under these conditions, the model predicted an ester yield of 83.31 wt.%, which could be experimentally validated in an independent assay (83.25 ± 1.11 wt.% ester yield after 16 h reaction, Figure 3). Figure 3 shows that the reaction equilibrium was reached after 23 h reaction (86.56 wt.% ester yield), after which the ester yield remains practically constant (86.84 wt.% after 48 h reaction). This result is very close to that reported by Wancura et al. [7] (85.08%) using methanol, deacidified cattle tallow, and liquid Eversa.

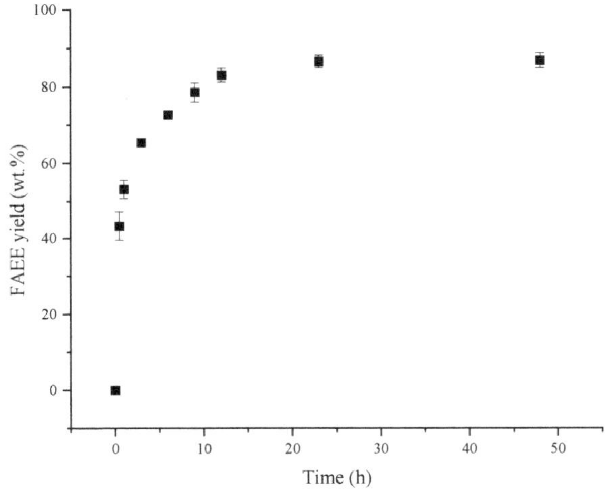

Figure 3. Profile of fatty acid ethyl ester (FAEE) yield (wt.%) vs. time for the esterification/transesterification of SODD saponifiable phase (SODD-SP) (15 g) with ethanol (8.10 g). Reactions conditions: up to 48 h reaction; 35 °C; molar ratio of 3.64:1 (ethanol:SODD-SP) and 8.36 wt.% enzyme concentration.

2.5. Performance of Liquid Eversa in Simultaneous Transesterification and Esterification of Fatty Material with Different Free Acidity

Under the best conditions described above, an experiment was carried out using SODD-SP (2.25% FFAs), crude SODD (17.18% FFAs), and hydrolyzed SODD (72.48% FFAs). Table 3 shows a mass percentage of the main components (FAEEs, FFAs, acylglycerides and tocopherols) for each step. Eversa similarly converted all substrates (with low, medium, and high FFAs content) in FAEEs (86.56, 76.85 and 80.02 wt.% ester yields, respectively); only a slightly higher ester yield (less than 10 wt.%) was observed for a substrate with low free acidity. However, for a highly acidic substrate, the product contained a higher content of FFAs and non-converted acylglycerides than the other substrates (low and medium acidity).

Table 3. Mass percentage of main components of producing FAEEs from low-to-high acidity SODD.

FAEEs Production (Reaction Step)				Caustic Treatment	
Inputs		Outputs		Outputs	
Component Crude SODD	Value (wt.%)	Component Crude FAEEs	Value (wt.%)	Component Crude FAEEs	Value (wt.%)
SI	91.27 ± 0.88	Saponifiable		Saponifiable	
FFAs	17.18 ± 0.26	FAEEs	76.85 ± 0.65	FAEEs	82.42 ± 0.78
Acyl-gly [a]	74.09	FFAs	5.08 ± 0.01	FFAs	1.08 ± 0.08
Toc-total	1.83	Acyl-gly	1.59	Acyl-gly	0.94
α-tocopherol	0.37 ± 0.001	Glycerol	0.04 ± 0.01	Glycerol	0.04 ± 0.004
β-tocopherol	0.09 ± 0.005	MAGs	0.73 ± 0.01	MAGs	0.42 ± 0.0
γ-tocopherol	1.01 ± 0.003	DAGs	0.79 ± 0.09	DAGs	0.45 ± 0.05
δ-tocopherol	0.36 ± 0.003	TAGs	0.03 ± 0.01	TAGs	0.03 ± 0.003
		Toc-total	1.39	Toc-total	0.41
		α-tocopherol	0.06 ± 0.01	α-tocopherol	0.002 ± 0.001
		β-tocopherol	0.07 ± 0.001	β-tocopherol	0.005 ± 0.001
		γ-tocopherol	0.99 ± 0.01	γ-tocopherol	0.19 ± 0.003
		δ-tocopherol	0.27 ± 0.001	δ-tocopherol	0.22 ± 0.01
SODD–SP		Crude FAEEs		Crude FAEEs	
SI	93.03 ± 1.08	Saponifiable		Saponifiable	

Table 3. *Cont.*

FAEEs Production (Reaction Step)				Caustic Treatment	
Inputs		Outputs		Outputs	
Component Crude SODD	Value (wt.%)	Component Crude FAEEs	Value (wt.%)	Component Crude FAEEs	Value (wt.%)
FFAs	2.25 ± 0.13	FAEEs	86.56 ± 0.31	FAEEs	90.83 ± 0.82
Acyl-gly [a]	90.78	FFAs	2.77 ± 0.08	FFAs	0.82 ± 0.08
Toc-total	1.33	Acyl-gly	1.89	Acyl-gly	1.03
α-tocopherol	0.34 ± 0.01	Glycerol	0.05 ± 0.01	Glycerol	N.d.
β-tocopherol	0.01 ± 0.003	MAGs	0.90 ± 0.05	MAGs	0.50 ± 0.02
γ-tocopherol	0.84 ± 0.01	DAGs	0.87 ± 0.23	DAGs	0.53 ± 0.09
δ-tocopherol	0.14 ± 0.001	TAGs	0.07 ± 0.03	TAGs	N.d.
		Toc-total	1.01	Toc-total	0.67
		α-tocopherol	0.20 ± 0.05	α-tocopherol	0.06 ± 0.02
		β-tocopherol	0.01 ± 0.001	β-tocopherol	0.01 ± 0.002
		γ-tocopherol	0.67 ± 0.02	γ-tocopherol	0.49 ± 0.03
		δ-tocopherol	0.13 ± 0.001	δ-tocopherol	0.11 ± 0.002
Hydrolyzed SODD		Crude FAEEs		Crude FAEEs	
SI [b]	96.24	Saponifiable		Saponifiable	
FFAs	72.48 ± 2.91	FAEEs	80.02 ± 0.11	FAEEs	88.83 ± 0.85
Acyl-gly	23.76	FFAs	8.07 ± 0.19	FFAs [c]	9.16 ± 0.001
Glycerol	N.d.	Acyl-gly	11.91	Acyl-gly	1.82
MAGs	8.02 ± 0.05	Glycerol	N.d.	Glycerol	0.73 ± 0.08
DAGs	7.12 ± 0.01	MAGs	2.41 ± 0.09	MAGs	1.09 ± 0.07
TAGs	8.62 ± 0.03	DAGs	7.64 ± 0.01	DAGs	N.d.
Toc-total	0.41	TAGs	1.86 ± 0.01	TAGs	N.d.
α-tocopherol	0.10 ± 0.003	Toc-total	0.21	Toc-total	0.29
β-tocopherol	0.10 ± 0.08	α-tocopherol	0.01 ± 0.005	α-tocopherol	0.06 ± 0.05
γ-tocopherol	0.06 ± 0.01	β-tocopherol	0.07 ± 0.003	β-tocopherol	0.07 ± 0.004
δ-tocopherol	0.15 ± 0.02	γ-tocopherol	0.03 ± 0.04	γ-tocopherol	0.04 ± 0.02
		δ-tocopherol	0.10 ± 0.005	δ-tocopherol	0.12 ± 0.01

SODD–soybean oil deodorizer distillate; FAEEs–fatty acid ethyl esters (by gas chromatography using EN-14103 method [39]); FFAs–free fatty acids (by gas chromatography using Agilent method [41]); SI–saponification index (by AOCS method Cd 3–25) [36]; Toc-total–sum of α-, β-, γ-, and δ-tocopherol (by liquid chromatography, AOCS method Ce 8–89 [40]); SODD-SP–SODD saponifiable phase; Acyl-gly–sum of monoglycerides, diglycerides, triglycerides, and free glycerol (by gas chromatography using ASTM D 6584 method [56]); [a] values indirectly calculated as (SI–FFAs); [b] values indirectly calculated as FFAs+total acylglycerides; [c] measured by AOCS method Ca 5a-40 [34]; reaction conditions: 24 h reaction; 35 °C; ethanol:SODD-SP molar ratio of 3.64:1, 8.36 wt.% enzyme, and 6.74 g of molecular sieves (only for FAEEs production from hydrolyzed SODD).

In general, even using raw material with different acidities, the Eversa performance was very close. In relation to the initial saponifiable matter (91–96 wt.% saponification index), the conversion of saponifiable matter to FAEEs was up to around 90%, showing the good performance of this enzyme for lowly and highly-acid raw materials, as already demonstrated with other fatty materials [1,4,7,8,10,11,13,14,19,55,57–59]. For example, Miranda et al., [5] reported FAEE yields around 90 wt.% after 48 h of reaction in the presence of 6.0% water, using liquid Eversa and refined soybean oil. After a caustic treatment, the ester yield could be increased to 98.2 wt.%. Other authors, using the same enzyme and methanol as acyl acceptor, obtained similar yields: 96% [8], 96.7 % [19], and 97.5 % [1].

Regarding the SODD as acyl donor, Facioli and Barrera-Arellano [30] reported 88% conversion of SODD to FAEEs; however, the enzyme used was the *Mucor miehei* immobilized lipase (Lipozyme[IM]). Wang et al., [29] and Du, Wang and Liu [28] reported ester yields of 97% and 95% (calculated as the percentage of methyl esters measured in relation to the theoretical methyl esters amount), respectively, using SODD, Novozym 435 (immobilized *Candida antarctica* lipase B) and methanol as acyl acceptor. This brief review shows that our results are very closed to those previously reported using other systems acyl donors-acyl acceptors-enzyme.

As an attempt to reduce residual FFAs in our product, a caustic treatment of the product was used. For the product from SODD-SP, the FAEEs content increased to 90.83 wt.%

and the non-converted acylglycerides and FFAs decreased to 1.03 and 0.82 wt.% (Table 3), respectively. Although our final product has still not met the values recommend to biodiesel for some parameters (such as FAEEs, min. 96.5 wt.%) [60], a deep study focused on this matter could adjust it as a biofuel and even recover other value-added compounds. A deep study of separating FAEEs, FFAs, tocopherols, and other compounds, as well as the economic analysis of these processes is in progress in our group.

3. Materials and Methods

3.1. Materials

Soybean oil deodorizer distillate (SODD) was supplied by COCAMAR (Maringá, PR, Brazil). Eversa® Transform 2.0 (Novozymes A/S, Bagsværd, DK), chromatography standards (α-, β-, γ- and δ-tocopherols, methyl heptadecanoate, monoolein, diolien, triolein, butanethiol, tricarpine, free fatty acids) and N-methyl-N-(trimethylsilyl) trifluoroacetamide (MSTFA) were purchased from Sigma Chem. Co. (St. Louis, MO, USA). Molecular sieves (3Å) were obtained from JT Baker (New Jersey, NJ, USA). All other chemicals were analytical grade and were used as received.

3.2. Characterization of the SODD

SODD was characterized in terms of acidity index by AOCS method Ca 5–40 [34], iodine index by AOCS method Cd 1–25 [35], saponification index by the AOCS method Cd 3–25 [36], density by AOCS method Cc 10a-25 [61] and humidity by the AOCS method Ca 2b-38 [37]. Viscosity at 25 and 40 °C was measured in a Brookfield Rheometer (Brookfield DV-III Ultra with TC-650 bath, Brookfield Brazil, Rio de Janeiro, RJ, Brazil) with a SC4–27 spindle.

3.3. SODD Saponification

The saponifiable matter from SODD was obtained by a cold saponification reaction according to the AOCS method Ca 6a-40 [62] with adaptations. SODD (100 mL), 10% alcohol solution (600 mL) and 50% KOH solution (100 mL) were added in an Erlenmeyer flask, and the mixture was stirred (in a magnetic stirrer) for 1 h at room temperature. Afterwards, 55 mL of ethyl ether were added, and the material was transferred to a separation funnel. Both phases were titrated with 0.1 M HCl for neutralization. The saponifiable phase was washed twice with hot distilled water (volume ratio 1:1), dried overnight in an oven at 60 °C, and used for enzymatic esterification/transesterification.

3.4. SODD Hydrolysis Reaction

The hydrolysis of SODD saponifiable matter was carried out at 37 °C in a thermostatically controlled closed reactor with mechanical stirring. The reaction medium was composed of 50 g of SODD, 200 g of distilled water (1:4 SODD/water mass ratio) and 2.5 g of free Eversa (5%, m/m_{SODD}). The reaction was monitored by measuring free fatty acids (FFAs) released in the reaction medium by gas chromatography [41]. For that, samples were withdrawn, immediately cooled in an ice bath, and centrifuged. The light phase (oily phase) was washed twice with hot distilled water (1:1 volume ratio), dried overnight in an oven at 60 °C, and used for FFAs analysis by gas chromatography. The heavy phase containing the enzyme was discarded. At the end of the reaction, the reaction product was washed twice with hot distilled water (1:1 volume ratio), dried overnight in an oven at 60 °C, and used for simultaneous enzymatic esterification and transesterification of FFAs and non-hydrolyzed acylglycerides (MAGs, DAGs, and TAGs).

3.5. Esterification/Transesterification Reaction Using SODD Saponifiable Phase (SODD-SP) and Ethanol

Firstly, a statistical design was performed to define the reaction conditions: ethanol: SODD-SP molar ratio, enzyme concentration and temperature (Table 4). The ethanol:SODD-

SP molar ratio (in terms of saponifiable material) was calculated based on the saponification index (SI) using Equations (2) and (3):

$$\text{SODD}-\text{SP (mol)} = m_{\text{SODD}-\text{SP}}(\text{g}) \times SI\left(\frac{\text{mg}_{\text{KOH}}}{\text{g}_{\text{SODD}-\text{SP}}}\right) \times \frac{1}{MM_{\text{KOH}}}\left(\frac{\text{mol}}{\text{g}}\right) \times 10^{-3}\left(\frac{\text{g}}{\text{mg}}\right) \quad (2)$$

$$\text{Ethanol : SODD}-\text{SP molar ratio}\left(\frac{\text{mol}_{\text{ethanol}}}{\text{mol}_{\text{SODD}-\text{SP}}}\right) = m_{\text{ethanol}}\ (\text{g}) \times \frac{1}{MM_{\text{ethanol}}}\left(\frac{\text{mol}}{\text{g}}\right) \times \frac{1}{\text{mol}_{\text{SODD}-\text{SP}}} \quad (3)$$

Table 4. Coded values of the input variables for statistical design.

Variables		−1.68	−1	0	+1	+1.68
Molar ratio (ethanol:SODD-SP)	X1	1.96:1	2.3:1	2.8:1	3.3:1	3.64:1
Enzyme concentration (wt.%)	X2	1.64	3	5	7	8.36
Temperature (°C)	X3	26.6	30	35	40	43.4

The data were analyzed and represented graphically using the software Statistica version 7.0 (Stat Soft), with a significance level of $\alpha = 0.05$. In the optimization stage, all tests were performed in closed flasks in an orbital shaker (Model MA832, Marconi, Piracicaba, SP, Brazil) at 250 rpm for 16 h. After defined the best conditions, the esterification/transesterification reactions were conducted in a batch reactor (50 mL working volume, thermostated and mechanically stirred at 2000 rpm) to construct the ester yield profile with the time (0–48 h) using SODD-SP. Samples were withdrawn, immediately cooled in an ice bath, and centrifuged. The light phase (oily phase) was washed twice with hot distilled water (1:1 volume ratio), dried overnight in an oven at 60 °C, and used for chromatography analyses (FFAs, FAEEs, MAGs, DAGs, TAGs, and free glycerol). The heavy phase containing the enzyme was discarded.

3.6. Caustic Treatment

A volume of 4% NaOH solution (*w/v*) was added to the esterification/transesterification product to reach 1.15 moles of base per mole of residual FFAs. The reaction mixture was stirred in a shaker (SL–222, Solab, Piracicaba, SP, Brazil) at 60 °C, 60 rpm, for 1 h. Then, the mixture was decanted for 10 min at 60 °C, and centrifuged at 8000 rpm for 10 min at 25 °C. The light phase (upper oily phase) was recovered, washed twice with hot distilled water (volume ratio 1:1), and dried overnight in an oven at 60 °C.

3.7. Tocopherol Quantification by Liquid Chromatography

Tocopherols were analyzed according to the AOCS method Ce 8–89 [40] with adaptations. The liquid chromatography system was a Waters E2695 chromatograph (Waters Co., Milford, MA, USA) equipped with UV detector (Photodiode Array Detector, Waters Co., Milford, MA, USA). The chromatographic separation was performed in a Luna® Silica 100 column (250 × 4.6 mm × 5 μm, Phenomenex INC., Torrance, CA, USA) at room temperature. The mobile phase was a mixture of n-hexane:isopropanol (98:2, *v/v*) at a flow rate of 1 mL/min, 20 μL injection volume, 12 min analysis time.

3.8. Quantification of Esters by Gas Chromatography

The yield of FAEEs (in wt.%) was determined by gas chromatography according to EN-14103 method [39], with modifications. An Agilent chromatograph (7890A, Agilent Technologies, Santa Clara, CA, USA) was used, equipped with a flame ionization detector (FID-250 °C) and a Rtx-Wax column (30 m × 0.25 mm × 0.25 μm, Restek Corporation, Bellefonte, PA, USA) at a temperature of 210 °C, with helium as carrier gas and methyl heptadecanoate as an internal standard. The samples were centrifuged at 9000 rpm for 10 min at 5 °C, the light phase was washed with hot distilled water and centrifuged (three times), and dried overnight in an over at 60 °C. For quantification, 50 mg of sample were diluted in 1 mL of methyl heptadecanoate solution (10 mg/mL, in heptane) and 1 μL was injected in the equipment.

3.9. Quantification of Glycerol, Triglycerides (TAGs), Diglycerides (DAGs), and Monoglycerides (MAGs) by Gas Chromatography

The content of free glycerol, TAGs, DAGs, and MAGs (in wt.%) was determined by gas chromatography in an Agilent chromatograph (7890A, Agilent Technologies, Santa Clara, CA, USA) equipped with a Select Biodiesel column (glycerides, UM + 2 m RG, 15 m × 0.32 mm × 0.1 μm, Agilent Technologies, Santa Clara, CA, USA) and a flame ionization detector. The temperature ramp was 50 °C for 1 min, heating to 180 °C at 15 °C/min, 230 °C at 7 °C/min and 380 °C at 10 °C/min, maintained for 10 min. The detector temperature was 380 °C and helium was used as the carrier gas. The calibration curves were constructed with diolein, monoolein and triolein standards, butanethiol and tricaprine as internal standards, and N-methyl-N-(trimethylsilyl) trifluoroacetamide (MSTFA) as derivatization reagent. Sample preparation, analysis and quantification were performed according to the ASTM D 6584 method [56].

3.10. Quantification of Free Fatty Acids (FFAs) by Gas Chromatography

FFAs were quantified by gas chromatography according to methodology adapted from the Agilent Catalog [41]. A gas chromatograph (7890A, Agilent Technologies, Santa Clara, CA, USA) was used, equipped with a flame ionization detector set at 250 °C, a split-splitless injector (250 °C, split ratio 40:1) and an Rtx-WAX column (30 m × 0.25 mm × 0.25 μm, Restek Corporation, Bellefonte, PA, USA). The oven temperature was set at 120 °C for 1 min, heating to 250 °C at 10 °C/min, and 250 °C for 5 min. Helium was used as carrier gas (42 cm/s, 24 psi at 120 °C, 1.8 mL/min). The samples were dissolved in dichloromethane at a concentration of 0.016 g/mL, and the standards were prepared in five different concentrations (0.25, 0.5, 1.0, 1.5 and 2.0 g/L) to adjust the calibration curve.

4. Conclusions

This study showed that liquid Eversa is a versatile lipase, serving multi-purposes, namely hydrolysis and simultaneous esterification and transesterification of feedstocks with low-to-high acidity. Besides that, soybean oil deodorizer distillate was shown to be an interesting source of unsaturated fatty acids (oleic, linoleic and linolenic acids), as well as a potential feedstock to produce fatty acid ethyl esters. Under the adopted conditions, products with a high content of free fatty acids (72.5 wt.%, by enzymatic hydrolysis) and fatty acid ethyl esters (up to 90 wt.%, by simultaneous esterification and transesterification, followed by caustic treatment) could be prepared.

Author Contributions: Conceptualization, P.W.T. and R.F.-L.; methodology, A.C.V.; A.B.M.C. and J.R.G.; formal analysis, P.W.T.; R.F.-L. and A.M.S.V.; investigation, A.C.V.; A.B.M.C. and J.R.G.; writing—original draft preparation, A.C.V. and A.B.M.C.; writing—review and editing, P.W.T.; R.F.-L.; supervision, P.W.T.; project administration, P.W.T. and A.M.S.V.; funding acquisition, P.W.T. and R.F.-L. All authors have read and agreed to the published version of the manuscript.

Funding: This research was funded by São Paulo Research Foundation (FAPESP), grant number 2016/10636–8; in part by Coordenação de Aperfeiçoamento de Pessoal de Nível Superior (CAPES), Finance Code 001; Conselho Nacional de Desenvolvimento Científico e Tecnológico (CNPq), grant number 315092/2020-3; Ministerio de Ciencia e Innovación (Spanish Government), grant number CTQ2017-86170-R; and Consejo Superior de Investigaciones Científicas (CSIC), grant number AEP045.

Acknowledgments: The authors thank COCAMAR (Maringá, PR, Brazil) for providing the soybean oil deodorizer distillate (SODD).

Conflicts of Interest: The authors declare no conflict of interest.

Appendix A

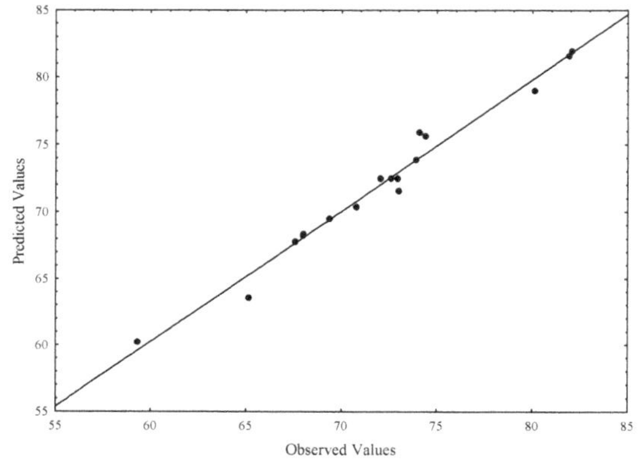

Figure A1. Correlation between experimental (**observed values**) and fitted (**predicted values**) ester yields (wt.%) of the esterification/transesterification with SODD saponifiable phase and ethanol using Eversa Transform 2.0 as biocatalyst.

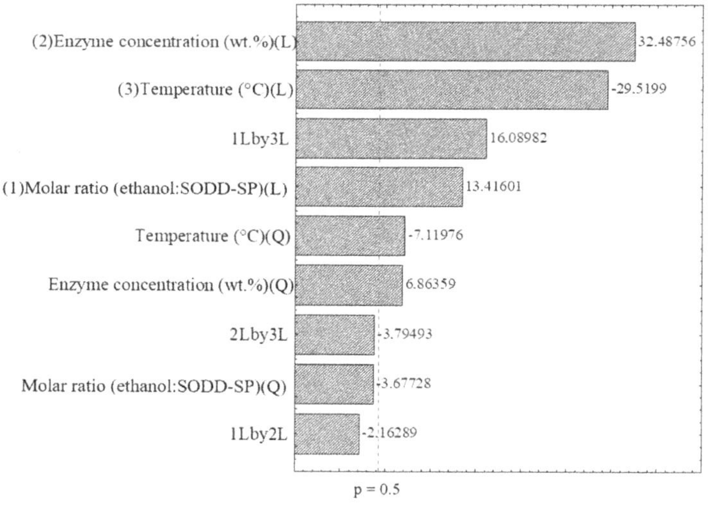

Figure A2. Pareto chart with the effects of the parameters and their interactions for the experimental design of the esterification/transesterification with SODD saponifiable phase and ethanol using Eversa Transform 2.0 as biocatalyst.

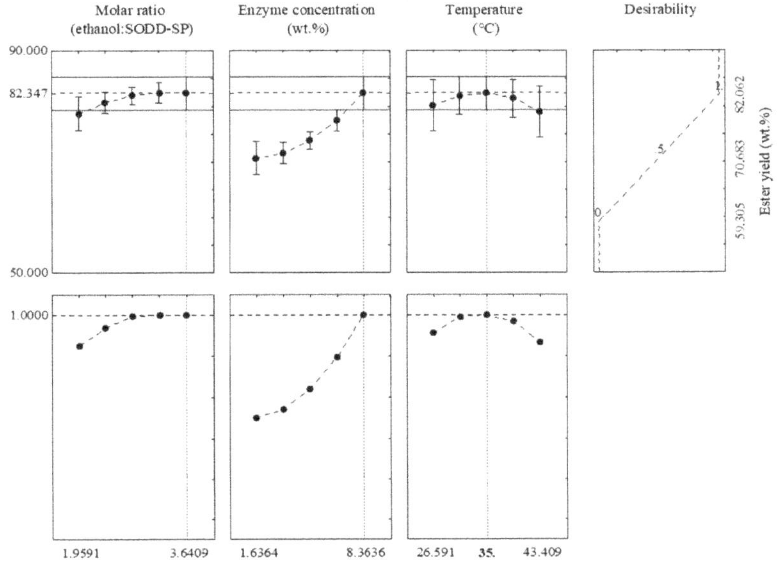

Figure A3. Profiles for predicted values and desirability of the esterification/transesterification with SODD saponifiable phase and ethanol using Eversa Transform 2.0 as biocatalyst.

Table A1. Studies of biodiesel production using Eversa® Transform lipase (liquid or immobilized) as biocatalyst.

Acyl Donor	Acyl Acceptor	Reaction Conditions	Biocatalyst Form	Yield (%)	Reference
Soy oil	Ethanol	6:1 alcohol: oil (molar ratio) 12 U esterification/g oi 40 °C, 1500–1700 rpm, and 24 h	Magnetic CLEAs [a]	98.9	[5]
Glyceril trioleate	Fusel oil	5:1 alcohol:oil (molar ratio) 2 wt.% enzyme 35 °C, 250 rpm, and 24 h	Liquid	>97%	[6]
Deacidified cattle tallow	Methanol	4.5:1 alcohol: oil (molar ratio) 1.0 wt.% enzyme 35 °C, 300 rpm, 6.0 wt.% water, and 8 h	Liquid	85.08	[7]
Soy oil	Methanol	550 kg oil and 2.2 kg methanol 0.2 wt.% enzyme, 45 °C, 20.0 wt% water, 100 ppm NaOH, and 24 h	Liquid	96	[8]
Castor oil	Methanol	6:1 alcohol: oil (molar ratio) 5.0 wt.% enzyme, 35 °C, 5.0 wt.% water, and 8 h	Liquid	83	[9]
Residual oil from a poultry industry	Methanol	100 g oil and 1.5 eqv. alcohol 0.3 wt.% enzyme, 45 °C, 1.5 wt.% water, 250 rpm, and 24 h	Liquid (NS 40116 trademark)	90.61	[10]
Cotton seed oil	Methanol	6:1 alcohol: oil (molar ratio) 5 wt.% enzyme, 35 °C, 6 wt.% water, 250 rpm, and24 h	Liquid	98.5	[11]

Table A1. *Cont.*

Acyl Donor	Acyl Acceptor	Reaction Conditions	Biocatalyst Form	Yield (%)	Reference
Sunflower oil	Ethanol	1:4 alcohol: oil (molar ratio) 4.1 mL hexane, 10 wt.% enzyme, 40 °C, 150 rpm, and 3 h	Immobilized on Sepabeads	99	[12]
Oleic acid	Methanol	3.44:1 alcohol:acid (molar ratio) 11.98% enzyme, 35.25 °C, and 2.5 h	Liquid	96.73	[13]
CTO [b] of the kraft pulping process	Methanol	1.5:1 alcohol: oil (molar ratio) 1 wt.% enzyme, 500 rpm, 40 °C, and 16 h	Liquid	96.57	[14]
Castor oil	Methanol	6:1 alcohol: oil (molar ratio) 5 wt.% enzyme, 5 wt.% water, 750 rpm, 35 °C, and 8 h	Liquid	94.21	[15]
Castor oil	Methanol	6:1 alcohol: oil (molar ratio) 5 w.t% enzyme, 5 wt.% water, 750 rpm, 35 °C, and 8 h	Liquid	Not informed	[16]
Bleached sardine oil	Ethanol	8:1 alcohol: oil (molar ratio) 60 U enzyme, 10 wt.% water, 25 °C, and 4 h	Liquid	93.98	[17]
Castor oil	Methanol	6:1 alcohol: oil (molar ratio) 10 wt.% enzyme, 750 rpm, 35 °C, and 8 h	Liquid	94.21	[18]
Soy oil	Methanol	1.5 eqv. methanol 1 wt.% enzyme, 2.5 wt.% water, 250 rpm, 35 °C, and 16 h	Liquid	96.7	[19]
Soy oil	Methanol	1.5 eqv. methanol 0.2 wt.% enzyme, 3 wt.% water, 500 rpm, 35 °C, and 24 h	Liquid	97.5	[1]
SODD-SP [c]	Ethanol	3.64:1 ethanol:SODD-SP (molar ratio) 8.36 wt.% enzyme 35 °C and 48 h	Liquid	90.83	This study

[a] Crosslinked enzyme aggregates, [b] Crude tall oil, [c] Soybean oil deodorizer distillate saponifiable phase.

Table A2. Research focused on the production of biodiesel by the enzymatic route from deodorizer distillates (DD) of vegetable oils.

DD Source Oil	Alcohol	Reaction Conditions	Biocatalyst	Yield (%)	Reference
Soy	Methanol	Methanol:DD 2.3:1 (molar ratio), 53.6 °C, and 2 h	Lipozyme IM [a]	88	[30]
Soy	Methanol	Methanol:DD 3.6:1 (molar ratio), 40 °C, and 24 h	Novozym 435 [b]	97	[29]
Soy	Methanol	Methanol:DD 3.9:1 (molar ratio), 40 °C, and 24 h	Novozym 435	95	[28]
Palm	Ethanol Methanol	2 g of alcohol added in two steps to 8 g of DD, 60 °C, and 2.5 h	Novozym 435 Lipozyme RM-IM [c] Lipozyme TL-IM [d]	93	[27]

Table A2. *Cont.*

DD Source Oil	Alcohol	Reaction Conditions	Biocatalyst	Yield (%)	Reference
Rapeseed	Ethanol	Ethanol:DD 4:1 (molar ratio), 40 °C, and 30 h	Lipase from *Rhizopus oryzae* immobilized on hydrophobic macroporous resin NKA [e]	98.23	[26]
Rapeseed	Methanol	Methanol:DD 167 μL:2 g, 34 °C, and 6 h	Lipase from *Rhizopus oryzae*	98.16	[25]

[a] Lipozyme IM - *Mucor miehei* lipase immobilized on a macroporous ion exchange resin, [b] *Candida antarctica* lipase immobilized on acrylic resin, [c] Lipozyme RM-IM-*Rhizomucor miehei* lipase immobilized on a macroporous ion exchange resin, [d] Lipozyme TL-IM-*Thermomyces lanuginosus* lipase immobilized on a macroporous ion exchange resin, [e] NKA-Neurokinin A.

Table A3. Analysis of variance (ANOVA) for ester yield (wt. %) from SODD saponifiable phase (SODD-SP) (response variable) as function of the independent variables (ethanol:SODD-SP molar ratio, enzyme concentration and temperature).

Factor	SS *	DF *	MS *	$F_{Calculatted}$	*p*-Value
Model	527.9898	9	58.6655	24.4483	0.000176
(1) Molar ratio (ethanol:SODD-SP)(L)	38.0718	1	38.0718	179.989	0.005510
Molar ratio (ethanol:SODD-SP)(Q)	2.8603	1	2.8603	13.522	0.066644
(2) Enzyme concentration (wt.%)(L)	223.2497	1	223.2497	1055.441	0.000946
Enzyme concentration (wt.%)(Q)	9.9646	1	9.9646	47.109	0.020575
(3) Temperature (°C)(L)	184.3259	1	184.3259	871.424	0.001146
Temperature (°C)(Q)	10.7223	1	10.7223	50.691	0.019162
1L by 2L	0.9895	1	0.9895	4.678	0.163034
1L by 3L	54.7595	1	54.7595	258.882	0.003841
2L by 3L	3.0462	1	3.0462	14.402	0.062951
Lack of Fit	12.0879	5	2.4176	11.429	0.082403
Pure Error	0.4230	2	0.2115		
Total SS	549.5108	16			

* DF: degree of freedom; SS: sum of squares; MS: mean square; R-squared = 0.9692; adjusted R-squared = 0.9295; predicted R-squared = 0.8890.

References

1. Nielsen, P.M.; Rancke-Madsen, A.; Holm, H.C.; Burton, R. Production of biodiesel using liquid lipase formulations. *J. Am. Oil Chem. Soc.* **2016**, *93*, 905–910. [CrossRef]
2. Lv, L.; Dai, L.; Du, W.; Liu, D. Effect of water on lipase NS81006-catalyzed alcoholysis for biodiesel production. *Process Biochem.* **2017**, *58*, 239–244. [CrossRef]
3. Pedersen, A.T.; Nordblad, M.; Nielsen, P.M.; Woodley, J.M. Batch production of FAEE-biodiesel using a liquid lipase formulation. *J. Mol. Catal. B Enzym.* **2014**, *105*, 89–94. [CrossRef]
4. Monteiro, R.R.C.; Arana-Peña, S.; da Rocha, T.N.; Miranda, L.P.; Berenguer-Murcia, Á.; Tardioli, P.W.; dos Santos, J.C.S.; Fernandez-Lafuente, R. Liquid lipase preparations designed for industrial production of biodiesel. Is it really an optimal solution? *Renew. Energy* **2021**, *164*, 1566–1587. [CrossRef]
5. Miranda, L.P.; Guimarães, J.R.; Giordano, R.C.; Fernandez-Lafuente, R.; Tardioli, P.W. Composites of crosslinked aggregates of Eversa® transform and magnetic nanoparticles. Performance in the ethanolysis of soybean oil. *Catalysts* **2020**, *10*, 817. [CrossRef]
6. Monroe, E.; Shinde, S.; Carlson, J.S.; Eckles, T.P.; Liu, F.; Varman, A.M.; George, A.; Davis, R.W. Superior performance biodiesel from biomass-derived fusel alcohols and low grade oils: Fatty acid fusel esters (FAFE). *Fuel* **2020**, *268*, 117408. [CrossRef]
7. Wancura, J.H.C.; Rosset, D.V.; Tres, M.V.; Oliveira, J.V.; Mazutti, M.A.; Jahn, S.L. Production of biodiesel catalyzed by lipase from *Thermomyces lanuginosus* in its soluble form. *Can. J. Chem. Eng.* **2018**, *96*, 2361–2368. [CrossRef]
8. Mibielli, G.M.; Fagundes, A.P.; Bender, J.P.; Oliveira, J.V. Lab and pilot plant FAME production through enzyme-catalyzed reaction of low-cost feedstocks. *Bioresour. Technol. Rep.* **2019**, *5*, 150–156. [CrossRef]
9. Andrade, T.A.; Errico, M.; Christensen, K.V. Investigation of the use of ceramic membranes in recovering liquid enzymes for castor oil transesterification. *Chem. Eng. Trans.* **2019**, *74*, 769–774. [CrossRef]
10. Coppini, M.; Magro, J.D.; Martello, R.; Valério, A.; Zenevicz, M.C.; De Oliveira, D.; Oliveira, J.V. Production of methyl esters by enzymatic hydroesterification of chicken fat industrial residue. *Braz. J. Chem. Eng.* **2019**, *36*, 923–928. [CrossRef]
11. Anderson, S.; Walker, T.; Moser, B.; Drapcho, C.; Zheng, Y.; Bridges, W. Evaluation of dominant parameters in lipase transesterification of cottonseed oil. *Trans. ASABE* **2019**, *62*, 467–474. [CrossRef]

12. Remonatto, D.; de Oliveira, J.V.; Guisan, J.M.; de Oliveira, D.; Ninow, J.; Fernandez-Lorente, G. Production of FAME and FAEE via alcoholysis of sunflower oil by Eversa lipases immobilized on hydrophobic supports. *Appl. Biochem. Biotechnol.* **2018**, *185*, 705–716. [CrossRef] [PubMed]
13. Nguyen, H.C.; Huong, D.T.M.; Juan, H.-Y.; Su, C.-H.; Chien, C.-C. Liquid lipase-catalyzed esterification of oleic acid with methanol for biodiesel production in the presence of superabsorbent polymer: Optimization by using response surface methodology. *Energies* **2018**, *11*, 1085. [CrossRef]
14. Adewale, P.; Vithanage, L.N.; Christopher, L. Optimization of enzyme-catalyzed biodiesel production from crude tall oil using Taguchi method. *Energy Convers. Manag.* **2017**, *154*, 81–91. [CrossRef]
15. Andrade, T.A.; Errico, M.; Christensen, K.V. Transesterification of castor oil catalyzed by liquid enzymes: Optimization of reaction conditions. *Comput. Aided Chem. Eng.* **2017**, *40*, 2863–2868. [CrossRef]
16. Andrade, T.A.; Errico, M.; Christensen, K.V. Evaluation of reaction mechanisms and kinetic parameters for the transesterification of castor oil by liquid enzymes. *Ind. Eng. Chem. Res.* **2017**, *56*, 9478–9488. [CrossRef]
17. He, Y.; Li, J.; Kodali, S.; Balle, T.; Chen, B.; Guo, Z. Liquid lipases for enzymatic concentration of n-3 polyunsaturated fatty acids in monoacylglycerols via ethanolysis: Catalytic specificity and parameterization. *Bioresour. Technol.* **2017**, *224*, 445–456. [CrossRef]
18. Andrade, T.A.; Errico, M.; Christensen, K.V. Castor oil transesterification catalysed by liquid enzymes: Feasibility of reuse under various reaction conditions. *Chem. Eng. Trans.* **2017**, *57*, 913–918. [CrossRef]
19. Remonatto, D.; Santin, C.M.T.; de Oliveira, D.; Di Luccio, M.; de Oliveira, J.V. FAME production from waste oils through commercial soluble lipase Eversa® catalysis. *Ind. Biotechnol.* **2016**, *12*, 254–262. [CrossRef]
20. Sherazi, T.H.S.; Mahesar, A.S. Vegetable oil deodorizer distillate: A rich source of the natural bioactive components. *J. Oleo Sci.* **2016**, *65*, 957–966. [CrossRef]
21. Naz, S.; Tufail, S.; Sherazi, H.A.; Talpur, F.N.; Mahesar, S.A.; Kara, H. Rapid determination of free fatty acid content in waste deodorizer distillates using single bounce-attenuated total reflectance-FTIR spectroscopy. *J. AOAC Int.* **2012**, *95*, 1570–1573. [CrossRef] [PubMed]
22. Gunawan, S.; Kasim, N.S.; Ju, Y.H. Separation and purification of squalene from soybean oil deodorizer distillate. *Sep. Purif. Technol.* **2008**, *60*, 128–135. [CrossRef]
23. Saini, R.K.; Keum, Y.S. Tocopherols and tocotrienols in plants and their products: A review on methods of extraction, chromatographic separation, and detection. *Food Res. Int.* **2016**, *82*, 59–70. [CrossRef]
24. De Camargo, A.C.; Franchin, M.; Shahidi, F. Tocopherols and tocotrienols: Sources, analytical methods, and effects in food and biological systems. *Encycl. Food Chem.* **2019**, 561–570. [CrossRef]
25. Zeng, L.; He, Y.; Jiao, L.; Li, K.; Yan, Y. Preparation of biodiesel with liquid synergetic lipases from rapeseed oil deodorizer distillate. *Appl. Biochem. Biotechnol.* **2017**, *183*, 778–791. [CrossRef] [PubMed]
26. Su, F.; Li, G.; Zhang, H.; Yan, Y. Enhanced performance of *Rhizopus oryzae* lipase immobilized on hydrophobic carriers and its application in biorefinery of rapeseed oil deodorizer distillate. *Bioenergy Res.* **2014**, *7*, 935–945. [CrossRef]
27. Corrêa, I.N.S.; Lorena de Souza, S.; Catran, M.; Bernardes, O.L.; Portilho, M.F.; Langone, M.A.P. Enzymatic biodiesel synthesis using a byproduct obtained from palm oil refining. *Enzyme Res.* **2011**, *2011*, 1–8. [CrossRef]
28. Du, W.; Wang, L.; Liu, D. Improved methanol tolerance during Novozym 435-mediated methanolysis of SODD for biodiesel production. *Green Chem.* **2007**, *9*, 173–176. [CrossRef]
29. Wang, L.; Du, W.; Liu, D.; Li, L.; Dai, N. Lipase-catalyzed biodiesel production from soybean oil deodorizer distillate with absorbent present in tert-butanol system. *J. Mol. Catal. B Enzym.* **2006**, *43*, 29–32. [CrossRef]
30. Facioli, N.L.; Barrera-Arellano, D. Optimisation of enzymatic esterification of soybean oil deodoriser distillate. *J. Sci. Food Agric.* **2001**, *81*, 1193–1198. [CrossRef]
31. Yin, X.; You, Q.; Ma, H.; Dai, C.; Zhang, H.; Li, K.; Li, Y. Biodiesel production from soybean oil deodorizer distillate enhanced by counter-current pulsed ultrasound. *Ultrason. Sonochem.* **2015**, *23*, 53–58. [CrossRef]
32. Yin, X.; Duan, X.; You, Q.; Dai, C.; Tan, Z.; Zhu, X. Biodiesel production from soybean oil deodorizer distillate usingcalcined duck eggshell as catalyst. *Energy Convers. Manag.* **2016**, *112*, 199–207. [CrossRef]
33. Campestre. Óleo de Soja Refinado. Ficha Ténica. Available online: https://www.campestre.com.br/oleos-vegetais/oleo-de-soja/oleo-de-soja-ficha-tecnica/ (accessed on 13 November 2021).
34. AOCS. Ca 5a-40. Free Fatty Acids in Crude and Refined Fats and Oils. In *Official Methods and Recommended Practices of the AOCS*; AOCS Press: Champaign, IL, USA, 1990.
35. AOCS. Cd 1-25. Wijs Method for Iodine Value. In *Official Methods and Recommended Practices of the AOCS*; AOCS: Champaign, IL, USA, 1990.
36. AOCS. Cd 3-25. Saponification Value. In *Official Methods and Recommended Practices of the AOCS*; AOCS: Champaign, IL, USA, 1990.
37. AOCS. Ca 2b-38. Moisture and Volatile Matter in Butter, Fats, Margarines, and Oils, Hot Plate. In *Official Methods and Recommended Practices of the AOCS*; AOCS: Champaign, IL, USA, 2004.
38. AOCS. Ce 2-66. Preparation of Methyl Esters of Fatty Acids. In *Official Methods and Recommended Practices of the AOCS*; AOCS: Champaign, IL, USA, 2004.
39. Duvekot, C. Determination of Total FAME and Linolenic Acid Methyl Esters in Biodiesel According to EN-14103. Available online: https://www.agilent.com/cs/library/applications/5990-8983EN.pdf (accessed on 18 June 2020).

40. AOCS. Ce 8-89. Tocopherols and Tocotrienols in Vegetable Oils and Fats by HPLC. In *Official Methods and Recommended Practices of the AOCS*; AOCS: Champaign, IL, USA, 2004.
41. Agilent Technologies. The Essential Chromatography and Spectroscopy Catalog, 2011/2012. Available online: https://www.agilent.com/en/promotions/catalog (accessed on 1 May 2021).
42. Sahu, S.; Ghosh, M.; Bhattacharyya, D.K. Isolation of the unsaponifiable matter (squalene, phytosterols, tocopherols, γ-oryzanol and fatty alcohols) from a fatty acid distillate of rice bran oil. *Grasas Aceites* **2018**, *69*, 1–8. [CrossRef]
43. Liu, Y.; Wang, L. Biodiesel production from rapeseed deodorizer distillate in a packed column reactor. *Chem. Eng. Process. Process Intensif.* **2009**, *48*, 1152–1156. [CrossRef]
44. Knothe, G. Structure indices in FA chemistry. How relevant is the iodine value? *J. Am. Oil Chem. Soc.* **2002**, *79*, 847–854. [CrossRef]
45. Barros, A.C.; WUST, E.; Meier, H.F. Estudo da viabilidade técnico-científica da produção de biodiesel a partir de resíduos gordurosos. *Eng. Sanit. Ambient.* **2008**, *13*, 255–262. [CrossRef]
46. Brock, J.; Nogueira, M.R.; Zakrzevski, C.; De Castilhos Corazza, F.; Corazza, M.L.; Vladimir De Oliveira, J. Experimental measurements of viscosity and thermal conductivity of vegetable oils. *Ciênc. Tecnol. Aliment. Camp.* **2008**, *28*, 564. [CrossRef]
47. Gunawan, S.; Ju, Y.H. Vegetable oil deodorizer distillate: Characterization, utilization and analysis. *Sep. Purif. Rev.* **2009**, *38*, 207–241. [CrossRef]
48. Kong, W.; Baeyens, J.; De Winter, K.; Urrutia, A.R.; Degrève, J.; Zhang, H. An energy-friendly alternative in the large-scale production of soybean oil. *J. Environ. Manag.* **2019**, *230*, 234–244. [CrossRef] [PubMed]
49. Kasim, N.S.; Gunawan, S.; Yuliana, M.; Ju, Y.H. A simple two-step method for simultaneous isolation of tocopherols and free phytosterols from soybean oil deodorizer distillate with high purity and recovery. *Sep. Sci. Technol.* **2010**, *45*, 2437–2446. [CrossRef]
50. Maniet, G.; Jacquet, N.; Richel, A. Recovery of sterols from vegetable oil distillate by enzymatic and non-enzymatic processes. *Comptes Rendus Chim.* **2019**, *22*, 347–353. [CrossRef]
51. Kapoor, M.; Gupta, M.N. Lipase promiscuity and its biochemical applications. *Process Biochem.* **2012**, *47*, 555–569. [CrossRef]
52. Paques, F.W.W.; Macedo, G.A. Lipases de látex vegetais: Propriedades e aplicações industriais. *Quim. Nova* **2006**, *29*, 93–99. [CrossRef]
53. Cavalcanti-Oliveira, E.D.A.; Da Silva, P.R.; Ramos, A.P.; Aranda, D.A.G.; Freire, D.M.G. Study of soybean oil hydrolysis catalyzed by *Thermomyces lanuginosus* lipase and its application to biodiesel production via hydroesterification. *Enzyme Res.* **2011**, *2011*, 618692. [CrossRef] [PubMed]
54. Vescovi, V.; Rojas, M.J.; Baraldo, A.; Botta, D.C.; Santana, F.A.M.; Costa, J.P.; Machado, M.S.; Honda, V.K.; Giordano, R.L.C.; Tardioli, P.W. Lipase-catalyzed production of biodiesel by hydrolysis of waste cooking oil followed by esterification of free fatty acids. *J. Am. Oil Chem. Soc.* **2016**, *93*, 1615–1624. [CrossRef]
55. Mibielli, G.M.; Fagundes, A.P.; Bohn, L.R.; Cavali, M.; Bueno, A.; Bender, J.P.; Oliveira, J.V. Enzymatic production of methyl esters from low-cost feedstocks. *Biocatal. Agric. Biotechnol.* **2020**, *24*, 101558. [CrossRef]
56. McCurry, J.D.; Wang, C.-X. Analysis of Glycerin and Glycerides in Biodiesel (B100) Using ASTM D6584 and EN14105. Available online: https://www.agilent.com/cs/library/applications/5989-7269CHCN.pdf (accessed on 18 June 2020).
57. Cesarini, S.; Diaz, P.; Nielsen, P.M. Exploring a new, soluble lipase for FAMEs production in water-containing systems using crude soybean oil as a feedstock. *Process Biochem.* **2013**, *48*, 484–487. [CrossRef]
58. Wancura, J.H.C.; Tres, M.V.; Jahn, S.L.; Oliveira, J.V. Lipases in liquid formulation for biodiesel production: Current status and challenges. *Biotechnol. Appl. Biochem.* **2020**, *67*, 648–667. [CrossRef] [PubMed]
59. Vargas, M.; Niehus, X.; Casas-Godoy, L.; Sandoval, G. Lipases as biocatalyst for biodiesel production. In *Lipases and Phospholipases. Methods in Molecular Biology*; Sandoval, G., Ed.; Humana Press: New York, NY, USA, 2018; Volume 1835, pp. 377–390. ISBN 978-1-4939-8672-9.
60. ANP (Agência Nacional do Petróleo, Gás Natural e Biocombustíveis. Resolução). ANP N° 45/2014. Available online: https://ubrabio.com.br/2014/08/26/resolucao-anp-no-45-2014/ (accessed on 2 December 2021).
61. AOCS. Cc 10a-25. Specific Gravity of Oils and Liquid Fats. In *Official Methods and Recommended Practices of the AOCS*; AOCS: Champaign, IL, USA, 1990.
62. AOCS. Ca 6a-40. Unsaponifiable Matter in Fats and Oils, Except Marine Oils. In *Official Methods and Recommended Practices of the AOCS*; AOCS: Champaign, IL, USA, 1990.

catalysts
MDPI

Article

Simplified Method to Optimize Enzymatic Esters Syntheses in Solvent-Free Systems: Validation Using Literature and Experimental Data

Ronaldo Rodrigues de Sousa [1,2], Ayla Sant'Ana da Silva [1,2], Roberto Fernandez-Lafuente [3,4,*] and Viridiana Santana Ferreira-Leitão [1,2,*]

1 Biocatalysis Laboratory, National Institute of Technology, Ministry of Science, Technology, and Innovations, Rio de Janeiro 20081-312, Brazil; ronaldo.rodrigues@int.gov.br (R.R.d.S.); ayla.santana@int.gov.br (A.S.d.S.)
2 Department of Biochemistry, Federal University of Rio de Janeiro, Rio de Janeiro 21941-909, Brazil
3 Biocatalysis Department, ICP-CSIC, Campus UAM-CSIC, 28049 Madrid, Spain
4 Center of Excellence in Bionanoscience Research, External Scientific Advisory Academics, King Abdulaziz University, Jeddah 21589, Saudi Arabia
* Correspondence: rfl@icp.csic.es (R.F.-L.); viridiana.leitao@int.gov.br (V.S.F.-L.)

Abstract: The adoption of biocatalysis in solvent-free systems is an alternative to establish a greener esters production. An interesting correlation between the acid:alcohol molar ratio and biocatalyst (immobilized lipase) loading in the optimization of ester syntheses in solvent-free systems had been observed and explored. A simple mathematical tool named Substrate-Enzyme Relation (SER) has been developed, indicating a range of reaction conditions that resulted in high conversions. Here, SER utility has been validated using data from the literature and experimental assays, totalizing 39 different examples of solvent-free enzymatic esterifications. We found a good correlation between the SER trends and reaction conditions that promoted high conversions on the syntheses of short, mid, or long-chain esters. Moreover, the predictions obtained with SER are coherent with thermodynamic and kinetics aspects of enzymatic esterification in solvent-free systems. SER is an easy-to-handle tool to predict the reaction behavior, allowing obtaining optimum reaction conditions with a reduced number of experiments, including the adoption of reduced biocatalysts loadings.

Keywords: immobilized lipases; solvent-free reactions; enzymatic esterification; esters

Citation: Sousa, R.R.d.; Silva, A.S.d.; Fernandez-Lafuente, R.; Ferreira-Leitão, V.S. Simplified Method to Optimize Enzymatic Esters Syntheses in Solvent-Free Systems: Validation Using Literature and Experimental Data. *Catalysts* **2021**, *11*, 1357. https://doi.org/10.3390/catal11111357

Academic Editors: Evangelos Topakas, David D. Boehr and Roland Wohlgemuth

Received: 25 October 2021
Accepted: 8 November 2021
Published: 12 November 2021

1. Introduction

Solvent-free systems (SFS) are becoming popular for enzymatic esterifications. These systems have many advantages because the reaction media is formed only by the reactants, increasing the volumetric productivity of the process and avoiding complex downstream and hazardous wastes [1–3]. The adoption of solvent-free systems may contribute to achieving the feasibility of biocatalytic ester syntheses on a large scale in both technical and economic aspects, in consonance with the principles of Green Chemistry.

Immobilized lipases have been utilized successfully for esterification reactions in SFS [4–9]. Enzyme immobilization enables enzyme recovery and reuse, associated with the possibility of improving enzyme stability, activity, selectivity, or specificity [10–12]. Moreover, it may enlarge the window of operating conditions (reducing inhibitions or inactivation by chemicals) and be coupled to the purification processes [13–16]. Immobilized lipases are extensively studied for esters syntheses [10,12,17–20], potentially addressing demands in many different sectors such as energy and transport [21–23], food industries [24–26], cosmetics and personal care [19,27], and chemical industries [28–30].

Molar ratio and biocatalyst loading are two of the main parameters studied in solvent-free enzymatic esterifications because this reaction is thermodynamically controlled and, thus, the concentration of the catalyst determines the conversion rate [31–33]. The reaction media and the experimental conditions have major influences on these aspects, and, in solvent-free systems, the medium is determined by the substrates and the molar ratio of the substrates [1]. The evaluation of molar ratio in SFS gains additional importance, considering that the use of a surplus reagent is required to obtain high conversions in the synthesis of the other substrate, and the surplus reagent, i.e., its nature and its quantity in the system, will define critical physicochemical characteristics of the reaction media in the different steps of the reaction [1,31,34–36]. A dynamic environment is presented in SFS esterification; initially, the media is formed exclusively by the mixture of the substrates, and at the end, it will be formed by the ester product, the co-product water (if is not removed), the remaining excess substrate, and some traces of the minority substrate [1,37]. The final reaction media is generally more hydrophobic than the initial system. The elimination or capture of the formed water is a way to shift the reaction thermodynamic equilibrium towards synthesis [37,38]. However, even if water is eliminated in the reaction medium when the enzyme activity is very high, the accumulation of water inside the biocatalyst particle (when the enzyme activity is very high) can generate a water phase in the enzyme environment, adversely affecting the enzyme performance [39,40]. The use of very hydrophobic supports or ultrasounds may reduce these adverse effects [41–48]. On the other hand, biocatalyst loading in the reactor defines the kinetics aspects of the reaction, such as reaction rate and the occurrence of inhibition/inactivation, with a substantial impact on reaction time, productiveness, and process costs [42,49–52]. As immobilized lipases are still expensive incomes, the definition of an optimized biocatalyst loading is a critical parameter for any applied biocatalytic process [44,49–52].

The classical approach of evaluating independent variables once a time is still commonly used for enzymatic esterification reaction optimization, even in recent studies [7–9,53–56]. However, it has been shown that correlation among some of the studied variables makes the independent optimization incomplete, and, thus, statistical tools, such as response surface methodology (RSM), became popular for optimization studies [49,57–64]. Although interactions between variables may be discovered using RSM, the molar ratio of reagents and biocatalyst loading are generally considered independent variables. Nonetheless, in a previous study, we have found an interesting correlation between both variables for the synthesis of octyl octanoate catalyzed by Novozym 435 in a SFS [6]. Our findings led us to develop a simple mathematic tool, that we named SER (Substrate-Enzyme Relation) that correlates the mass of reagents and immobilized lipases, as described in Equation (1):

$$\mathrm{SER} = \frac{\text{m alcohol} - \text{m acid}}{\text{m biocatalyst}} \tag{1}$$

where "m alcohol" and "m acid" are the masses (grams) of alcohol and acid in the reaction, respectively, and "m biocatalyst" is the mass of immobilized lipase (enzyme + support). SER is a dimensionless number that expresses a certain reaction condition—a mass of reagents and biocatalyst in the system. This reaction condition will generate a conversion result in the reaction, i.e., the percentual degree of conversion of the reagents (in general carboxylic acid) into products (ester). Considering that lipases are specific catalysts that catalyze only esterification/hydrolysis reactions, the conversion of the reagents means the yield of the reaction. Thus, a SER number may be associated with a conversion result. In our first approach, the study of the interaction between molar ratio and biocatalyst loading in the reactor and their influence on the process performance resulted in high conversions (above 96.0%), using relatively low enzyme loading (1.5% *wt*/*wt* acid mass) and 30% stoichiometric excess of alcohol [6]. SER possible fundamentals were based on the hypothesis of a shift of chemical equilibrium by using stoichiometric excess of alcohol until an estimated level in which the yields are not further improved because lipase suffers inhibition or inactivation. The simple combination of the masses of reagents and biocatalyst enabled us to obtain practical information about the reaction thermodynamics and kinetics, using the outcome to establish a range of reaction conditions in which high conversions could be achieved (in this case, SER between 0 and 65).

Thus, considering that SER can be an easy-to-handle tool to predict the enzymatic esterification reaction optimization, this study aims to validate its applicability in synthesizing aliphatic esters in SFS mediated by immobilized lipases. We applied this mathematic tool for lipase-catalyzed esterification reactions previously described in the literature and using experimental data obtained in this study (39 different examples).

2. Results

2.1. SER Validation from Literature and Experimental Data

Table 1 shows the SER outcomes calculated utilizing available data in the selected publications and the correlation with the optimal range suggested by our previous work. We observe that high conversions (above 80%) were associated, in most cases, with intermediate positive values of SER, between 0 and 65, as observed by Sousa and co-workers (2020) [6] for the octyl octanoate synthesis. In addition, conversion higher than 90% were observed within this SER range for a variety of cases—92–93% on butyl formate and octyl formate with SER 35–36 [8]; 92–94% on isobutyl propionate with SER 20 [4] and SER 13 [65]; 92% on butyl octanoate with SER 0.5 [7]; 96–99% on octyl octanoate with SER 4–9 [6]; 95–98% on cetyl tetradecanoate, cetyl hexadecanoate, cetyl stearate, and cetyl oleate with SER 4 [66].

Table 1. Enzymatic syntheses studies of aliphatic esters in solvent-free systems with calculated SER and its respective conversion results.

Acid	Alcohol	Lipase	Immobilization Support	T (°C)	Conversion Response Reported (%)	SER Calculated Range	SER with the Highest Conversion	Ref.
			Good level of agreement					
Metanoic Acid	*n*-Butanol	Novozym 435®	Lewatit VP OC 1600	40	55.0–93.0%	35 to 39	36 (93.0%)	[8]
Metanoic Acid	*n*-Octanol	Novozym 435®	Lewatit VP OC 1600	40	-	-	35 (92.0%)	[8]
Propionic Acid	Isobutanol	Fermase CALB 10000	Polyglycidemethaacrylate	60	71.8–94.2%	0 to 20	13 (94.2%) ***	[65]
Propionic Acid	Isobutanol	Novozym 435®	Lewatit VP OC 1600	40	63.8–92.5%	0 to 30	20 (92.5%)	[4]
Butanoic Acid	*n*-Butanol	Novozym 435®	Lewatit VP OC 1600	37	41.0–48.6%	−3 to −1	−1 (48.6%) *	[67]
Pentanoic Acid	Ethanol	Lipase from S. *simulans*	$CaCO_3$	37	29.0–51.0%	−20 to −4	−4 (51.0%) **	[68]
Pentanoic Acid	Ethanol	Novozym 435®	Lewatit VP OC 1600	50	40.0–69.0%	−50 to −21	−33 (69.0%) ***	[55]
Octanoic Acid	*n*-Butanol	Novozym 435®	Lewatit VP OC 1600	37	-	-	−4 (38.0%) *	[67]
Octanoic Acid	*n*-Butanol	Novozym 435®	Lewatit VP OC 1600	50	85.0–89.0%	−24 to 27	27 (89.0%)	Data not published
Octanoic Acid	*n*-Butanol	Lipozyme RM IM®	Duolite ES 562	50	68.2–89.0%	−12 to 14	14 (89.0%)	Data not published
Octanoic Acid	*n*-Butanol	Novozym 435®	Lewatit VP OC 1600	60	76.5–92.5%	−14 to 9	0.5 (92.5%)	[7]
Octanoic Acid	*n*-Octanol	Novozym 435®	Lewatit VP OC 1600	65	74.7–96.1%	−19 to 207	9 (96.1%)	[6]
Octanoic Acid	*n*-Octanol	Lipozyme RM IM®	Duolite ES 562	65	19.6–99.0%	−10 to 171	4 (99.0%)	Data not published
Decanoic Acid	*n*-Propanol	Fermase CALB 10,000	Polyglycidemethaacrylate	60	19.0–83.8%	−96 to 8	1 (83.8%)	[9]
Dodecanoic Acid	*n*-Butanol	Novozym 435®	Lewatit VP OC 1600	37	28.0–35.0%	−10 to −3	−10 (35.0%)	[67]
Dodecanoic Acid	Hexadecan-1-ol	Lipozyme RM IM®	Duolite ES 562	70	67.0–98.1%	4 to 32	4 (98.1%)	[66]
Tetradecanoic Acid	Isopropanol	Novozym 435®	Lewatit VP OC 1600	60	7.0–87.7%	7 to 60	15 (87.7%)	[5]
Tetradecanoic Acid	Hexadecan-1-ol	Lipozyme RM IM®	Duolite ES 562	70	65.9–97.3%	4 to 33	4 (97.3%)	[66]
Hexadecanoic Acid	Hexadecan-1-ol	Lipozyme RM IM®	Duolite ES 562	70	61.8–97.1%	4 to 33	4 (97.1%)	[66]
Octadecanoic Acid	Hexadecan-1-ol	Lipozyme RM IM®	Duolite ES 562	70	60.5–95.8%	4 to 33	4 (95.8%)	[66]
Octadec-9-enoic Acid	*n*-Octanol	Lipase from *R. miehei*	Poly(ehtylene)-g-co-hydroxyethyl methaacrylate	37	-	-	56 (82.0%)	[69]
			Intermediate level of agreement					
Propionic Acid	*n*-Butanol	Novozym 435®	Lewatit VP OC 1600	45	81.2–92.7%	100 to 600	100 (92.7%)	[70]
Propionic Acid	*n*-Butanol	Novozym 435®	Lewatit VP OC 1600	60	35.0–92.0%	39 to 160	85 (92.0%) ***	[56]
Butanoic Acid	*n*-Butanol	Novozym 435®	Lewatit VP OC 1600	53.9	-	-	91 (99.6%)	[64]
Decanoic Acid	*n*-Octanol	Novozym 435®	Lewatit VP OC 1600	50	92.0–96.0%	−1 to 0	−0.4 (96.0%)	[71]
DecanoicAcid	*n*-Octanol	Lipozyme RM IM®	Duolite ES 562	50	88.8–95.1%	−1 to 0	−0.2 (95.1%)	[71]
Dodecanoic Acid	*n*-Octanol	Novozym 435®	Lewatit VP OC 1600	50	93.9–96.3%	−10 to −1.5	−1.5 (96.3%)	[71]
Dodecanoic Acid	*n*-Octanol	Lipozyme RM IM®	Duolite ES 562	50	86.7–97.7%	−10 to −1.5	−4 (97.7%)	[71]
Octadec-9-enoic Acid	*n*-Butanol	Lipase from *R. oryzae*	$CaCO_3$	37	18.0–81.0%	−54 to −5	−5 (81.0%)	[72]
			Low level of agreement					
Etanoic Acid	*n*-Butanol	Lipase from *R. oryzae*	Celite 545	37	12.0–61.0%	−7.5 to 9	1 (61.0%) **	[73]
Etanoic Acid	Isopentanol	Lipase from S. *simulans*	$CaCO_3$	37	2.0–64.0%	−26 to 35	−7 (64.0%) **	[74]
Etanoic Acid	Isopentanol	Novozym 435®	Lewatit VP OC 1600	30	46.7–68.4%	3 to 10	3 (68.4%)	[60]
Etanoic Acid	*n*-Hexanol	Lipase from S. *simulans*	$CaCO_3$	37	21.0–43.0%	3 to 14	3 (43.0%) **	[68]
Dodecanoic Acid	Ethanol	Fermase CALB 10,000	Polyglycidemethaacrylate	60	67.0–92.4%	−31 to −2	−15 (92.4%)	[75]
TetradecanoicAcid	Isopentanol	Novozym 435®	Lewatit VP OC 1600	60	82.0–97.0%	−197 to 17	−99 (97.0%) ***	[76]
Hexadecanoic Acid	Isopropanol	Novozym 435®	Lewatit VP OC 1600	75	33.5–88.0%	−35 to −10	−10 (88.0%)	[57]
Octadecanoic Acid	Ethanol	Novozym 435®	Lewatit VP OC 1600	60	66.0–92.0%	−65 to 27	−19 (92.0%)	[77]
Octadecanoic Acid	*n*-Butanol	Novozym 435®	Lewatit VP OC 1600	60	61.0–92.0%	−74 to 104	−48 (92.0%)	[77]
Octadec-9-enoic Acid	Ethanol	Lipozyme®	Duolite A568	40	72.0–99.0%	−41 to −4	−41 (99.0%)	[1]

* Included the use of molecular sieves or vacuum pressure to remove the water. ** With the addition of water. *** Included activation by microwave or ultrasound.

SER optimum range seems to be applicable for different carboxylic acid and alcohols of different chain lengths, as showed by the first section of Table 1, but some important aspects should be highlighted. For short-chain acids and alcohols, as studied by Aljawish et al. (2019) [8] and Kuperkar et al. (2014) [4], optimum SER was found in the middle of the proposed range, between 20 and 36, due to the use of significant molar excess of alcohol for reducing the strong inhibition potential of short-chain acids. As short-chain alcohols also promote inhibition of lipases and may cause damages in their hydration layers [1,57,72,76], the formed water in the system helps to attenuate this problem, resulting in high conversions. Jaiswal & Rathod (2017) [65] did not use the same level of a stoichiometric excess of isobutanol, which resulted in SER equals 13, but the reaction was assisted by microwave. Similarly, the formed water remained in the system. Contrarily, for long-chain acid and alcohols, as reported by Arnaldos et al. (2018) [66], optimum SER was found to be close to 0, as shown by Table 1. Since long-chain acids or alcohols have less potential to affect the biocatalysts' hydration layer, it is essential to remove the formed water due to the possible water accumulation in lipase vicinity in such a hydrophobic environment, forming a diffusional barrier for the substrate and lipase active sites [2,37,78]. In these cases, it is possible to adopt biocatalyst loadings lower than those used for esterification with short-chain reagents, which is coherent with the reduced probability of substrate enzyme inhibition when using long-chain reagents [1,79,80]. The same rationale can be extended for mid-chain acids and alcohols, as suggested by the SER in our previous study [6] and the experimental data collected in this work, as shown in the first section of Table 1. Studies in which water was not removed, include own data, showed conversions slightly lower [5]. Data from Ghamgui et al. (2004) [72] and Sousa et al. (2021) [71] show that mid and long-chain acids and alcohols seem to have optimum SER close to 0, although within a broader range of values including slightly negative outcomes. These studies showed optimum conditions using discrete stoichiometric excess of alcohol, associated with low biocatalyst loadings, which resulted in a low SER number.

It is plausible to compare different studies of solvent-free enzymatic esterification with immobilized lipases because (i) chemical equilibrium in an esterification reaction, using monofunctional acids and alcohols, is not dependent on the (bio)catalyst adopted [31,81,82]; (ii) the theoretical equilibrium constant should be the same for the same type of reaction independently of the reagents, even though slight variations may occur due to effects of the solvation of different reagents and products and the ionization of the different carboxylic acids, as observed in experimental data [31,82]. Furthermore, external mass transfer limitations are generally neglected in immobilized lipase-catalyzed esterification reactions at a lab-scale [1,5,10,83,84].

Considering the many different aspects that define the kinetics and yields of enzymatic esterifications in SFS using immobilized lipases, it is counterintuitive to think that a simple mathematical equation is effective in predicting the behavior of the reaction with different reagents. The numerator of SER represents, in terms of mass, the quantity of exceeding reagent. Indirectly, the analysis of its nature coupled with its quantity may give an idea if the solvation of the second reagent by the exceeding reagent will be favored or not. Solvation of reagents and products are associated with their respective thermodynamic activities, which govern the equilibrium equation of the reaction [31,81,85–88]. The denominator of SER is the mass of biocatalyst that provides information about the reaction rate: in the absence of harmful interferences, increasing biocatalyst loadings in the reactor will linearly increase the reaction rate. The stoichiometric excess of one of the reagents (the numerator of SER) and the biocatalyst loading (denominator of SER) are linked by the ability of the exceeding reagent to affect the enzyme properties, causing modifications in the essential hydration layer of the enzyme or generating enzyme inhibition. An intermediate outcome is obtained by a relation between a proper biocatalyst loading and an excess of reagent that potentially leads the system to shift the equilibrium towards synthesis without noticeable inhibition or inactivation of enzymes. If a highly hydrophilic reagent is present in the system, a sizeable molar excess of the second reagent or adopting high biocatalyst loading is required to counterbalance its deleterious effect using to compensate for the possible reaction rate reduction, as observed in the study of Vadgama and co-workers (2015) [5]. This condition is associated with an intermediate SER outcome, depending on the nature of the reagents and biocatalysts involved. The proposition of a different range of values that correspond to high conversions is a way to deal with the variability of reaction conditions that include the nature of reagents and biocatalysts and different methods of activation and reactional strategies. SER should be understood as a general tool, not a mathematical simulator of the reaction.

The data of short-chain esterifications lead us to hypothesize that the optimum range established for mid-chain esterifications should be shifted for higher SER values. The synthesis of butyl propionate carried out by Dai and co-workers (2014) [70] is an interesting case because only conditions with high SER were evaluated. Although the lowest SER number tested—the closest to the previously defined optimum range—resulted in the highest conversion, this result was obtained using a similar strategy than Aljawish et al. (2019) [8] and Kuperkar et al. (2014) [4] maintaining the formed water in the media. The reduction in the reaction rate seems clear in this case due to the long reaction times (30 h) compared to the synthesis of isobutyl propionate carried out by Kuperkar and co-workers (10 h) [4]. The study of Rahman and co-workers (2017) [64] reinforces the relevance of water maintenance in the esterification of short-chain acids and alcohols, and although only one condition was tested, the optimum SER value in this study (91) is above the supposed optimum range.

The studies of short-chain esterifications classified as low level of agreement with our previous work also reinforce this hypothesis, because conversions below 80.0% with SER close to 0 (supposedly inside the proposed optimum range) were obtained [68,73,74,89]. However, three studies of the same research group [68,73,74] adopted specific conditions in the reactions related to the addition of a considerable quantity of water (10% or more) to the system. As potent inhibition caused by highly polar acids (as etanoic or pentanoic acid) and short-chain alcohols are expected in these cases, the authors tried to avoid this using stoichiometric excess of some of the reagents. Thus, the addition of the water (acting as a co-solvent) reduced the acid concentration and may have restored the hydration layer of the enzymes; however, as water is a product of the reaction, thermodynamic equilibrium was shifted to hydrolysis, as discussed by the authors. Moreover, the syntheses studied by Karra-Châabouni et al. (2006) [68] were carried out only in equimolar substrate conditions at different biocatalyst loadings, indicating why the SER trend is not followed in this case.

The use of microwave irradiation can be included in the same set of specific reaction conditions. Bhavsar & Yadav (2018) [55,56] studied the syntheses of *n*-butyl propionate and ethyl valerate—two pairs of short-chain substrates—assisted by microwave. The first study shows SER value equals 85 as the optimum; the lowest conversions observed were associated with SER value 39 and 170 (respectively, 52.0% and 35.0%), indicating a range of optimum reaction conditions with different values than the initially suggested SER trend. However, when the condition corresponding to SER 85 was evaluated under conventional heating (60 °C), the reaction presented a 72.0% of conversion after 8 h, not so far from the expected trend. The second study evaluates reaction conditions corresponding to negative SER outcomes (between −50 and −21). Besides the use of microwave—which affects the reaction media behavior and immobilized lipase performance differently than conventional heating—the reaction time was too short (40 min), suggesting that the equilibrium may not have been achieved. Comparing the results obtained in optimum reaction conditions under microwave and conventional heating indicates that the reaction presented a low conversion (53.1%) with SER −33. Then, the adoption of microwave in SFS brings additional difficulties for analyzing the data. However, utilizing the data obtained using conventional heating in the same studies, we observe an acceptable level of agreement with the SER trend.

The most remarkable exceptions are the studies involving long-chain acids (above C12) and short-chain alcohols (below C7) [1,57,76,77]. High conversions were obtained with negative SER results far from 0, within a range of −10 and −48. The reason is that the SER equation, as firstly established, does not consider the molar masses of the different involved reagents. When a long-chain acid reacts with short-chain alcohol, the subtraction of utilized masses of alcohol (even using an excess of alcohol) and acid generates a negative SER number due to the accentuated differences in molar masses. For instance, the highest conversion obtained by Pereira and co-authors [77] for butyl stearate synthesis (92.0%) was obtained with acid:alcohol molar ratio 1:2, which corresponded to the SER value of −48. Yadav and Thorat (2012) [76] obtained a high yield on isopentyl tetradecanoate with a very negative SER value, −99. The authors indicated the equimolar ratio of reagents with a low biocatalyst loading of Novozym 435® (0.38% *wt/wt*) as the optimum condition for this synthesis, but the system was submitted to microwave. The same reaction conditions under conventional heating resulted in a low 56.0% of conversion. Considering these cases, we have elements to hypothesize that, for esterifications between long-chain acids and short-chain alcohols, SER negative outcomes also follow the same trends as the positive ones, keeping the same rationale about extreme values.

Summarizing: (i) SER optimum range (0 to 65) is also observed for other mid-chain esters syntheses besides octyl octanoate mediated by Novozym 435®; (ii) SER optimum range seems to be shifted for higher values (20 to 100) in short-chain esters syntheses and lower values in long-chain esters syntheses (−65 to 0); (iii) the trend is not followed either when the reaction is assisted by microwave or ultrasound, nor when water is added to the media.

Few studies evaluated a wide range of SER values at a constant temperature [6,9,76,77]. The highest conversion observed in these studies were achieved not only by the manipulation of molar ratio and biocatalyst loadings at a constant temperature, but with the application of some additional strategy to obtain the maximum yield, as the continuous removal of the water from the media [6,7,9,66], or different methods of activation, as microwave irradiation [55,56,76]. Therefore, we believe that SER will only indicate a range of conditions in which thermodynamics and kinetics aspects have a convergence towards high conversions but not predicting the exact condition that will result in the maximum conversion. The effective use of SER to optimize molar ratios and biocatalyst loadings implies that temperature and stirring rate are enough to promote a proper reaction performance.

The work of Santos and co-authors (2007) [67] illustrates well the limitations caused by stirring and temperature. The authors adopted stoichiometric excesses of acid for esterifications of *n*-butanol with butanoic, octanoic, or dodecanoic acid, with high biocatalyst loadings (5, 10, and 15% *wt/wt*) of Novozym 435®, at 37 °C and 150 rpm of stirring rate. Besides the unfavorable relation between these variables using SER as a parameter, 37 °C seems inadequate in the esterification of dodecanoic acid for a proper mixture of reagents since the dodecanoic acid melting point is 43 °C; its use in stoichiometric excess may limit the diffusion of *n*-butanol inside the biocatalyst particle. Moreover, the stirring rate (150 rpm) could not be enough to properly mix biocatalyst and substrates, considering that most cited studies used 200 rpm or higher stirring rates in these reactions [4,7–9,72–74,77]. Moreover, high biocatalyst loadings were adopted, and the poor conversions after 72 h (below 50.0%) of reaction may be associated with biocatalyst particle aggregation, increasing the diffusional limitations on the system. Partial inactivation of the immobilized lipase may also be suggested in the esterification of butanoic acid by reducing pH in the enzyme environment. To reinforce the hypothesis of diffusional limitations caused by high biocatalyst loadings in that stirring rate, we obtained 73.0% of conversion in butyl octanoate synthesis using 2.0% Novozym 435® and an acid:alcohol molar ratio of 1:2 (SER equals to 2), at 30 °C and 150 rpm, in just 3 h. Similar issues may be suggested for the results obtained by Güvenç and co-workers (2007) [89] in the esterification of etanoic acid and *n*-pentanol, where high biocatalyst loadings (6% and 10%, which corresponds to SER 3 and 10, respectively), temperatures of 35 °C, and 150 rpm of stirring rate were employed.

As we can see in Table 1, most of studies used Novozym 435® as biocatalyst. Its immobilization support material is a moderately hydrophobic resin [20]. The same feature can be observed in Lipozyme RM IM®, whose support is Duolite ES 562® [18]. By adopting proper reaction conditions, reactions mediated by these biocatalysts achieve high conversions. Thus, it is possible to observe that moderately hydrophobic materials are enough to promote a proper partitioning of the water in the media, keeping the hydration layer of enzymes and simultaneously avoiding the occurrence of hydrolysis of formed esters.

2.2. Thermodynamics and Kinetics Aspects

SER contains a term indirectly correlated with thermodynamics (difference between substrates masses) and another term indirectly correlated to kinetics (biocatalyst mass in the system). An exploration of these topics is required to check how SER is coherent with the phenomena involved in enzymatic esterification. Chemical equilibrium is established by the relation of products and reagents concentrations, or more accurately, its thermodynamics activities (in non-ideal systems) [31–34]. Substrates thermodynamic activities should increase to favor the esterification, and this increase will not happen if the reagents are very well solvated (by the solvent or by the exceeding reagent in SFS); contrarily, the thermodynamic activities of the products should decrease to avoid hydrolysis [1,31,36,90]. The quantity and the nature of the exceeding reagent can provide simple predictions about solvation and thermodynamic activities, helping obtain a good overview of the reaction. For instance, in a reaction between short-chain alcohol (C3) and a long-chain acid (C14) with a molar ratio acid:alcohol 1:3, the mutual solubility of the substrates will not be favored—highly polar alcohol with a non-polar acid. Due to the differences in polarity, the alcohol will not solvate the acid completely, although some degree of solvation may affect the final conversion due to its excess. In the case of a reduction in alcohol quantity that corresponds to a discrete stoichiometric excess, it is expected that the alcohol will be unable to solvate a more significant percentage of the acid, which means that the acid conversion will be favored. The opposite situation—the increase of alcohol concentration in the system—may increase the probability of damages in the hydration layer of lipases or inhibition [1,4,72,76,91]. Not only surplus reagent can cause these damages, but its probability is higher than that of the limiting reagent. SER helps visualize these situations by a simple way. A discrete stoichiometric excess of the mentioned alcohol that is unable to solvate the acid is associated with low number in the SER numerator; similarly, discrete stoichiometric excess of alcohol that reduces the probability of damages on immobilized lipases or inhibition is also associated with low numbers.

Adopting a significant stoichiometric excess of alcohol in this hypothetical case may cause inhibition, with a consequential reduction in the reaction rate. Then, an increase in the biocatalyst loading may be adopted to compensate for this reduction. Mathematically, an increase in the biocatalyst loading means an increase in the SER denominator, reducing the outcome of SER. From a thermodynamic perspective, a discrete stoichiometric excess of alcohol seems adequate to shift the chemical equilibrium, avoiding excessive solvation of acid in this case and with a reduced probability of inhibition on lipases. These effects will depend on the amounts and the nature of the surplus reagent in the media. The combination of this condition, expressed by SER numerator, associated with low biocatalyst loading, expressed by SER denominator as the mass of biocatalyst, will result in a low SER outcome. As shown in Table 1, this observation is coherent with high conversions observed in the literature (and own data) for the esterification of long-chain acids and short-chain alcohols [1,72].

Water plays an essential role in enzymatic esterification, affecting both thermodynamic and kinetics aspects. For example, in a C14 acid and C3 alcohol esterification, the formed water is expected to be solubilized mainly in the alcohol. However, some water may remain in the vicinity of the enzyme if the immobilization support of the enzyme is moderately hydrophobic [92–94]. The ester formed is always more apolar than both substrates, and its concentration will be growing along with the reaction time. Partitioning of the water between the reaction media and the catalytic phase may occur, driving to favor the lipases hydrolytic activity. The effects of an excess of the alcohol (indicated by an increase in SER numerator) may cause a complete strip-off of the hydration layer of the enzyme with probable enzyme inhibition/inactivation. As the remaining quantity of alcohol will decrease along with the reaction, the essential hydration layer of the immobilized enzyme may be less affected at the end than at the beginning of the reaction, depending on the biocatalyst loading and the excess of alcohol in the system (indicated by SER denominator and numerator, respectively). Most of the water formed may be bounded to the biocatalyst, forming a water layer around the enzymes, affecting enzyme activity and stability. Also, the water layer may produce a partition of the ester from the active site of the enzyme, considering the ester hydrophobicity; then, diffusional limitation caused by the water layer may avoid the access of the ester into the active site of the enzyme, preventing likely inhibitions by the reaction product.

Further information of the reaction media may be predicted in qualitative terms, and SER can be used to a quantitative understanding of these phenomena. Overall considerations about the behavior of the reaction from a chemical equilibrium and kinetics perspective are summarizing in Table 2.

Table 2. Predictions of the behavior of enzymatic esterifications in different conditions of molar ratio using stoichiometric excess of reagents.

Reaction Condition	Initial Condition of the Media	Effect on Biocatalyst	Condition on Equilibrium	Possible Optimization Path
Large stoichiometric excess of a highly polar reagent	Possible strip-off of essential water on enzymes pH acid (if exceeding reagent is the acid) Low log P; low viscosity	Possible reduction in the enzymatic activity	Accumulation of the water on the organic phase pH slightly acid Low log P; increased viscosity	Adoption of high biocatalyst loadings Control of the water activity Fractioned additions of the exceeding reagent
Discrete excess of a highly polar reagent	pH slightly acid (if the surplus reagent is the acid) Limited solubility/poor solvation (if limiting reagent is non-polar) Intermediate log P; intermediate viscosity (if limiting reagent have a long-chain)	-	pH neutral (or slightly acid) Intermediate log P; considerable viscosity	Possibility of adopting reduced biocatalyst loadings
Discrete excess of a non-polar reagent	pH slightly acid (if the surplus reagent is the acid) Limited solubility/poor solvation (if limiting reagent is highly-polar) Intermediate log P; intermediate viscosity (if limiting reagent have a long-chain)	-	pH neutral (or slightly acid) Intermediate log P; considerable viscosity	Possibility of adopting reduced biocatalyst loadings
Large stoichiometric excess of a non-polar reagent	pH acid (or slightly acid) High log P; high viscosity (if surplus reagent have a long-chain)	Possible reduction in the enzymatic activity	Accumulation of the water on catalytic phase pH neutral (or slightly acid) High log P; high viscosity	Adoption of high biocatalyst loadings Control of the water activity Increase the temperature

For different reasons, we can observe that large stoichiometric excesses of one of the reagents seem to be unfavorable to enzymatic esterifications since it increases the solvation of the limiting reagent requires the use of high biocatalyst loading or other strategies to avoid inhibition of lipases. Nevertheless, some high conversion results showed in Table 1 were attained with a large stoichiometric excess of one of the reagents [4,8,64], diminishing the inhibition potential of the utilized acid by increasing its solvation. Vadgama et al. (2015) [5] did not observe deleterious effects on Novozym 435® above 4% biocatalyst loading using a substrate molar ratio acid:alcohol of 1:15. This stoichiometric alcohol excess was applied to dissolve the tetradecanoic acid, increasing its solvation. These authors, however, did not explore different molar ratios. Besides the thermodynamic and kinetics issues, this excessive amount of alcohol must be removed after the reaction completion, bringing additional difficulties for the downstream processes.

When varied molar ratios and biocatalyst loadings in short-term evaluations (limited reaction time), the conversion reductions of the limiting reagent may be associated exclusively with a reaction rate reduction. In these cases, it is possible to observe which concentration of surplus reagent (correlated to the biocatalyst loading in the system) is not favorable for the enzymatic activity, indicating some potential inhibition. For this reason, SER may be helpful to indicate in which conditions (or a range of conditions) inhibition by exceeding reagent may arise. Another essential aspect to emphasize is the possibility of aggregation of biocatalyst particles when high biocatalyst loadings are adopted. The aggregation will reduce the reaction rates due to increased substrate diffusion limitations [83,95]. SER values close to zero (positive or negative) indicate, in some cases, the employment of high biocatalyst loadings in the system, increasing the possibility of aggregation of the biocatalysts. This phenomenon is favored when immobilization support has affinity by water or when high water content is present in the system [91]. For example, Novozym 435®, the most widely studied lipase in these studies, is immobilized in Lewatit VP OC 1600, an organic carrier composed of poly(methyl methacrylate) and divinylbenzene. This support material is moderately hydrophobic and susceptible to absorb some water [20]. Observations like this should be considered, mainly when highly hydrophilic reagents are present and high biocatalyst loadings are adopted.

To check the accuracy of these kinds of predictions with SER, a more detailed analysis of the reaction condition of a selected study—Parikh et al. (2019) [9]—was carried out, as showed in Table 3.

Table 3. SER and conversions results of the reaction conditions studied by Parikh et al. (2019) in the synthesis of propyl decanoate mediated by Fermase at 60 °C, in the solvent-free system, 300 rpm, and 10 h. The biocatalyst loading is showed based both on *wt/wt* of the total mass of reagents and based on *wt/wt* of acid.

Acid:Alcohol Molar Ratio	Biocatalyst Loading (% *wt/wt* Acid)	Biocatalyst Loading (% *wt/wt* Total)	SER	Conversion (%)
1:1	0.7	0.5	−97	19%
1:1	1.3	1.0	−48	47%
1:1	2.7	2.0	−24	59%
1:1	4.0	3.0	−16	62%
1:2	3.4	2.0	−9	71%
1:3	4.1	2.0	1	83%
1:4	4.8	2.0	8	78%
2:1	2.3	2.0	−35	36%
3:1	2.2	2.0	−40	24%

Table 3 shows that the intermediate SER values (−9 to 8) resulted in the highest conversions observed in the conditions tested. The equimolar substrate condition and biocatalyst loadings below 2.0% (*wt/wt* total mass) were unfavorable for the synthesis. Above 2.0% biocatalysts loading at equimolar condition, it is possible to observe that the 60% conversion was reached without shifting the equilibrium due to the lack of surplus reagent. The molar ratios 1:2 to 1:4 improved the yields and showed that the excess of *n*-propanol shifted the equilibrium, promoting the low acid solvation without generating an unfavorable partitioning of the water. Although this information is insufficient to indicate the most proper conditions for this synthesis, it is possible to discard the evaluation of molar ratios superior to 4, as well as the conditions with an excess of acid (SER negative) due to the increase of the viscosity and the decrease of pH. Considering the molar mass differences between decanoic acid and *n*-propanol, SER optimum condition seems to be close to 0 (positive or slightly negative values), implying that the evaluation of molar ratio conditions between 1:2 and 1:3 with biocatalyst loadings close to 2.0%. Another important observation is that slight differences in SER values seem to be associated with slight variations in conversion (the difference between SER 1 and 8, resulting in conversions of 83% and 78%), as observed initially by Sousa et al. (2020) [6]. This trend may be used to optimize the conditions, simultaneously reducing the molar ratio (avoiding the waste of alcohol) and the biocatalyst loading (reducing the costs of the process).

In this sense, important consequences can be extracted: (i) the interdependence of substrate molar ratio and biocatalyst loading; (ii) the possibility of using reduced biocatalyst loadings in the reactor in some cases. We emphasize that this kind of analysis is not trivial, or even feasible, to be made using equilibrium constant or kinetics parameters (Vmax, Km, Kcat, Ki), whose determinations are laborious, reinforcing the potential utility of SER as a tool for predicting enzymatic esterification behavior. Figure 1 shows, in a simplified way, the convergence between molar ratio (expressed as a mass ratio) and biocatalyst loading that may lead to high conversions in esterification reactions in SFS. This convergence may be numerically translated by SER, which can be used as a tool for designing experiments and optimizing reaction conditions.

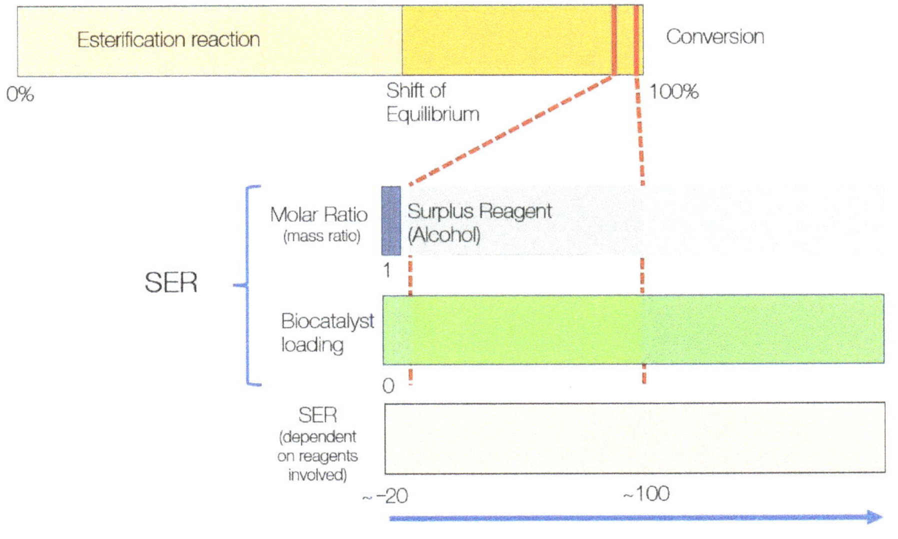

Figure 1. Graphical representation of the optimal reaction condition (molar ratio and biocatalyst loading) range for optimizing enzymatic esterification and its mathematical expression using SER.

2.3. Handling SER for Design Experiments

We discussed up to now the use of SER to analyze reaction conditions already established. However, the real interest is the use of SER to simplify the design of the process, thus predicting an optimum range of conditions and discarding exhaustive assays in extreme conditions.

When designing an optimization study, some additional information should be known to allow a reasonable prediction based on SER. The first one is the enzymatic activity of the biocatalyst used, which will establish a "starting point" of the biocatalyst loading studied and the reaction time. Enzymatic activity expresses the rate of the conversion of a given acid per time (usually per minute) per mass of biocatalyst (usually per gram); if we extrapolate this data for the acid used, we can obtain an estimation of the minimum quantity of biocatalyst is required to consume the acid within a specific time of the reaction. The second one is temperature, which should promote activation of the reaction without denaturation of the enzyme, taking into account the maximum temperature tolerated by the enzyme and the melting point/boiling point of the reagents. Moreover, we should establish the mass of the acid used and some bounds in which we want to find the biocatalyst loadings (considering its cost and the reaction time). Then, we should establish some SER values within the optimum estimated range (considering the nature of the reagents) and the respective mass of alcohol. The general procedure to design experiments with SER is given in Figure 2, and hypothetic examples of the application shown in Table 4.

The enzymatic activity of Novozym 435® is expressed as PLU (propyl dodecanoate units), which means that this data refers to the conversion of dodecanoic acid to esters in the conditions tested. It is not possible to extrapolate this data to the conversion of any other acid, and in this sense, we adopt an estimation that 260 PLU will be sufficient. This biocatalyst content is sufficient to convert 0.025 mols of dodecanoic acid to ester in 1.5 h. We can suggest that this biocatalyst loading will convert these different acids in esters in less than 3 h, considering the broad substrate specificity shown by Novozym 435® [20]. As the moles of hexadecanoic acid are much lower than 0.025, we suggested using minimum biocatalyst content (135 PLU). The temperature selection considers favoring the reaction kinetics, the melting of the hexadecanoic acid, and the fact that Novozym 435® is stable at high temperatures [20]. The selection of SER values was empiric, considering the analysis of the previous data carried out in this study, but we adopt the premise to make a coupled increase of molar ratio and enzyme loading. We already observed that the conditions that conciliate large stoichiometric excesses of reagents with low biocatalyst loadings resulted in a poor conversion due to the increased probability of enzyme inhibition. Moreover, we did not consider adopting an excess of acid to avoid potential issues related to the pH media.

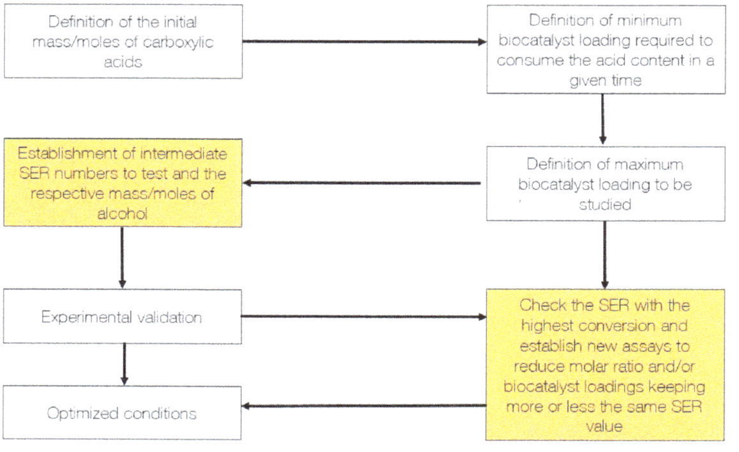

Figure 2. Fluxogram of steps required to design a set of experiments using SER.

Table 4. Hypothetical experimental design for enzymatic esterification using SER.

Synthesis (Product)	Pentyl Propionate	Ethyl Hexadecanoate	Dodecyl Hexanoate
Enzyme (commercial lipase)	Novozym 435® (7000 PLU)	Novozym 435® (7000 PLU)	Novozym 435® (7000 PLU)
Temperature	60 °C	70 °C	60 °C
Initial quantity of acid	1.85 g (0.025 mol)	1.85 g (0.0072 mol)	1.85 g (0.016 mol)
Minimum biocatalyst loading	0.037 g (260 PLU)	0.019 g (135 PLU)	0.037 g (260 PLU)
Maximum biocatalyst loading	0.065 g (455 PLU)	0.037 g (260 PLU)	0.065 g (455 PLU)
SER and the respective reaction conditions	SER 18 Molar ratio 1:1.1/Bioc. loading 2.0%	SER −43 Molar ratio 1:3.2/Bioc. loading 1.0%	SER 27 Molar ratio 1:0.9/Bioc. loading 2.0%
	SER 24 Molar ratio 1:1.5/Bioc. loading 3.5%	SER −30 Molar ratio 1:3.9/Bioc. loading 1.0%	SER 38 Molar ratio 1:1.1/Bioc. loading 2.0%
	SER 39 Molar ratio 1:2/Bioc. loading 2.0%	SER −18 Molar ratio 1:3.6/Bioc. loading 2.0%	SER 42 Molar ratio 1:1.4/Bioc. loading 3.5%
	SER 42 Molar ratio 1:1.5/Bioc. loading 2.0%	SER −1 Molar ratio 1:5.4/Bioc. loading 2.0%	SER 68 Molar ratio 1:1.7/Bioc. loading 3.5%
	SER 74 Molar ratio 1:3/Bioc. loading 3.5%	SER 4 Molar ratio 1:6/Bioc. loading 2.0%	SER 75 Molar ratio 1:2/Bioc. loading 3.5%

Bioc. loading = Biocatalyst loading, expressed as % *wt*/*wt* of the acid mass.

One may argue against using the mass of biocatalyst instead of using the enzymatic activity in SER calculation. Some reasons support our decision. The first one is that enzymatic activity data are expressed by many different methodologies, which may make the comparison between different studies unfeasible. Although enzymatic activity is indeed a more precise measure of catalytic activity than the mass of biocatalyst (which considers a significant proportion of immobilization support mass), the enzymatic activity is measured in units per mass (as U g^{-1}), which means a direct correlation between mass and quantity of active enzymes in the media. In an industrial context, it is reasonable to think that frequent enzymatic activity measures must be carried out as part of quality control of raw materials, giving the relation between the mass of biocatalyst and enzymatic activity. The second and most important reason is the influence of the immobilization support on the partition of the water in the system, which impacts the equilibrium position depending on the reaction media [1,10,90,96] and the influence of the immobilization support on kinetics aspects [42,92,97,98]. The third reason is that the measure of the mass is practical; indirectly, the mass of the biocatalyst in the system may bring valuable information about the operational costs, productivity, design of the reaction system, stirring rate, downstream processes, among others.

It is noteworthy that if the enzymatic esterification study aims to understand the mutual interaction between reaction variables, an extensive set of assays is required to generate accurate data that support the results. On the other hand, if the aim is to obtain optimized conditions for a specific synthesis, SER helps find the optimum molar ratio of reagents and enzyme loading. However, the reaction yield can still be increased by using different methods for shifting the reaction equilibrium, activating the reaction, and designing the reaction system [19,42,99–102].

Some experiments were carried out with different SER in various syntheses to obtain additional pieces of evidence of the relation between molar ratio and biocatalyst loading, as shown in Table 5. As a result, we can observe a general trend—variations in molar ratio and biocatalyst loading that generate discrete differences in SER resulted in similar conversions. In these cases, we increase simultaneously molar ratio (the stoichiometric excess of alcohol) and biocatalyst loading.

Table 5. Enzymatic syntheses of ethyl, butyl, and octyl dodecanoate, butyl octanoate, and butyl decanoate in different reaction conditions, respective SER, and conversion results.

Ethyl Dodecanoate		**Butyl Dodecanoate**		**Octyl Dodecanoate**		**Butyl Octanoate**		**Butyl Decanoate**	
SER	**Conversion**	**SER**	**Conversion**	**SER**	**Conversion**	**SER**	**Conversion**	**SER**	**Conversion**
13 (1:5.2) (1.5%)	89.0%	8 (1:3) (1.5%)	88.9%	10 (1:1.7) (1.5%)	90.7%	1 (1:1.9) (1.5%)	87.4%	7 (1:2.6) (1.5%)	89.5%
19 (1:6.4) (2.5%)	85.2%	27 (1:3.7) (2.5%)	92.1%	16 (1:2.1) (2.5%)	90.9%	13 (1:2.7) (2.5%)	90.3%	20 (1:3.5) (2.5%)	91.9%

The molar ratio acid:alcohol and biocatalyst loading (*wt*/*wt* mass of acid) used in each synthesis are described below the SER value. All syntheses were mediated by Novozym 435®, in closed stoppered flasks for 24 h, under 50 °C, 200 rpm, in a solvent-free system.

It is possible to obtain similar results when we carried out an equivalent reduction in keeping the same (or similar) resultant SER, which means the possibility of attaining advantageous conditions of molar ratio and biocatalyst loading. The variation of the conversion result with SER was also checked with esterification of decanoic acid and *n*-butanol mediated by Novozym 435® at 65 °C, using the SER equals 20 in two different conditions. The first one was equivalent to molar ratio acid:alcohol 1:3 and biocatalyst loading 1.5%, and the second one was molar ratio acid:alcohol 1:3.5 and biocatalyst loading 2.5%. Very similar conversions were observed: 93.5% and 93.6% after 24 h. Although both conditions did not represent a significant change in the system, we emphasize that this kind of analysis is helpful to reduce the biocatalyst loading. As the biocatalyst cost represents the highest cost item in a biocatalytic process, reduced biocatalyst loadings in the reactor are a way to overcome the main obstacle for a broader application of biocatalysis in the production of commodities and chemical specialties [49,50,52,83]. We also observed a discrete increase in the conversion of decanoic acid by increasing the temperature from 50 °C (Table 5) to 65 °C.

To illustrate how changes in biocatalyst loading impact SER, we can take as an example the synthesis of medium-chain alcohol (C8) and acid (C6). If we use molar ratio acid:alcohol 1:2 and biocatalyst loading (*wt*/*wt* of acid mass) of 5%, resultant SER will be 25; if we use the same molar ratio with biocatalyst loading 10%, a not considerable variation will occur—SER value becomes 12. In practical terms, this discrete difference can be associated with a common observation present in enzymatic esterification studies—an increase in biocatalyst loading after a certain level did not increase the conversion in a specific time [4,5,8,67,100]. Although the biocatalyst does not affect the equilibrium position (i.e., the yield of the reaction), this observation refers to the conversion obtained after a short interval of time, which turns evident the alterations in the reaction rate. As already mentioned, discrete differences in SER (inside the range of intermediate values) seem to be associated with marginal differences in the reaction conversion. Both conditions show high biocatalyst loadings, and it seems evident that there is an unnecessary excess of biocatalyst in the system—as biocatalysts are high-cost incomes, it is necessary to recycle them for obtaining a cost-effective biocatalytic process. However, if we adopt a biocatalyst loading of 1%, the resultant SER will be 124, which is an extreme value. In this case, it is possible to evaluate a coupled reduction of the enzyme loading and the stoichiometric excess generating a more favorable condition. SER will remain equals to 25 if we use biocatalyst loading 2% and a molar ratio 1:1.3. We reduce the potential inhibitor effect of the reagent (keeping their effect on the shift of equilibrium), coupling these effects with a proportional reduction of biocatalyst loading using SER as a guide. However, the reduction of biocatalyst loading is limited by the minimum quantity of enzyme required to process the reagents within a given time. This kind of rationale using SER as a guide shows how interesting this tool could be for economic purposes, considering the strict correlation between raw materials quantities (reagents and biocatalyst) or, in other words, between different operational costs terms.

2.4. Limitations of SER

The first explicit limitation of SER is related to the temperature of the reaction. It is well-known that temperature influences the equilibrium constant, but Flores and co-workers (2000) [81] found that the equilibrium position variation of a reaction between octadec-9-enoic acid acid and *n*-butanol was negligible between 35 and 50 °C. As already pointed out, we are dealing with reaction conditions (molar ratio and biocatalyst loading) at a constant temperature in a given synthesis to reduce the complexity of our analysis. The temperature probably produces some SER response variation due to the thermodynamics and kinetics temperature effects that should be further investigated. In here, we considered that the temperature was enough to promote a proper activation of the reaction without causing the enzyme denaturation.

The SER formulae, as proposed, did not consider the mass molar between reagents. The lack of this parameter in SER calculation may lead to inaccurate conclusions when a reaction between reagents with substantial differences in chain length (for example, a long-chain acid and short-chain alcohol) is evaluated. For example, using 5% of biocatalyst loading, an SER equals 20 is equivalent to the molar ratio acid:alcohol 1:10 in a reaction between tetradecanoic acid (C14) and ethanol (C2), but the same SER equals 20 is equivalent to the molar ratio acid:alcohol 1:1.1 in a reaction between butanoic acid (C4) and decanol (C10). We can manipulate the SER formulae to consider the molar mass of reagents keeping the outcome as a dimensionless number. Moreover, this alteration (proposed in Equation (2)) will avoid the outcome equals to zero, which was considered non-applicable when SER was first described [6], which is another explicit limitation of the equation. However, this alteration will produce specific ranges for every combination of reagents and, although noticeable gain in precision of the analyses may be obtained, general trends and predictions may be offset.

$$\mathrm{SER} = \frac{((\mathrm{m/M})\ \mathrm{alcohol} - (\mathrm{m/M})\ \mathrm{acid})(\mathrm{M\ alcohol})}{\mathrm{m\ biocatalyst}} \tag{2}$$

Another expected limitation of the SER equation is that the same outcome can be associated with more than one experimental condition (molar ratio and biocatalyst loading), even though both variables proportionality is maintained. However, there is no guarantee that this trend will be observed in reactions between short-chain acids and long-chain alcohols.

As the SER was described for synthesizing an aliphatic ester in SFS, this study focuses on applying the SER for other aliphatic molecules in the same condition. We can suppose that applying a mathematic relation like SER (as it is, or with modifications) could help predict enzymatic esterifications with alicyclic reagents, polyfunctional reagents (like ethylene glycol, glycerol, acrylic acid, for example), using free lipases, or in the presence of solvents.

Table 6 summarizes which information can be extracted from SER and its limitations, showing the advantages and the disadvantages of its use. Further studies should be performed to address all these issues, considering the practical applicability of SER in predicting different kinds of esterification reactions in SFS.

Table 6. Advantages and disadvantages of SER.

Advantages
Predictability of reaction' s behavior
Easy to handle
Reduced number of experiments to achieve high yields
Offer a range of reaction conditions that achieve high yields
Possibility to reduce the biocatalyst loading keeping high yields
Disadvantages
One SER number may be associated with more than one reaction condition
Optimum range may be shifted depending on the nature of the reagents
Temperature effects are not considered
Differences in molar ratio of reagents are not considered
Differences in esterification activity of lipases are not considered

3. Materials and Methods

3.1. Data Obtained from the Literature

A bibliographic survey was carried out to collect relevant enzymatic esterification studies in literature in which SER could be appropriately calculated and compared with our previous work [6]. To obtain a direct comparison and to facilitate the SER validation, only monofunctional and aliphatic reagents were considered. The following criteria were adopted for their selection: (i) ester syntheses should involve monofunctional carboxylic acids and alcohols in SFS using immobilized lipases; (ii) the selected studies should provide a detailed description of the quantities of reagents and biocatalysts (enzyme + support) used in the assays, allowing SER calculation. Studies that simultaneously fulfilled these criteria were considered for further analysis. However, we did not differentiate studies that applied a one-by-one variable optimization approach or RSM. In addition, studies that synthesize aliphatic esters from mixtures of different fatty acids or alcohols were not considered to facilitate the validation of SER.

As SER does not include the temperature effects, the data utilized for SER calculation were studied only at a constant temperature. For example—if a given study evaluated three different molar ratios and three different enzyme loadings at two different temperatures, only the set of data in which higher conversions were reported was used for SER application. In RSM studies, the maximum amount of data at a constant temperature was selected. The same rationale was applied when different stirring rates were evaluated. We considered that the selected temperature and stirring rate were high enough to favor the reaction but not so high to generate deleterious effects on immobilized lipases. Figure 3 shows the schematic procedure used to collect and calculate the data in both cases.

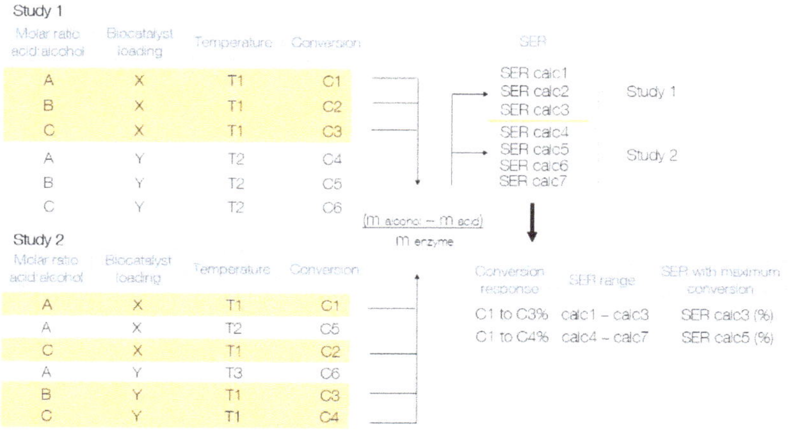

Figure 3. Schematic representation of collecting data from two hypothetical studies of enzymatic ester syntheses for SER calculation and compilation of obtained data.

The molar ratio and biocatalyst loading data extracted from the studies were compiled in a sheet, converted to masses (g) of alcohol, carboxylic acids, and biocatalysts when presented in mol or enzymatic activity per gram. The mass of biocatalyst considers the mass of the whole immobilized enzymes, accounting for protein and immobilization supports masses. The data collected were displayed in a Table, including the description of reagents (in IUPAC names) and immobilized lipase adopted. The respective SER outcomes were calculated as described in Equation (1) for each data of mass of reagents and biocatalyst. These SER outcomes were correlated to the conversion result obtained for each reaction condition. Thus, a range of SER numbers and conversion results were obtained. However, in some cases, only one condition of molar ratio and enzyme loading was available for calculations.

The range of SER numbers and the respective conversion results for the different ester syntheses were compared to the data reported by our group [6] to check if similar conclusions could be obtained—conversions above 80.0% but frequently around 90.0%, are associated with intermediate values of SER, i.e., between 0 and 65. Similarly, conversions below 80.0% could be associated with extreme values of SER, i.e., negative values or SER higher than 100. Therefore, we divided the data into three different sections: studied with (i) a good level of agreement, with optimum range between 0 and 65; (ii) an intermediate level of agreement, whose optimum SER are deviated from the estimated optimum range but still delimited between -10 and 100; (iii) and a low level of agreement, in which high conversions were observed outside the range or low conversions (using the maximum conversion observed as reference) were observed inside the range. In these cases, the likely reasons for the discrepancies were discussed. For the purpose of this text, in the discussion of the results the term "extreme values of SER" were considered as SER numbers largely outside the proposed optimum range (for example, ≥ -25 or ≤ 120); on other hand, the term "intermediate values of SER" were considered as SER within the proposed optimum range, or more specifically in a narrower range (for example, 5 or 30).

3.2. Data Obtained Experimentally

Part of the data used in this study were obtained experimentally by our group. Esterification reactions in a solvent-free system were carried out in 150 mL stoppered closed flasks under orbital stirring (200 rpm). Different molar ratios of carboxylic acid and alcohols were mixed, starting from 2.0 g of acid. For the syntheses of octyl decanoate and octyl dodecanoate, molar ratio acid:alcohol 1:1.3 were evaluated using different enzyme loadings of Novozym 435® and Lipozyme RM IM®, varying from 1.5% to 10.0% (*wt*/*wt* acid). To synthesize butyl octanoate, molar ratios acid:alcohol varied from 1:1 to 1:3 with enzyme loading 2.0% *wt/wt* acid. The synthesis of octyl octanoate mediated by Lipozyme RM IM® evaluated molar ratios acid:alcohol varied from 1:1 to 1:3 and enzyme loadings from 2.5% to 4.0% *wt/wt* acid. Syntheses of dodecanoates—ethyl dodecanoate, butyl dodecanoate, and octyl dodecanoate—and butyl decanoate mediated by Novozym 435® were carried out in the reaction conditions described in Table 6, aiming to check slight variations in reaction condition using SER as a parameter. These systems were studied considering the previous experience of our group in medium-chain esters syntheses [6]. The immobilized lipases used, Novozym 435® and Lipozyme RM IM®, were provided by Novozymes Latin American (Araucária, Brazil). Reactions between octanoic acid and *n*-butanol or *n*-octanol were carried out at 65 °C for 3 h; reactions between decanoic acid and *n*-butanol or *n*-octanol, and dodecanoic acid with ethanol, *n*-butanol, and *n*-octanol were carried out at 50 °C for 24 h. All reagents were of analytical grade. The titration of the samples was used to monitor the progress of the reaction by the reduction of the acidity index. The samples were collected at the beginning and the end of the reactions. The conversions of the esterification reactions were calculated by the percentual reduction observed between the final and initial acidity index.

4. Conclusions

This study shows that SER is an empiric mathematical relation that presents a reasonable agreement with the literature results for solvent-free enzymatic syntheses of aliphatic esters. SER offers a simple way to conciliate the thermodynamics and kinetics aspects towards high conversions in enzymatic ester syntheses, providing a helpful overview of the reaction variables. In combination with other physicochemical parameters, a general prediction of the reaction's behavior can be obtained using SER. Its main advantage is to indicate a range of conditions in which the equilibrium shift towards synthesis may be attained without inhibition of immobilized lipases. By analyzing the application of the SER trend which indicated a range of reaction conditions correspondent to 0 from 65 as an optimum in a mid-chain ester synthesis, we observed an expanded range of −20 from 100 as potentially able to cover a variety of cases—short, mid, and long-chain esters syntheses—taken into account certain exceptions when esterifications are carried under ultrasound or microwaves, or with addition of water. A fast and simple methodology to design experiments using SER was proposed, by defining some boundaries of temperature, reaction time, and maximum biocatalyst loading, and after the determination of molar ratio using arbitrary SER values inside the estimated optimum range for the esterification reaction, considering the nature of the reagents involved. Although some limitations of the SER application are evident, its use brings attractive features—the possibility of achieving high yields with a reduced number of experiments, the possibility of understanding the practical implications of thermodynamics and kinetics on the syntheses, and the obtaining of more realistic reaction conditions for scaling-up (reduced biocatalyst loadings). Specific studies are required to evaluate how robust this mathematical tool is in various cases and an accurate definition of the optimum SER range and extreme values for each case. SER was applied successfully for the synthesis of octyl octanoate and its application can be extended to the optimization of other similar syntheses, considering the analyses carried out in this study with an extensive set of data.

Author Contributions: This article was performed as collaborative research between R.R.d.S. and A.S.d.S. For R.R.d.S., the work included: conceptualization, methodology, formal analysis, data curation, visualization, and writing—original, draft preparation; A.S.d.S. performed conceptualization, methodology, review, and corrections; V.S.F.-L. and R.F.-L., performed review, corrections, main supervision, project administration, and funding acquisition. All authors have read and agreed to the published version of the manuscript.

Funding: INT Biocatalysis Laboratory was funded by Ministry of Science, Technology, and Innovations from the Brazilian Government. Brazilian National Council for Scientific and Technological Development (CNPq) grant number 306619/2019-9. ICP-CSIC Biocatalysis Department was funded by Ministry of Science and Innovation from the Spanish Government (project number CTQ2017-86170-R), and CSIC for the project AEP045. The data presented in this study are available within the article.

Acknowledgments: We gratefully recognize the support from the Ministry of Science, Technology, and Innovations from the Brazilian Government for PIBITI grants, Ministry of Science and Innovation from the Spanish Government (project number CTQ2017-86170-R), and CSIC for the project AEP045.

Conflicts of Interest: The authors declare no conflict of interest.

References

1. Sandoval, G.; Condoret, J.S.; Monsan, P.; Marty, A. Esterification by Immobilized Lipase in Solvent-Free Media: Kinetic and Thermodynamic Arguments. *Biotechnol. Bioeng.* **2002**, *78*, 313–320. [CrossRef] [PubMed]
2. Yahya, A.R.M.; Anderson, W.A.; Moo-young, M. Ester synthesis in lipase catalyzed reactions. *Enzym. Microb. Technol.* **1998**, *23*, 438–450. [CrossRef]
3. Chowdhury, A.; Mitra, D. A kinetic study on the Novozyme 435-catalyzed esterification of free fatty acids with octanol to produce octyl esters. *Biotechnol. Prog.* **2015**, *31*, 1494–1499. [CrossRef] [PubMed]
4. Kuperkar, V.V.; Lade, V.G.; Prakash, A.; Rathod, V.K. Synthesis of isobutyl propionate using immobilized lipase in a solvent free system: Optimization and kinetic studies. *J. Mol. Catal. B Enzym.* **2014**, *99*, 143–149. [CrossRef]
5. Vadgama, R.N.; Odaneth, A.A.; Lali, A.M. Green synthesis of isopropyl tetradecanoate in novel single phase medium Part I: Batch optimization studies. *Biotechnol. Rep.* **2015**, *8*, 133–137. [CrossRef]
6. Sousa, R.R.; Pazutti, L.V.B.; Dalmaso, G.Z.L.; Siqueira, D.F.; Silva, A.S.; Ferreira-Leitão, V.S. A practical approach to obtain high yield lipase-mediated synthesis of octyl octanoate with Novozym 435. *Biocatal. Biotransform.* **2020**, *38*, 293–303. [CrossRef]
7. Sose, M.T.; Bansode, S.R.; Rathod, V.K. Solvent free lipase catalyzed synthesis of butyl octanoate. *J. Chem. Sci.* **2017**, *129*, 1755–1760. [CrossRef]
8. Aljawish, A.; Heuson, E.; Bigan, M.; Froidevaux, R. Lipase catalyzed esterification of formic acid in solvent and solvent-free systems. *Biocatal. Agric. Biotechnol.* **2019**, *20*, 101221. [CrossRef]
9. Parikh, D.T.; Lanjekar, K.J.; Rathod, V.K. Kinetics and thermodynamics of lipase catalysed synthesis of propyl decanoate. *Biotechnol. Lett.* **2019**, *41*, 1163–1175. [CrossRef]
10. Adlercreutz, P. Immobilisation and application of lipases in organic media. *Chem. Soc. Rev.* **2013**, *42*, 6406–6436. [CrossRef]
11. Sheldon, R.A.; van Pelt, S. Enzyme immobilisation in biocatalysis: Why, what and how. *Chem. Soc. Rev.* **2013**, *42*, 6223–6235. [CrossRef]
12. Thangaraj, B.; Solomon, P.R. Immobilization of Lipases—A Review. Part I: Enzyme Immobilization. *ChemBioEng Rev.* **2019**, *6*, 157–166. [CrossRef]
13. Mateo, C.; Palomo, J.M.; Fernandez-Lorente, G.; Guisan, J.M.; Fernandez-Lafuente, R. Improvement of enzyme activity, stability and selectivity via immobilization techniques. *Enzym. Microb. Technol.* **2007**, *40*, 1451–1463. [CrossRef]
14. Rodrigues, R.C.; Ortiz, C.; Berenguer-Murcia, Á.; Torres, R.; Fernández-Lafuente, R. Modifying enzyme activity and selectivity by immobilization. *Chem. Soc. Rev.* **2013**, *42*, 6290–6307. [CrossRef] [PubMed]
15. Garcia-Galan, C.; Berenguer-Murcia, Á.; Fernandez-Lafuente, R.; Rodrigues, R.C. Potential of different enzyme immobilization strategies to improve enzyme performance. *Adv. Synth. Catal.* **2011**, *353*, 2885–2904. [CrossRef]
16. Barbosa, O.; Ortiz, C.; Berenguer-Murcia, Á.; Torres, R.; Rodrigues, R.C.; Fernandez-Lafuente, R. Strategies for the one-step immobilization-purification of enzymes as industrial biocatalysts. *Biotechnol. Adv.* **2015**, *33*, 435–456. [CrossRef] [PubMed]
17. Fernandez-Lafuente, R. Lipase from Thermomyces lanuginosus: Uses and prospects as an industrial biocatalyst. *J. Mol. Catal. B Enzym.* **2010**, *62*, 197–212. [CrossRef]
18. Rodrigues, R.C.; Fernandez-Lafuente, R. Lipase from *Rhizomucor miehei* as a biocatalyst in fats and oils modification. *J. Mol. Catal. B Enzym.* **2010**, *66*, 15–32. [CrossRef]
19. Khan, N.R.; Rathod, V.K. Enzyme catalyzed synthesis of cosmetic esters and its intensification: A review. *Process Biochem.* **2015**, *50*, 1793–1806. [CrossRef]
20. Ortiz, C.; Ferreira, M.L.; Barbosa, O.; Dos Santos, J.C.S.; Rodrigues, R.C.; Berenguer-Murcia, Á.; Briand, L.E.; Fernandez-Lafuente, R. Novozym 435: The "perfect" lipase immobilized biocatalyst? *Catal. Sci. Technol.* **2019**, *9*, 2380–2420. [CrossRef]
21. Nielsen, P.M.; Brask, J.; Fjerbaek, L. Enzymatic biodiesel production: Technical and economical considerations. *Eur. J. Lipid Sci. Technol.* **2008**, *110*, 692–700. [CrossRef]
22. Szczesna Antczak, M.; Kubiak, A.; Antczak, T.; Bielecki, S. Enzymatic biodiesel synthesis—Key factors affecting efficiency of the process. *Renew. Energy* **2009**, *34*, 1185–1194. [CrossRef]
23. Pourzolfaghar, H.; Abnisa, F.; Daud, W.M.A.W.; Aroua, M.K. A review of the enzymatic hydroesterification process for biodiesel production. *Renew. Sustain. Energy Rev.* **2016**, *61*, 245–257. [CrossRef]

24. Neta, N.S.; Teixeira, J.A.; Rodrigues, L.R. Sugar Ester Surfactants: Enzymatic Synthesis and Applications in Food Industry. *Crit. Rev. Food Sci. Nutr.* **2015**, *55*, 595–610. [CrossRef] [PubMed]
25. Aravindan, R.; Anbumathi, P.; Viruthagiri, T. Lipase applications in food industry. *Indian J. Biotechnol.* **2007**, *6*, 141–158.
26. Hasenhuettl, G.L. Synthesis and commercial preparation of food emulsifiers. In *Food Emulsifiers and Their Applications*, 2nd ed.; Springer: Cham, Switzerlands, 2008; pp. 11–37. [CrossRef]
27. Ansorge-Schumacher, M.B.; Thum, O. Immobilised lipases in the cosmetics industry. *Chem. Soc. Rev.* **2013**, *42*, 6475–6490. [CrossRef]
28. Bolina, I.C.A.; Gomes, R.A.B.; Mendes, A.A. Biolubricant Production from Several Oleaginous Feedstocks Using Lipases as Catalysts: Current Scenario and Future Perspectives. *Bioenergy Res.* **2021**. [CrossRef]
29. Cecilia, J.A.; Plata, D.B.; Saboya, R.M.A.; de Luna, F.M.T.; Cavalcante, C.L.; Rodríguez-Castellón, E. An overview of the biolubricant production process: Challenges and future perspectives. *Processes* **2020**, *8*, 257. [CrossRef]
30. Straathof, A.J.J. Transformation of biomass into commodity chemicals using enzymes or cells. *Chem. Rev.* **2014**, *114*, 1871–1908. [CrossRef]
31. Halling, P.J. Solvent selection for biocatalysis in mainly organic systems: Predictions of effects on equilibrium position. *Biotechnol. Bioeng.* **1990**, *35*, 691–701. [CrossRef]
32. Kasche, V. Mechanism and yields in enzyme catalysed equilibrium and kinetically controlled synthesis of β-lactam antibiotics, peptides and other condensation products. *Enzym. Microb. Technol.* **1986**, *8*, 4–16. [CrossRef]
33. Kasche, V.; Haufler, U.; Riechmann, L. Equilibrium and Kinetically Controlled Synthesis with Enzymes: Semisynthesis of Penicillins and Peptides. In *Methods in Enzymology*; Academic Press: Cambridge, MA, USA, 1987; Volume 136, pp. 280–292. ISBN 9780121820367.
34. Halling, P.J. Thermodynamic predictions for biocatalysis in nonconventional media: Theory, tests, and recommendations for experimental design and analysis. *Enzym. Microb. Technol.* **1994**, *16*, 178–206. [CrossRef]
35. Hari Krishna, S.; Karanth, N.G. Lipases and Lipase-Catalyzed Esterification Reactions in Nonaqueous Media. *Catal. Rev.* **2002**, *44*, 499–591. [CrossRef]
36. Castillo, E.; Torres-Gavillan, A.; Sandoval, G.; Marty, A. Thermodynamical Methods for Optimization of Lipase-Catalyzed Reactions. In *Lipases and Phospholipases*; Humana Press: Totowa, NJ, USA, 2012; Volume 861, pp. 383–400.
37. Colombié, S.; Tweddell, R.J.; Condoret, J.S.; Marty, A. Water activity control: A way to improve the efficiency of continuous lipase esterification. *Biotechnol. Bioeng.* **1998**, *60*, 362–368. [CrossRef]
38. Castillo, E.; Dossat, V.; Marty, A.; Stéphane Condoret, J.; Combes, D. The role of silica gel in lipase-catalyzed esterification reactions of high-polar substrates. *JAOCS J. Am. Oil Chem. Soc.* **1997**, *74*, 77–85. [CrossRef]
39. Dossat, V.; Combes, D.; Marty, A. Continuous enzymatic transesterification of high oleic sunflower oil in a packed bed reactor: Influence of the glycerol production. *Enzym. Microb. Technol.* **1999**, *25*, 194–200. [CrossRef]
40. Marty, A.; Dossat, V.; Condoret, J.S. Continuous operation of lipase-catalyzed reactions in nonaqueous solvents: Influence of the production of hydrophilic compounds. *Biotechnol. Bioeng.* **1997**, *56*, 232–237. [CrossRef]
41. Séverac, E.; Galy, O.; Turon, F.; Pantel, C.A.; Condoret, J.S.; Monsan, P.; Marty, A. Selection of CalB immobilization method to be used in continuous oil transesterification: Analysis of the economical impact. *Enzym. Microb. Technol.* **2011**, *48*, 61–70. [CrossRef]
42. Martins, A.B.; Schein, M.F.; Friedrich, J.L.R.; Fernandez-Lafuente, R.; Ayub, M.A.Z.; Rodrigues, R.C. Ultrasound-assisted butyl acetate synthesis catalyzed by Novozym 435: Enhanced activity and operational stability. *Ultrason. Sonochem.* **2013**, *20*, 1155–1160. [CrossRef]
43. Paludo, N.; Alves, J.S.; Altmann, C.; Ayub, M.A.Z.; Fernandez-Lafuente, R.; Rodrigues, R.C. The combined use of ultrasound and molecular sieves improves the synthesis of ethyl butyrate catalyzed by immobilized Thermomyces lanuginosus lipase. *Ultrason. Sonochem.* **2014**, *22*, 89–94. [CrossRef]
44. Martins, A.B.; Friedrich, J.L.R.; Cavalheiro, J.C.; Garcia-Galan, C.; Barbosa, O.; Ayub, M.A.Z.; Fernandez-Lafuente, R.; Rodrigues, R.C. Improved production of butyl butyrate with lipase from *Thermomyces lanuginosus* immobilized on styrene-divinylbenzene beads. *Bioresour. Technol.* **2013**, *134*, 417–422. [CrossRef] [PubMed]
45. Fallavena, L.P.; Antunes, F.H.F.; Alves, J.S.; Paludo, N.; Ayub, M.A.Z.; Fernandez-Lafuente, R.; Rodrigues, R.C. Ultrasound technology and molecular sieves improve the thermodynamically controlled esterification of butyric acid mediated by immobilized lipase from *Rhizomucor miehei*. *RSC Adv.* **2014**, *4*, 8675–8681. [CrossRef]
46. Graebin, N.G.; Martins, A.B.; Lorenzoni, A.S.G.; Garcia-Galan, C.; Fernandez-Lafuente, R.; Ayub, M.A.Z.; Rodrigues, R.C. Immobilization of lipase B from *Candida antarctica* on porous styrene-divinylbenzene beads improves butyl acetate synthesis. *Biotechnol. Prog.* **2012**, *28*, 406–412. [CrossRef]
47. Poppe, J.K.; Garcia-Galan, C.; Matte, C.R.; Fernandez-Lafuente, R.; Rodrigues, R.C.; Ayub, M.A.Z. Optimization of synthesis of fatty acid methyl esters catalyzed by lipase B from *Candida antarctica* immobilized on hydrophobic supports. *J. Mol. Catal. B Enzym.* **2013**, *94*, 51–56. [CrossRef]

48. Alves, J.S.; Garcia-Galan, C.; Schein, M.F.: Silva, A.M.; Barbosa, O.; Ayub, M.A.Z.; Fernandez-Lafuente, R.; Rodrigues, R.C. Combined effects of ultrasound and immobilization protocol on butyl acetate synthesis catalyzed by CALB. *Molecules* **2014**, *19*, 9562–9576. [CrossRef]
49. Martins, A.B.; Graebin, N.G.; Lorenzoni, A.S.G.; Fernandez-Lafuente, R.; Ayub, M.A.Z.; Rodrigues, R.C. Rapid and high yields of synthesis of butyl acetate catalyzed by Novozym 435: Reaction optimization by response surface methodology. *Process Biochem.* **2011**, *46*, 2311–2316. [CrossRef]
50. Aguieiras, E.C.G.; Ribeiro, D.S.; Couteiro, P.P.; Bastos, C.M.B.; de Queiroz, D.S.; Parreira, J.M.; Langone, M.A.P. Investigation of the Reuse of Immobilized Lipases in Biodiesel Synthesis: Influence of Different Solvents in Lipase Activity. *Appl. Biochem. Biotechnol.* **2016**, *179*, 485–496. [CrossRef] [PubMed]
51. Mulalee, S.; Srisuwan, P.; Phisalaphong, M. Influences of operating conditions on biocatalytic activity and reusability of Novozym 435 for esterification of free fatty acids with short-chain alcohols: A case study of palm fatty acid distillate. *Chin. J. Chem. Eng.* **2015**, *23*, 1851–1856. [CrossRef]
52. Serrano-Arnaldos, M.; Montiel, M.C.; Ortega-Requena, S.; Máximo, F.; Bastida, J. Development and economic evaluation of an eco-friendly biocatalytic synthesis of emollient esters. *Bioprocess Biosyst. Eng.* **2019**, *43*, 495–505. [CrossRef]
53. Aguieiras, E.C.G.; de Barros, D.S.N.; Sousa, H.; Fernandez-Lafuente, R.; Freire, D.M.G. Influence of the raw material on the final properties of biodiesel produced using lipase from *Rhizomucor miehei* grown on babassu cake as biocatalyst of esterification reactions. *Renew. Energy* **2017**, *113*, 112–118. [CrossRef]
54. Alves, M.D.; Cren, É.C.; Mendes, A.A. Kinetic, thermodynamic, optimization and reusability studies for the enzymatic synthesis of a saturated wax ester. *J. Mol. Catal. B Enzym.* **2016**, *133*, S377–S387. [CrossRef]
55. Bhavsar, K.V.; Yadav, G.D. Process intensification by microwave irradiation in immobilized-lipase catalysis in solvent-free synthesis of ethyl valerate. *Mol. Catal.* **2018**, *461*, 34–39. [CrossRef]
56. Bhavsar, K.V.; Yadav, G.D. Microwave assisted solvent-free synthesis of *n*-butyl propionate by immobilized lipase as catalyst. *Biocatal. Agric. Biotechnol.* **2018**, *14*, 264–269. [CrossRef]
57. Garcia, T.; Sanchez, N.; Martinez, M.; Aracil, J. Enzymatic synthesis of fatty esters Part II. Optimization studies. *Enzym. Microb. Technol.* **1999**, *25*, 591–597. [CrossRef]
58. Hari Krishna, S.; Divakar, S.; Prapulla, S.G.; Karanth, N.G. Enzymatic synthesis of isoamyl acetate using immobilized lipase from *Rhizomucor miehei*. *J. Biotechnol.* **2001**, *87*, 193–201. [CrossRef]
59. Chang, S.-W.; Shaw, J.-F.; Yang, K.-H.; Shih, I.-L.; Hsieh, C.-H.; Shieh, C.-J. Optimal lipase-catalyzed formation of hexyl laurate. *Green Chem.* **2005**, *7*, 547. [CrossRef]
60. Güvenç, A.; Kapucu, N.; Kapucu, H.; Aydoğan, Ö.; Mehmetoğlu, Ü. Enzymatic esterification of isoamyl alcohol obtained from fusel oil: Optimization by response surface methodolgy. *Enzym. Microb. Technol.* **2007**, *40*, 778–785. [CrossRef]
61. Mahapatra, P.; Kumari, A.; Kumar, V. Enzymatic synthesis of fruit flavor esters by immobilized lipase from *Rhizopus oligosporus* optimized with response surface methodology. *J. Mol. Catal. B Enzym.* **2009**, *60*, 57–63. [CrossRef]
62. Richetti, A.; Leite, S.G.F.; Antunes, O.A.C.; Souza, A.L.F. De Optimization of 2-ethylhexyl palmitate Production Using Lipozyme RM IM as Catalyst in a Solvent-Free System. *Appl. Biochem. Biotechnol.* **2010**, *160*, 2498–2508. [CrossRef]
63. Barros, D.; Azevedo, A.A.; Cabral, J.; Fonseca, L. Optimization Of Flavor Esters Synthesis By *Fusarium Solani Pisi* Cutinase. *J. Food Biochem.* **2012**, *36*, 275–284. [CrossRef]
64. Rahman, I.N.A.; Manan, F.M.A.; Marzuki, N.H.C.; Mahat, N.A.; Attan, N.; Keyon, A.S.A.; Jamalis, J.; Aboul-Enein, H.Y.; Wahab, R.A. A statistical approach for optimizing the high yield green production of the flavor ester butyl butyrate. *J. Teknol.* **2017**, *79*, 141–151. [CrossRef]
65. Jaiswal, K.S.; Rathod, V.K. Acoustic cavitation promoted lipase catalysed synthesis of isobutyl propionate in solvent free system: Optimization and kinetic studies. *Ultrason. Sonochem.* **2018**, *40*, 727–735. [CrossRef]
66. Serrano-Arnaldos, M.; Bastida, J.; Máximo, F.; Ortega-Requena, S.; Montiel, C. One-Step Solvent-Free Production of a Spermaceti Analogue Using Commercial Immobilized Lipases. *ChemistrySelect* **2018**, *3*, 748–752. [CrossRef]
67. Santos, J.C.; Bueno, T.; Rós, P.C.M.; de Castro, H.F. Lipase-catalyzed Synthesis of Butyl Esters by Direct Esterification in Solvent-Free System. *J. Chem. Technol. Biotechnol.* **2007**, *82*, 956–961. [CrossRef]
68. Karra-chaâbouni, M.; Ghamgui, H.; Bezzine, S.; Rekik, A.; Gargouri, Y. Production of flavour esters by immobilized *Staphylococcus simulans* lipases in a solvent-free system. *Process Biochem.* **2006**, *41*, 1692–1698. [CrossRef]
69. Rocha, J.M.S.; Gil, M.H.; Garcia, F.A.P. Optimisation of the enzymatic synthesis of *n*-octyl oleate with immobilised lipase in the absence of solvents. *J. Chem. Technol. Biotechnol.* **1999**, *74*, 607–612. [CrossRef]
70. Dai, W.C.; Chiu, S.J.; Huang, D.Y.; Juan, H.Y.; Chen, C.Y.; Chen, S.S.; Su, C.H.; Li, S.Y. Lipase-catalyzed synthesis of butyl propionate in solvent-free system: Optimization by response surface methodology. *J. Taiwan Inst. Chem. Eng.* **2014**, *45*, 2233–2237. [CrossRef]
71. Sousa, R.R.; Costa, M.; Pinto, C.; Cristina, E.; Aguieiras, G.; Cipolatti, E.P.; Andrade, E.; Ayla, M.; Ana, S.; Pinto, J.C.; et al. Comparative performance and reusability studies of lipases on syntheses of octyl esters with an economic approach. *Bioprocess Biosyst. Eng.* **2021**. [CrossRef] [PubMed]
72. Ghamgui, H.; Karra-Chaabouni, M.; Gargouri, Y. 1-Butyl oleate synthesis by immobilized lipase from *Rhizopus oryzae*: A comparative study between n-hexane and solvent-free system. *Enzym. Microb. Technol.* **2004**, *35*, 355–363. [CrossRef]

73. Ben Salah, R.; Ghamghui, H.; Miled, N.; Mejdoub, H.; Gargouri, Y. Production of butyl acetate ester by lipase from novel strain of *Rhizopus oryzae*. *J. Biosci. Bioeng.* **2007**, *103*, 368–372. [CrossRef] [PubMed]
74. Ghamgui, H.; Karra-chaabouni, M.; Bezzine, S.; Miled, N.; Gargouri, Y. Production of isoamyl acetate with immobilized *Staphylococcus simulans* lipase in a solvent-free system. *Enzym. Microb. Technol.* **2006**, *38*, 788–794. [CrossRef]
75. Gawas, S.D.; Jadhav, S.V.; Rathod, V.K. Solvent Free Lipase Catalysed Synthesis of Ethyl Laurate: Optimization and Kinetic Studies. *Appl. Biochem. Biotechnol.* **2016**, *180*, 1428–1445. [CrossRef] [PubMed]
76. Yadav, G.D.; Thorat, P.A. Microwave assisted lipase catalyzed synthesis of isoamyl myristate in solvent-free system. *J. Mol. Catal. B Enzym.* **2012**, *83*, 16–22. [CrossRef]
77. Pereira, G.N.; Holz, J.P.; Giovannini, P.P.; Oliveira, J.V.; de Oliveira, D.; Lerin, L.A. Enzymatic esterification for the synthesis of butyl stearate and ethyl stearate. *Biocatal. Agric. Biotechnol.* **2018**, *16*, 373–377. [CrossRef]
78. Goldberg, M.; Thomas, D.; Legoy, M.D. Water activity as a key parameter of synthesis reactions: The example of lipase in biphasic (liquid/solid) media. *Enzym. Microb. Technol.* **1990**, *12*, 976–981. [CrossRef]
79. Nordblad, M.; Adlercreutz, P. Effects of acid concentration and solvent choice on enzymatic acrylation by *Candida antarctica* lipase B. *J. Biotechnol.* **2008**, *133*, 127–133. [CrossRef]
80. Bélafi-Bakó, K.; Badr, A.K.; Ehrenstein, U.; Gubicza, L. Kinetics of Ethyl Acetate Formation by Lipase in Organic Solvent and Solvent-Free System. *Chem. Pap.* **2003**, *57*, 278–281.
81. Flores, M.V.; Sewalt, J.J.W.; Janssen, A.E.M.; Van Der Padt, A. The Nature of Fatty Acid Modifies the Equilibrium Position in the Esterification Catalyzed by Lipase. *Biotechnol. Bioeng.* **2000**, *67*, 364–371. [CrossRef]
82. Paiva, A.L.; Balcão, V.M.; Malcata, F.X. Kinetic and mechanisms of reactions catalyzed by immobilized lipases. *Enzym. Microb. Technol.* **2000**, *27*, 187–204. [CrossRef]
83. Foresti, M.L.; Pedernera, M.; Ferreira, M.L.; Bucalá, V. Kinetic modeling of enzymatic ethyl oleate synthesis carried out in biphasic systems. *Appl. Catal. A Gen.* **2008**, *334*, 65–72. [CrossRef]
84. Lopresto, G.C.; Calabrò, V.; Woodley, J.M.; Tufvesson, P. Enzymatic kinetic study on the enzymatic esterification of octanoic acid and hexanol by immobilized *Candida antarctica* lipase B. *J. Mol. Catal. B. Enzym.* **2014**, *110*, 64–71. [CrossRef]
85. Eggers, D.K.; Blanch, H.W.; Prausnitz, J.M. Extractive catalysis: Solvent effects on equilibria of enzymatic reactions in two-phase systems. *Enzym. Microb. Technol.* **1989**, *11*, 84–89. [CrossRef]
86. Valivety, R.H.; Johnston, G.A.; Suckling, C.J.; Halling, P.J. Solvent effects on biocatalysis in organic systems: Equilibrium position and rates of lipase catalyzed esterification. *Biotechnol. Bioeng.* **1991**, *38*, 1137–1143. [CrossRef] [PubMed]
87. Janseen, A.E.M.; Van der Padt, A.; Van Sonsbeek, H.M.; Van't Riet, K. The effect of organic solvents on the equilibrium position of enzymatic acylglycerol synthesis. *Biotechnol. Bioeng.* **1993**, *41*, 95–103. [CrossRef]
88. Janssen, A.E.M.; Van Der Padt, A.; Van't Riet, K. Solvent Effects on Lipase-Catalyzed Esterification of Glycerol and Fatty Acids. *Biotechnol. Bioeng.* **1993**, *42*, 953–962. [CrossRef]
89. Güvenç, A.; Kapucu, N.; Mehmetoğlu, Ü. The production of isoamyl acetate using immobilized lipases in a solvent-free system. *Process Biochem.* **2002**, *38*, 379–386. [CrossRef]
90. Bucalá, V.; Briozzo, M.; Foresti, M.L.; Ferreira, M.L.; Trubiano, G.; Bottini, S. Influence of liquid-liquid equilibria on the modelling of solvent-free ethyl oleate synthesis. In Proceedings of the 2nd Mercosur Congress on Chemical Engineering, Rio de Janeiro, Brazil, 14–18 August 2005; pp. 1–10.
91. Garcia, S.; Vidinha, P.; Arvana, H.; Gomes Da Silva, M.D.R.; Ferreira, M.O.; Cabral, J.M.S.; Macedo, E.A.; Harper, N.; Barreiros, S. Cutinase activity in supercritical and organic media: Water activity, solvation and acid-base effects. *J. Supercrit. Fluids* **2005**, *35*, 62–69. [CrossRef]
92. Garcia, T.; Sanchez, N.; Martinez, M.; Aracil, J. Enzymatic synthesis of fatty esters. Part I. Kinetic approach. *Enzym. Microb. Technol.* **1999**, *25*, 584–590. [CrossRef]
93. Martins, A.B.; Da Silva, A.M.; Schein, M.F.; Garcia-Galan, C.; Záchia Ayub, M.A.; Fernandez-Lafuente, R.; Rodrigues, R.C. Comparison of the performance of commercial immobilized lipases in the synthesis of different flavor esters. *J. Mol. Catal. B Enzym.* **2014**, *105*, 18–25. [CrossRef]
94. Barbosa, O.; Ortiz, C.; Torres, R.; Fernandez-Lafuente, R. Effect of the immobilization protocol on the properties of lipase B from *Candida antarctica* in organic media: Enantiospecifc production of atenolol acetate. *J. Mol. Catal. B Enzym.* **2011**, *71*, 124–132. [CrossRef]
95. Fernández-Lorente, G.; Betancor, L.; Carrascosa, A.V.; Palomo, J.M.; Guisan, J.M. Modulation of the selectivity of immobilized lipases by chemical and physical modifications: Release of omega-3 fatty acids from fish oil. *JAOCS J. Am. Oil Chem. Soc.* **2012**, *89*, 97–102. [CrossRef]
96. Foresti, M.L.; Ferreira, M.L. Solvent-free ethyl oleate synthesis mediated by lipase from *Candida antarctica* B adsorbed on polypropylene powder. *Catal. Today* **2005**, *107–108*, 23–30. [CrossRef]
97. Condoret, J.S.; Vankan, S.; Joulia, X.; Marty, A. Prediction of water adsorption curves for heterogeneous biocatalysis in organic and supercritical solvents. *Chem. Eng. Sci.* **1997**, *52*, 213–220. [CrossRef]
98. Eby, J.M.; Peretti, S.W. Performance in synthetic applications of a yeast surface display-based biocatalyst. *RSC Adv.* **2015**, *5*, 30425–30432. [CrossRef]

99. Tacias-Pascacio, V.G.; Virgen-Ortíz, J.J.; Jiménez-Pérez, M.; Yates, M.; Torrestiana-Sanchez, B.; Rosales-Quintero, A.; Fernandez-Lafuente, R. Evaluation of different lipase biocatalysts in the production of biodiesel from used cooking oil: Critical role of the immobilization support. *Fuel* **2017**, *200*, 1–10. [CrossRef]
100. Khan, N.R.; Rathod, V.K. Microwave mediated lipase-catalyzed synthesis of *n*-butyl palmitate and thermodynamic studies. *Biocatal. Agric. Biotechnol.* **2020**, *29*, 101741. [CrossRef]
101. Baum, S.; Mueller, J.J.; Hilterhaus, L.; Eckstein, M.; Thum, O.; Liese, A. The bubble column reactor: A novel reactor type for cosmetic esters. In *Applied Biocatalysis: From Fundamental Science to Industrial Applications*; Wiley: Hoboken, NJ, USA, 2016; pp. 343–366. [CrossRef]
102. Sousa, R.R.; Silva, A.S.; Fernandez-Lafuente, R.; Ferreira-Leitão, V.S. Solvent-free esterifications mediated by immobilized lipases: A review from thermodynamic and kinetic perspectives. *Catal Sci. Technol.* **2021**, *11*, 5696–5711. [CrossRef]

catalysts

Article

Kinetic and Structural Properties of a Robust Bacterial L-Amino Acid Oxidase

Simone Savino [1,2], J. Daniël-Moráh Meijer [2], Henriëtte J. Rozeboom [2], Hugo L. van Beek [1] and Marco W. Fraaije [2,*]

[1] GECCO Biotech, Nijenborgh 4, 9747 AG Groningen, The Netherlands; s.savino@rug.nl (S.S.); h.l.van.beek@gecco-biotech.com (H.L.v.B.)
[2] Molecular Enzymology Group, Groningen Biomolecular Sciences and Biotechnology Institute, University of Groningen, Nijenborgh 4, 9747 AG Groningen, The Netherlands; j.d.m.meijer@rug.nl (J.D.-M.M.); h.j.rozeboom@rug.nl (H.J.R.)
* Correspondence: m.w.fraaije@rug.nl

Abstract: L-Amino acid oxidase (LAAO) is a flavin adenine dinucleotide (FAD)-dependent enzyme active on most proteinogenic L-amino acids, catalysing their conversion to α-keto acids by oxidative deamination of the substrate. For this oxidation reaction, molecular oxygen is used as the electron acceptor, generating hydrogen peroxide. LAAO can be used to detect L-amino acids, for the production of hydrogen peroxide as an oxidative agent or antimicrobial agent, and for the production of enantiopure amino acids from racemates. In this work, we characterised a previously reported LAAO from the bacterium *Pseudoalteromonas luteoviolacea*. The substrate scope and kinetic properties of the enzyme were determined, and the thermostability was evaluated. Additionally, we elucidated the crystal structure of this bacterial LAAO, enabling us to test the role of active site residues concerning their function in catalysis. The obtained insights and ease of expression of this thermostable LAAO provides a solid basis for the development of engineered LAAO variants tuned for biosensing and/or biocatalysis.

Keywords: flavin-dependent oxidase; L-amino acids; deracemisation; biocatalysis; crystal structure

Citation: Savino, S.; Meijer, J.D.-M.; Rozeboom, H.J.; van Beek, H.L.; Fraaije, M.W. Kinetic and Structural Properties of a Robust Bacterial L-Amino Acid Oxidase. *Catalysts* **2021**, *11*, 1309. https://doi.org/10.3390/catal11111309

Academic Editors: Evangelos Topakas, David D. Boehr and Roland Wohlgemuth

Received: 12 October 2021
Accepted: 27 October 2021
Published: 28 October 2021

1. Introduction

LAAOs are flavin-dependent enzymes present in many organisms [1]. Until recently, representatives were only known and isolated from eukaryotes. LAAOs have been isolated from various snakes and are often a major component in venom [2] and given these venomous secretions their typical yellow colour [3]. Studies have revealed that these proteins consist of three structural elements: a substrate-binding domain, an FAD domain, and a helical domain [4]. LAAOs catalyse the oxidative deamination of L-amino acids to α-keto acids, and substrate ranges have been determined for different LAAOs [5].

LAAOs harbour a tightly bound FAD cofactor which assists in catalysis. The mechanism can be divided in two discrete catalytic steps (Figure 1). Initially, the amino acid substrate is oxidised to the imino acid, which spontaneously reacts with water to form the α-keto acid and ammonium. This results in the formation of the flavin adenine dinucleotide (FAD) cofactor in the reduced state ($FADH_2$). Regeneration of the reduced FAD by molecular oxygen leads to the production of hydrogen peroxide. It is the formation of this reactive oxygen species that has been the focus of studies concerning the role of LAAOs as inducers of oxidative stress [6]. The oxidative damage induced by LAAO activity from snake venom leads to cytotoxicity, antibacterial activity, and the disruption of platelet aggregation [7–9]. Recent biochemical studies have revealed that LAAOs are not only present in the venom of snakes but are also produced in other eukaryotes and even bacteria [1].

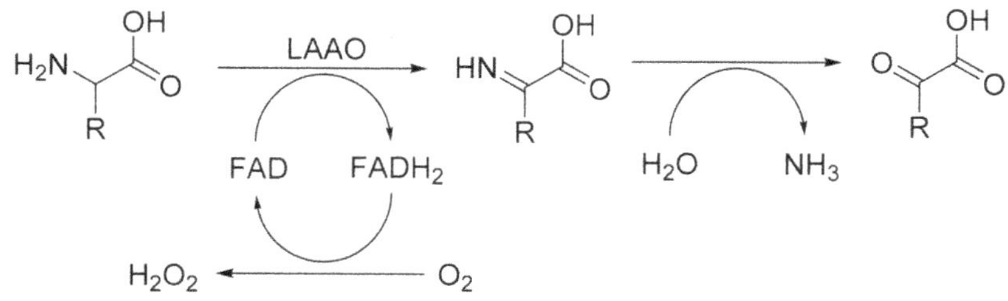

Figure 1. The catalytic mechanism of LAAO: oxidation of amino acids into imino acids and the subsequent reoxidation of the FAD cofactor by reduction of dioxygen into hydrogen peroxide. The hydration of the initial product, imino acid, is a spontaneous reaction.

LAAOs are regarded as valuable biotechnological tools. Applications include their use as biosensors to detect amino acids [10], as antibacterial agents, and as biocatalysts [11]. As a biocatalytic tool, LAAO can be applied to deracemise amino acids, generating enantiopure D-amino acids. This approach takes advantage of an almost exclusive activity of LAAO towards L-amino acids, leaving D-amino acids untouched. Through in situ reduction of the imino acid intermediate produced in the first step of the reaction using a chemical reductant, enantiopure D-amino acids can be produced [12]. LAAO can also be considered for alternative biocatalytic routes. A more recent example of the use of an LAAO in biocatalysis was described by Busch et al. [13], where it was used for the in situ generation of α-ketoglutarate using L-glutamate.

LAAO from the marine bacterium *Pseudoalteromonas luteoviolacea* (*Pl*-LAAO, 653 amino acids, 74 kDa) was shown to display antibacterial activity [14]. The reported broad substrate range of this enzyme makes it an attractive biocatalyst. We decided to investigate its biochemical features in more detail, which included the elucidation of its molecular structure. The obtained structure was analysed and compared with a recently reported structure of an ancestrally reconstructed LAAO, AncLAAO [15,16]. The determined structure of *Pl*-LAAO was also used to design, create, and analyse several *Pl*-LAAO mutants that revealed the role of specific active site residues.

2. Results

2.1. Expression and Purification

The final yield of *Pl*-LAAO was approximately 200 mg/L of *Escherichia coli* culture in terrific broth (TB). This shows that this recombinant bacterial LAAO is well expressed as His-tagged protein, in contrast to previous attempts to express bacterial LAAOs [17]. It also outperforms the expression levels of recently reported ancestral reconstructed LAAOs (40–80 mg/L) [16]. *Pl*-LAAO was purified using immobilised metal affinity chromatography (IMAC) as shown by SDS-PAGE (Supplementary Figure S1). The UV–Vis absorbance spectrum of purified *Pl*-LAAO shows absorbance features that are typical for a flavin-containing protein, in line with its yellow appearance (Supplementary Figure S2). The extinction coefficient of the FAD-bound enzyme was determined to be 11.4 mM^{-1} cm^{-1} at 455 nm and used to quantify *Pl*-LAAO. For crystallisation experiments, *Pl*-LAAO was further purified using size exclusion chromatography, yielding a single peak corresponding to a monodispersed sample of dimeric *Pl*-LAAO (Supplementary Figure S3). This fits the observation of Andreo-Vidal et al. [14] that *Pl*-LAAO is only active as an oligomeric protein and is in accordance with the oligomeric states previously reported for bacterial LAAOs [17].

2.2. Thermostability

To establish optimal conditions for studying and applying *Pl*-LAAO, the thermostability of *Pl*-LAAO was tested in several buffers and at different pH values. Interestingly, the apparent melting temperatures at pH 5–6.5 (50 mM Tris-HCl) were found to be relatively high: 82–84 °C (Supplementary Figure S4a). At higher pH, the stability decreased (T_m = 60 °C in 50 mM Tris-HCl, pH 7.5). The addition of salt fully rescued the stability of *Pl*-LAAO at pH 7.5 (T_m = 85 °C in 500 mM NaCl, 50 mM Tris-HCl, pH 7.5) (Supplementary Figure S4b).

To verify that the enzyme was not only retaining its structure at high temperatures but also remained functional, its residual activity was measured after incubation at various temperatures for 10 min. The remaining activity of samples in 500 mM NaCl, 50 mM Tris-HCl, pH 7.5 was measured at elevated temperatures (Supplementary Figure S4c). Incubation at 70 and 80 °C only led to partial inactivation (>60% remaining activity), but at 90 °C, the residual activity dropped to about 10%. Residual activity was also monitored at 37 °C for an extended time period (Supplementary Figure S4d). When using 500 mM NaCl, 50 mM Tris-HCl, pH 7.5, this resulted in slow inactivation, with 50% remaining activity after 56 h. Interestingly, these thermostability data reveal that this extant LAAO is significantly more stable when compared with ancestrally reconstructed LAAOs [15,16]. The most stable ancestral LAAO lost most of its activity in 10 min at 70 °C (<20% remaining activity) and in 10 h at 30 °C (about 30% remaining activity).

2.3. Kinetic Properties

Next, the kinetic parameters of *Pl*-LAAO were determined for all 20 naturally occurring amino acids (Table 1, Supplementary Figure S5). No significant activity was detected in the case of proline and cysteine. Gratifyingly, oxidase activity was observed for all other L-amino acids. The highest rates were found for L-leucine and L-glutamine. This is in full agreement with the initial characterisation of this bacterial LAAO that only reported activities on L-amino acids at concentrations of 2 mM [14]. We have found that *Pl*-LAAO is most efficient with non-charged bulky L-amino acids (L-leucine, L-methionine, L-phenylalanine, L-glutamine, L-tryptophan) with k_{cat} values of 36–136 s^{-1} and K_m values ranging from 0.34 to 2.4 mM. There are also other L-amino acids with high k_{cat} values (L-glutamate, L-isoleucine, L-arginine, L-valine, L-histidine, L-alanine, L-lysine) but they suffer from relatively high K_m values. Other L-amino acids (mainly small and polar amino acids) displayed poor catalytic efficiencies. The substrate specificity resembles those of the recently reported ancestral LAAOs [15,16] which were described to be most efficient with L-glutamine, L-phenylalanine, and L-methionine. The kinetic parameters of the most stable AncLAAO (AncLAAO-N4) for L-methionine (k_{cat} = 34.2 s^{-1}, K_m = 0.76 mM [16]) were found to be similar to *Pl*-LAAO (k_{cat} = 63 s^{-1}, K_m = 0.42 mM, Table 1).

2.4. Structure Elucidation

The crystal structure of *Pl*-LAAO was solved at 2.8 Å resolution. The crystallographic data and the model refinement values are reported in Supplementary Table S1. The content of the asymmetric unit is the physiological dimer of *Pl*-LAAO, with both molecules binding one FAD (Figure 2a). The structure of the *Pl*-LAAO monomer reveals an overall conserved folding with respect to the recently reported crystal structure of the ancestral LAAO, AncLAAO-N5 [16], with an rmsd value of 0.77 Å when the monomers of each protein are superimposed. This is in line with the 71% sequence identity between both protein sequences. Two main differences are (i) the identity of surface residues, specifically some secondary structure elements exposed to the solvent and involved in the dimerisation interface (Supplementary Figure S6), and (ii) the number of intra-chain salt bridges (20 in *Pl*-LAAO and 17 in AncLAAO). These differences in structural features may explain the relatively high stability of *Pl*-LAAO.

Table 1. Steady state kinetic parameters for amino acid substrates, ordered from highest to lowest k_{cat}/K_m values. For L-glutamate, L-aspartate, and L-tyrosine, the individual kinetic parameters, k_{cat} and K_m, could be determined because of the relatively high K_m values in combination with relatively low solubility of some of these substrates.

Substrate	k_{cat} (s^{-1})	K_m (mM)	k_{cat}/K_m (s^{-1} mM^{-1})	V_{max} (U/mg)
L-leucine	72	0.40	180	58
L-methionine	63	0.42	150	51
L-phenylalanine	52	0.34	150	42
L-glutamine	136	2.4	56	110
L-tryptophan	36	1.5	25	29
L-glutamate	>30	>6.9	4.4	24
L-isoleucine	25	11	2.3	20
L-arginine	43	25	1.7	35
L-valine	9.4	9.5	0.99	7.5
L-histidine	5.3	11	0.48	4.3
L-alanine	14	48	0.30	11.3
L-lysine	7.2	57	0.13	5.8
L-serine	0.21	14	0.015	0.17
L-tyrosine	>0.75	>10	0.075	0.60
L-asparagine	0.46	8.7	0.053	0.37
L-aspartate	>0.023	>3.8	0.0061	0.018
L-threonine	0.05	14	0.0035	0.040
L-glycine	0.01	30	0.0003	0.0081

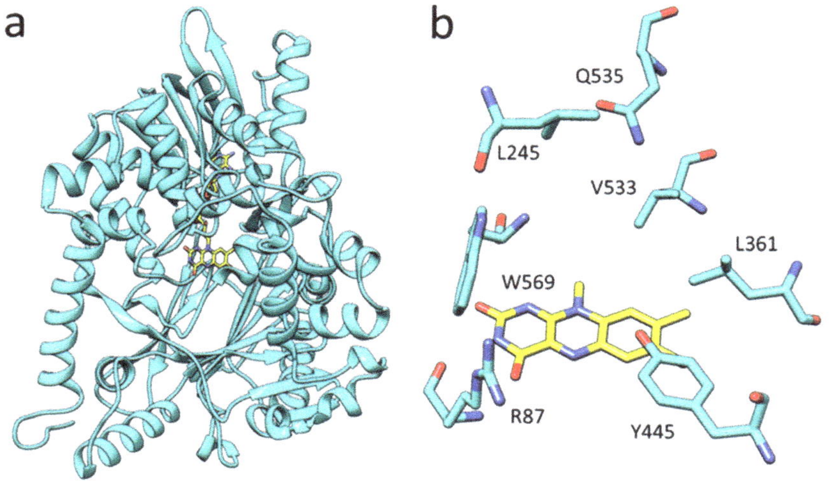

Figure 2. (**a**) The overall structure of *Pl*-LAAO monomer in cyan ribbons and the FAD cofactor highlighted in yellow sticks. (**b**) Close-up of the active site, with the isoalloxazine moiety of the flavin cofactor in yellow and active site residues in cyan.

In one of the monomers, a glycerol molecule occupies the pocket that is next to the redox active part of the bound FAD, identifying the active site at the *Re*-face of the flavin cofactor (Supplementary Figure S7). Comparison with the L-phenylalanine-complexed structure of AncLAAO confirms the location of the substrate binding pocket. There is a tunnel leading from the solvent exposed surface to the substrate binding pocket (Supplementary Figure S8). A number of hydrophobic residues line the substrate binding pocket of *Pl*-LAAO (L245, L361, V533, W569) and account for the preference of *Pl*-LAAO for bulky hydrophobic amino acids (Figure 2b). There is one polar residue (Q535) located in the back of the binding pocket which explains the relatively good activity and affinity of *Pl*-LAAO for L-glutamine. The residues R87 and Y445, positioned close to the isoalloxazine moiety of the flavin cofactor, bind the carboxyl group of amino acid substrates. Their essential role in catalysis was confirmed by generating and testing R87M and Y445F *Pl*-LAAO mutants, which displayed drastically reduced activity on L-phenylalanine (<2% activity).

3. Discussion

We have found that *Pl*-LAAO can be overexpressed in *E. coli*, yielding large amounts of FAD-containing, soluble, active enzyme. Surprisingly, such high levels of expression do not seem to affect the viability of the expression host, even though it is active on abundantly present L-amino acids. It was discovered that *Pl*-LAAO is quite thermotolerant and shows a higher tolerance to temperature when compared with other microbial and ancestral reconstructed LAAOs [16]. The relatively high temperature tolerance and preference for high salt concentrations may reflect the conditions at which the oxidase is active in real life. It has been found that it is secreted by the original bacterial host, contributing to the antimicrobial activity of *Pseudoalteromonas luteoviolacea* [14]. Substrate profiling of *Pl*-LAAO revealed that it is active on most L-amino acids. Yet, it shows the highest activity and affinity towards bulky hydrophobic L-amino acids and L-glutamine. No activity or very poor activity is observed with small amino acids.

To understand the structural basis for the stability and the substrate preferences of *Pl*-LAAO we set out to determine its crystal structure. The obtained structure sheds light on the relatively high stability of this enzyme with respect to close homologs, the recently reported ancestral LAAOs [16]. While the core of the protein is structurally highly similar to the ancestral LAAOs, the surface residues are not conserved, especially in some specific regions (Supplementary Figure S6). The residues involved in the dimerisation of LAAO differ by 26 substitutions between *Pl*-LAAO and AncLAAO. The latter enzyme accumulates asparagine and lysine residues in hotspots (N520-N522-N524 and K341-K345-K346) which may prevent complementarity of charges between the two monomers, lowering the stability of the dimeric protein. Moreover, the difference in the number of intra-chain salt bridges might contribute to the difference in thermostability.

By inspecting the *Pl*-LAAO structure and, specifically, the substrate binding site, being rather large and hydrophobic, it becomes clear why a specific set of L-amino acids is preferred, similar to the ancestral LAAO. Comparing *Pl*-LAAO with the structures of other flavin-dependent amino acid oxidases provides some clues on how evolution has generated several solutions for oxidising amino acids with the help of a flavin cofactor. Snake venom LAAOs, for example LAAO from the Malayan pit viper, also have well-described substrate scopes, typically with low K_m values for many amino acids [2,18] and its crystal structure has been reported [19]. Both *Pl*-LAAO (or AncLAAO) and venom LAAO use the same configuration of residues for binding and positioning of their substrates (Figure 3a,b). In both types of LAAOs, an arginine and a tyrosine residue interact with the carboxyl group of the amino acid substrate. This array of H-bonds together with the side chain of a tryptophan, which defines one boundary of the active site, orient the L-amino acids for hydride transfer to the N5 atom of the FAD cofactor. The other part of the substrate binding pocket, interacting with the side chain of the amino acid substrate, shows many differences, explaining the different substrate preferences (Figure 3a,b). Intriguingly, D-amino acid oxidase (DAAO) from yeast, acting on mirror-image substrates and having quite a different protein structure, also contains a tyrosine and arginine couple that binds the carboxylate of the amino acid substrate (Figure 3c). For DAAO, the binding pocket accommodating the substrate amino acid side chain is orthogonal compared to LAAO, allowing binding of the stereoisomeric substrates, D-amino acids [20]. The structural dissimilarities indicate that LAAOs and DAAOs independently evolved and converged to perform the same reaction, oxidation of amino acids, using a similar active site arrangement. The structural insights into the basis of substrate binding in *Pl*-LAAO and other amino acid oxidases is valuable to understand their catalytic mechanism and can inform future enzyme engineering efforts towards other substrates.

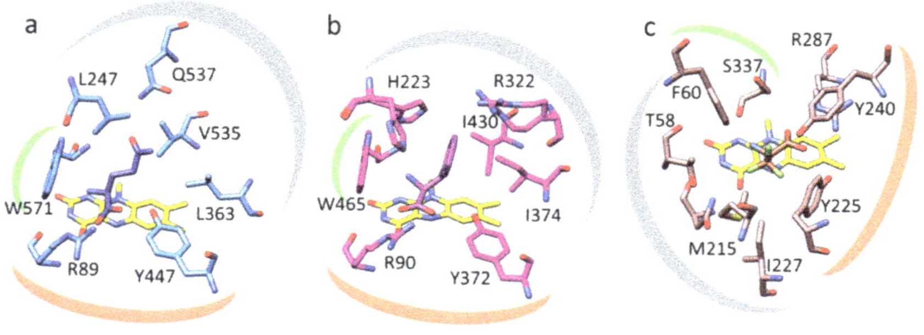

Figure 3. Structural comparison of active sites of amino acid oxidases. Active sites of (**a**) ancestral LAAO (7C4L), (**b**) LAAO from the Malayan pit viper (2IID), and (**c**) yeast DAAO (1C0L). In (**a**), residues R89 and Y447, highlighted by the orange mark, interact with the carboxyl group of the bound L-glutamine substrate (dark blue). In (**b**), the same role is exerted by R90 and Y372 for binding L-phenylalanine (purple). In (**c**), R287 and Y240 interact with the carboxyl moiety of the bound D-trifluoroalanine (dark brown). The green marks highlight the W571 and W465 in (**a**,**b**) respectively, which form part of the substrate binding pocket, close to the amine moiety of the bound substrate. In DAAO (**c**), a phenylalanine (F60) serves a similar role. The grey marks indicate the residues involved in binding of the substrate side chain in the three different amino acid oxidases.

4. Materials and Methods

4.1. Materials

The reagents used in this study were from Merck (Kenilworth, NJ, United States), Acros (Geel, Belgium), and BioRad (Hercules, CA, United Stated). Culture media components were from Difco (Franklin Lakes, NJ, United States) and were used to prepare TB medium (12 g/L tryptone, 24 g/L yeast extract, 5 g/L glycerol, 23 g/L 170 g/L KH_2PO_4, and 125 g/L K_2HPO_4). The crystallisation screens were purchased from Hampton Research (Aliso Viejo, CA, United States) and Molecular Dimensions (Sheffield, UK).

4.2. Protein Expression and Purification

E. coli NEB10β cells were transformed with pBAD-based plasmid containing the codon-optimised sequence coding for wild type *Pl*-LAAO including a His-tag for affinity chromatography located at the N-terminus of the *Pl*-LAAO sequence. A single colony was used to inoculate 10 mL LB preculture which was used, after overnight growth, to inoculate (1% *v/v*) 1 L TB_{amp}. When the culture reached an OD_{600} of 0.6, L-arabinose was added to a final concentration of 0.02% (*w/v*) to trigger protein expression. The flasks were moved from 37 to 24 °C and incubated for 2 days. The obtained cells were harvested by centrifugation (3153× *g*, 30 min) and the obtained cell pellet was resuspended in 50 mM Tris-HCl pH 7.5 with 100 mM NaCl (buffer A). Cells were sonicated for 20 min using 70% amplitude with 10 s on/ 5 s off intervals on a Sonic Vibra cell processor, and cell debris was separated from the supernatant via centrifugation (20,130× *g*, 30 min). The supernatant was collected, filtered, and applied to pre-equilibrated columns loaded with Ni-Sepharose. After a washing step with 3 CV of buffer A and the removal of not specifically bound proteins with buffer A containing 10 mM imidazole (3 CV), the enzyme was eluted using buffer B (buffer A + 500 mM imidazole). The collected sample was desalted using Vivaspin concentrators.

4.3. Determination of the Extinction Coefficient and Stability

In order to determine the extinction coefficient of *Pl*-LAAO, the absorbance spectra of the flavin-containing enzyme and of the FAD after protein denaturation and pellet removal were measured at 455 and 450 nm, respectively. The known extinction coefficient of FAD at 450 nm (11.3 mM^{-1} cm^{-1}) was used to calculate the extinction coefficient of *Pl*-LAAO.

The stability of *Pl*-LAAO was tested using the ThermoFAD protocol [21]. A thermocycler was programmed to run a temperature gradient from 25 to 99 °C on the samples composed of 12.5 µL of different buffers and 12.5 µL protein diluted to 1 mg/mL. The emission signal was measured using a SYBR filter (ThermoFisher, Waltham, MA, United States), taking advantage of the increase in FAD fluorescence caused by protein denaturation. The first derivative of the measured peaks provides apparent Tm values.

The residual activity of *Pl*-LAAO was measured after incubation for 10 min at different temperatures (30, 40, 50, 60, 70, 80, and 90 °C) and over time (56 h) at 37 °C. The activity was measured using L-phenylalanine as the substrate following the protocol below.

4.4. Mutagenesis

The rationally designed mutants of *Pl*-LAAO were obtained by Quick Change PCR using pairs of primers designed by AAscan (https://github.com/dmitryveprintsev/AAScan, (accessed on 27 October 2021)) to introduce the mutations R87M and Y445F. Successful introduction of mutations was confirmed by sequencing.

4.5. Kinetic Analysis

LAAO activity was measured at 37 °C and both the enzyme samples and the reaction mixtures were preincubated at this temperature. Substrates were added to start the reaction. The final reaction mixture contained 8 U of horseradish peroxidase, 0.1 mM amino antipyrine, 1 mM 3,5-dichloro-2-hydroxybenzene, 180–680 µL substrate at varying concentrations and 0.1 µM–0.5 nM *Pl*-LAAO in buffer A, for a total volume of 1 mL. Reaction rates of mutants were measured in the same setup, using a fixed concentration of 0.02 µM enzyme.

4.6. Crystallisation and Structure Determination

For crystallisation, wild type *Pl*-LAAO samples underwent further purification by size exclusion chromatography on a Superdex 200 10/300 column (Cytiva, Medemblik, The Netherlands) pre-equilibrated with 100 mM Tris-HCl pH 7.5 containing 150 mM NaCl on an ÄKTA purifier (Cytiva). The eluted peak was pooled and concentrated to 24 mg/mL for crystallisation experiments. Multiple hits were found, with crystals growing in conditions from several commercially available screens. The most easily reproducible crystals were obtained by sitting drop method in 20% (*w/v*) PEG 3350 and 0.1 M Tris-HCl pH 7.0, but also in HEPES pH 7.0 and Tris pH 8.0. The crystals were small cubes of a bright yellow colour, which were cryoprotected by adding 20% (*v/v*) glycerol and frozen in liquid nitrogen for synchrotron data collection at Diamond Light Source (Didcot, UK). The collected datasets were merged [22] and scaled [23]. The structure of PDB 7C4K [16] was used as a model for molecular replacement [24]. The obtained model underwent *REFMAC5* refinement and manual fitting with *COOT* [25] yielding the final structure deposited at the PDB with code 7OG2. Images of the *Pl*-LAAO structure and other models were made using Chimera [26]. The number of salt bridges present in the structure was calculated using the ESBRI tool available on the LSCF Bioinformatics Unit portal [27]. The *Pl*-LAAO structure was inspected in *PyMOL* [28] using the *CAVER* [29] plugin. This allowed analysis of the tunnel leading from the solvent accessible surface to the active site.

5. Conclusions

Due to the high expression level, ease of purification, high thermostability, and activity on many different L-amino acids, *Pl*-LAAO is an attractive enzyme for biocatalysis and biosensing. It can be used for the synthesis of enantiopure amino acids or related compounds. The high catalytic efficiency on a subset of L-amino acids allows it to be used for detection of amino acids. The determined crystal structure can be exploited for tailor-made variants for activity on specific amino acids.

Supplementary Materials: The following are available online at https://www.mdpi.com/article/10.3390/catal11111309/s1, Figure S1: SDS-PAGE of the LAAO purification process. Figure S2: UV–Vis absorbance spectrum of purified LAAO. Figure S3: Size exclusion chromatogram of LAAO sample before crystallisation experiments. The sample was collected in small fractions and pooled together, avoiding the shoulder on the right side of the peak. Figure S4: (A) pH-dependence of Tm of Pl-LAAO in 50 mM TRIS/Cl at different pH values, (B) salt concentration dependence of Tm of Pl-LAAO in 50 mM TRIS/Cl, pH 6.5, with different NaCl concentrations, (C) observed Pl-LAAO activity after 10 min incubation at different temperature, (D) observed Pl-LAAO activity upon incubation at 37 °C over time. For the last two figures, data are shown for samples in low salt buffer (blue line) and in high salt buffer (orange line) as comparison. Figure S5: Steady state kinetics for 18 of the 20 proteinogenic amino acids. All measurements were performed in duplicate. No activity was observed for proline and cysteine. Figure S6: Top, sequence alignment between *Pl*-LAAO and the homologue from ancestral sequence reconstruction, the structure of which is deposited under the PDB 7C4K. Non-conserved sites are highlighted by black bars. Bottom, the structure of 7C4K with non-conserved sites in dark blue. All of these are limited to surface-exposed portions of the protein, while the active site and the cofactor binding residues are fully retained. For sake of clarity, the portions with highest divergence are reported in rectangles of the same colour in both sequence alignment and structure representation. Figure S7: Focus on the electron density (in blue chicken wire) of FAD (in yellow carbons) and glycerol (in green carbons) present only in molecule B of the asymmetric unit. Figure S8: (a) Structure of LAAO with the tunnel (magenta) leading from the FAD cofactor to the solvent-accessible surface. (b) Side view of the tunnel, the entrance of which is enclosed in the black circle. A set of side chains, the motion of which is necessary to allow access to the active site, is highlighted. (c) A 180° rotated view of the same side view. Table S1: Crystallographic data collection and refinement statistics. The values for the high-resolution shell are shown in parenthesis.

Author Contributions: Conceptualisation, H.L.v.B. and M.W.F.; methodology, S.S., J.D.-M.M. and H.J.R.; writing—original draft preparation, S.S.; writing—review and editing, S.S., H.L.v.B. and M.W.F.; and funding acquisition, M.W.F. All authors have read and agreed to the published version of the manuscript.

Funding: This research was funded by NWO through a VICI grant to M.W.F.

Data Availability Statement: Data are contained within the article.

Conflicts of Interest: The authors declare no conflict of interest.

References

1. Kasai, K.; Ishikawa, T.; Nakamura, T.; Miura, T. Antibacterial properties of L-amino acid oxidase: Mechanisms of action and perspectives for therapeutic applications. *Appl. Microbiol. Biotechnol.* **2015**, *99*, 7847–7857. [CrossRef] [PubMed]
2. Guo, C.; Liu, S.; Yao, Y.; Zhang, Q.; Sun, M.Z. Past decade study of snake venom L-amino acid oxidase. *Toxicon* **2012**, *60*, 302–311. [CrossRef] [PubMed]
3. Izidoro, L.F.; Sobrinho, J.C.; Mendes, M.M.; Costa, T.R.; Grabner, A.N.; Rodrigues, V.D.M.; Da Silva, S.L.; Zanchi, F.B.; Zuliani, J.P.; Fernandes, C.F.C.; et al. Snake venom L-amino acid oxidases: Trends in pharmacology and biochemistry. *Biomed. Res. Int.* **2014**, *2014*, 196754. [CrossRef]
4. Pawelek, P.D.; Cheah, J.; Coulombe, R.; Macheroux, P.; Ghisla, S.; Vrielink, A. The structure of L-amino acid oxidase reveals the substrate trajectory into an enantiomerically conserved active site. *EMBO J.* **2000**, *19*, 4204–4215. [CrossRef] [PubMed]
5. Chen, W.M.; Sheu, F.S.; Sheu, S.Y. Novel L-amino acid oxidase with algicidal activity against toxic cyanobacterium *Microcystis aeruginosa* synthesized by a bacterium *Aquimarina* sp. *Enzyme Microb. Technol.* **2011**, *49*, 372–379. [CrossRef] [PubMed]
6. Samel, M.; Tõnismägi, K.; Rönnholm, G.; Vija, H.; Siigur, J.; Kalkkinen, N.; Siigur, E. L-Amino acid oxidase from *Naja naja oxiana* venom. *Comp. Biochem. Physiol. Part B* **2008**, *149*, 572–580. [CrossRef] [PubMed]
7. Zuliani, J.P.; Kayano, A.M.; Zaqueo, K.D.; Neto, A.C.; Sampaio, S.V.; Soares, A.M.; Stabeli, R.G. Snake venom L-amino acid oxidases: Some consideration about their functional characterization. *Protein Pept. Lett.* **2009**, *16*, 908–912. [CrossRef]
8. Li, Z.Y.; Yu, T.F.; Lian, E.C. Purification and characterization of L-amino acid oxidase from king cobra (*Ophiophagus hannah*) venom and its effects on human platelet aggregation. *Toxicon* **1994**, *32*, 1349–1358. [CrossRef]
9. Lee, M.L.; Tan, N.H.; Fung, S.Y.; Sekaran, S.D. Antibacterial action of a heat-stable form of L-amino acid oxidase isolated from king cobra (*Ophiophagus hannah*) venom. *Comp. Biochem. Physiol. Toxicol. Pharmacol.* **2010**, *153*, 237–242. [CrossRef]
10. Lata, S.; Pundir, C.S. L-amino acid biosensor based on L-amino acid oxidase immobilized onto NiH-CNFe/c-MWCNT/PPy/GC electrode. *Int. J. Biol. Macromol.* **2013**, *54*, 250–257. [CrossRef]
11. Kasai, K.; Nakano, M.; Ohishi, M.; Nakamura, T.; Miura, T. Antimicrobial properties of L-amino acid oxidase: Biochemical features and biomedical applications. *Appl. Microbiol. Biotechnol.* **2021**, *105*, 4819–4832. [CrossRef] [PubMed]

12. Huh, J.W.; Yokoigawa, K.; Esaki, N.; Soda, K. Total conversion of racemic pipecolic acid into the L-enantiomer by a combination of enantiospecific oxidation with D-amino acid oxidase and reduction with sodium borohydride. *Biosci. Biotechnol. Biochem.* **1992**, *56*, 2081–2082.
13. Busch, F.; Brummund, J.; Calderini, E.; Schürmann, M.; Kourist, R. Cofactor generation cascade for α-ketoglutarate and Fe(II)-dependent dioxygenases. *ACS Sustain. Chem. Eng.* **2020**, *8*, 8604–8612. [CrossRef]
14. Andreo-Vidal, A.; Sanchez-Amat, A.; Campillo-Brocal, J.C. The *Pseudoalteromonas luteoviolacea* L-amino acid oxidase with antimicrobial activity is a flavoenzyme. *Mar. Drugs* **2018**, *16*, 499. [CrossRef]
15. Nakano, S.; Kozuka, K.; Minamino, Y.; Karasuda, H.; Hasebe, F.; Ito, S. Ancestral L-amino acid oxidases for deracemization and stereoinversion of amino acids. *Commun. Chem.* **2020**, *3*, 181. [CrossRef]
16. Nakano, S.; Minamino, Y.; Hasebe, F.; Ito, S. Deracemization and stereoinversion to aromatic D-amino acid derivatives with ancestral L-amino acid oxidase. *ACS Catal.* **2019**, *9*, 10152–10158. [CrossRef]
17. Pollegioni, L.; Motta, P.; Molla, G. L-amino acid oxidase as biocatalyst: A dream too far? *Appl. Microbiol. Biotechnol.* **2013**, *97*, 9323–9341. [CrossRef]
18. Ponnudurai, G.; Chung, M.C.; Tan, N.H. Purification and properties of the L-amino acid oxidase from Malayan pit viper (*Calloselasma rhodostoma*) venom. *Arch. Biochem. Biophys.* **1994**, *313*, 373–378. [CrossRef] [PubMed]
19. Moustafa, I.M.; Foster, S.; Lyubimov, A.Y.; Vrielink, A. Crystal structure of LAAO from *Calloselasma rhodostoma* with an L-phenylalanine substrate: Insights into structure and mechanism. *J. Mol. Biol.* **2006**, *364*, 991–1002. [CrossRef] [PubMed]
20. Umhau, S.; Pollegioni, L.; Molla, G.; Diederichs, K.; Welte, W.; Pilone, M.S.; Ghilsa, S. The x-ray structure of D-amino acid oxidase at very high resolution identifies the chemical mechanism of flavin-dependent substrate dehydrogenation. *Proc. Natl. Acad. Sci. USA* **2000**, *97*, 12463–12468. [CrossRef]
21. Forneris, F.; Orru, R.; Bonivento, D.; Chiarelli, L.R.; Mattevi, A. ThermoFAD, a Thermofluor-adapted flavin ad hoc detection system for protein folding and ligand binding. *FEBS J.* **2009**, *276*, 2833–2840. [CrossRef] [PubMed]
22. McCoy, A.J.; Grosse-Kunstleve, R.W.; Adams, P.D.; Winn, M.D.; Storoni, L.C.; Read, R.J. Phaser crystallographic software. *J. Appl. Crystallogr.* **2007**, *40*, 658–674. [CrossRef]
23. Kabsch, W. XDS. *Acta Crystallogr. D Biol. Crystallogr.* **2010**, *66 Pt 2*, 125–132. [CrossRef] [PubMed]
24. Winn, M.D.; Ballard, C.C.; Cowtan, K.D.; Dodson, E.J.; Emsley, P.; Evans, P.R.; Keegan, R.M.; Krissinel, E.B.; Leslie, A.G.W.; McCoy, A.; et al. Overview of the CCP4 suite and current developments. *Acta Crystallogr. D Biol. Crystallogr.* **2011**, *67*, 235–242. [CrossRef] [PubMed]
25. Emsley, P.; Cowtan, K. Coot: Model-building tools for molecular graphics. *Acta Crystallogr. D Biol. Crystallogr.* **2004**, *60*, 2126–2132. [CrossRef] [PubMed]
26. Pettersen, E.F.; Goddard, T.D.; Huang, C.C.; Couch, G.S.; Greenblatt, D.M.; Meng, E.C.; Ferrin, T.E. UCSF Chimera—A visualization system for exploratory research and analysis. *J. Comput Chem.* **2004**, *25*, 1605–1612. [CrossRef]
27. Sarakatsannis, J.N.; Duan, Y. Statistical characterization of salt bridges in proteins. *Proteins* **2005**, *60*, 732–739. [CrossRef]
28. *The PyMOL Molecular Graphics System*, version 2.0; Schrödinger, LLC.: New York, NY, USA, 2015.
29. Chovancová, E.; Pavelka, A.; Benes, P.; Strnad, O.; Brezovsky, J.; Kozlikova, B.; Góra, A.; Sustr, V.; Klvana, M.; Medek, P.; et al. CAVER 3.0: A tool for the analysis of transport pathways in dynamic protein structures. *PLoS Comput. Biol.* **2012**, *8*, e1002708. [CrossRef]

catalysts

MDPI

Communication

Enzymatic Hydrogen Electrosynthesis at Enhanced Current Density Using a Redox Polymer

John C. Ruth [1], Fabian M. Schwarz [2], Volker Müller [2] and Alfred M. Spormann [1,3,*]

1 Department of Chemical Engineering, Stanford University, Stanford, CA 94305, USA; jruth@stanford.edu
2 Department of Molecular Microbiology and Bioenergetics, Johann Wolfgang Goethe University, 60438 Frankfurt am Main, Germany; f.schwarz@bio.uni-frankfurt.de (F.M.S.); vmueller@bio.uni-frankfurt.de (V.M.)
3 Department of Civil and Environmental Engineering, Stanford University, Stanford, CA 94305, USA
* Correspondence: spormann@stanford.edu

Abstract: High-temperature tolerant enzymes offer multiple advantages over enzymes from mesophilic organisms for the industrial production of sustainable chemicals due to high specific activities and stabilities towards fluctuations in pH, heat, and organic solvents. The production of molecular hydrogen (H_2) is of particular interest because of the multiple uses of hydrogen in energy and chemicals applications, and the ability of hydrogenase enzymes to reduce protons to H_2 at a cathode. We examined the activity of Hydrogen-Dependent CO_2 Reductase (HDCR) from the thermophilic bacterium *Thermoanaerobacter kivui* when immobilized in a redox polymer, cobaltocene-functionalized polyallylamine (Cc-PAA), on a cathode for enzyme-mediated H_2 formation from electricity. The presence of Cc-PAA increased reductive current density 340-fold when used on an electrode with HDCR at 40 °C, reaching unprecedented current densities of up to 3 $mA{\cdot}cm^{-2}$ with minimal overpotential and high faradaic efficiency. In contrast to other hydrogenases, *T. kivui* HDCR showed substantial reversibility of CO-dependent inactivation, revealing an opportunity for usage in gas mixtures containing CO, such as syngas. This study highlights the important potential of combining redox polymers with novel enzymes from thermophiles for enhanced electrosynthesis.

Keywords: hydrogen-dependent CO_2 reductase; redox polymer; cobaltocene; enzymatic electrosynthesis; hydrogen evolution; carbon monoxide

Citation: Ruth, J.C.; Schwarz, F.M.; Müller, V.; Spormann, A.M. Enzymatic Hydrogen Electrosynthesis at Enhanced Current Density Using a Redox Polymer. *Catalysts* **2021**, *11*, 1197. https://doi.org/10.3390/catal11101197

Academic Editors: Evangelos Topakas, David D. Boehr and Roland Wohlgemuth

Received: 29 August 2021
Accepted: 27 September 2021
Published: 30 September 2021

1. Introduction

Rising global temperatures and their resulting environmental effects are leading to an urgent transition towards renewable energies [1]. Converting sustainably generated energies to a chemical form is desirable and important for long-term energy storage and transport [2]. A variety of small, high-energy molecules are promising target products for this conversion, including hydrogen and formate, because only simple, two-electron reduction reactions are required for their production and their usability in existing technologies [3–5]. Hydrogen is particularly interesting due to its high energy content, low redox potential, and usage in a wide variety of industrial processes [4,6].

Enzymes, such as hydrogenases and formate dehydrogenases, are promising catalysts for conversion of electrical energy to storable chemical energy, due to their high selectivities and activities [3,7]. In particular, Hydrogen-Dependent CO_2 Reductase (HDCR) is an enzyme of interest for both enzymatic formate and hydrogen synthesis as a result of its high turnover rates and selectivity at high temperatures [8–11]. In vivo, HDCR catalyzes the simultaneous oxidation of H_2 to protons and reduction of CO_2 to formate in the methyl branch of the Wood-Ljungdahl pathway of the two acetogenic bacteria *Acetobacterium woodii* and *Thermoanaerobacter kivui*. HDCR-like gene clusters were also found in the genome of different organisms but biochemical evidence of purified HDCRs was so far only given for the two mentioned acetogens. HDCR from *Thermoanaerobacter kivui* is O_2-sensitive and contains a hydrogenase and a formate dehydrogenase subunit, linked by two 4Fe–4S cluster-containing electron transfer subunits [10]. This HDCR is a Fe-Fe hydrogenase with two Fe atoms in its active site, which has the highest measured hydrogen production activities among hydrogenases. Other Fe-Fe hydrogenases were previously shown to be irreversibly inactivated by both O_2 and carbon monoxide (CO) [12]. Purified HDCR from thermophilic *T. kivui* catalyzes hydrogen evolution from formate at a specific activity of 930 μmol hydrogen·min^{-1}·mg^{-1} at 60 °C [10]. This enzyme is stable above 37 °C and has its optimal activity between 60 and 70 °C. Interestingly, HDCR from the mesophilic acetogen *Acetobacterium woodii* is completely tolerant of CO, regaining full activity after an initial inhibition [8]. Reversible inactivation after CO addition to this Fe-Fe hydrogenase is unusual, as Fe-Fe hydrogenases are typically irreversibly inactivated by CO via binding to the distal Fe [12]. CO-insensitivity in enzymatic catalysts is desirable in many industrial processes, such as those involving syngas. While the HDCR of *A. woodii* has displayed reactivation after exposure to CO, CO sensitivity of the heat stable HDCR of *T. kivui*, which has significantly higher turnover rates and also contains an Fe-Fe hydrogenase, has not been investigated.

Many studies have investigated the ability of oxidoreductase enzymes to produce chemicals of interest in an electrochemical cell, in a process known as enzymatic electrosynthesis [3,7,13–19]. In such a system, enzymes receive electrons from an electrode either directly or through a chemical mediator, and then catalyze the reduction of their specific substrate to the specific product. Redox polymers have been shown to greatly enhance oxidoreductase activity at the electrode surface through enzyme immobilization and mediation of electron flux to the enzyme [3,20]. Recently characterized redox polymers for enzymatic hydrogen evolution have been either cobaltocene functionalized poly(allylamine) (Cc-PAA) or viologens, due to their low redox potential [3,18,21,22]. An advantage of viologens is the ability to choose from a wide variety of viologen derivatives with different redox potentials. However, many such systems require diffusion of the species for electron transfer to occur, and none have been tested with hydrogenases above ambient temperature. Cobaltocene-functionalized poly(allylamine) has previously been shown to effectively mediate electron transfer to both formate dehydrogenases and hydrogenases for reduction of their respective substrates (Figure 1) [18,23]. The polyallylamine backbone forms a hydrogel when crosslinked with ethylene glycol diglycidyl ether (EGDGE), allowing for enzyme immobilization. Electron transfer can then occur via the cobaltocene functional groups throughout the hydrogel. However, there has been very little research on the ability of redox polymers as mediators for hydrogen evolution by enzymes at elevated temperatures. Given the high specific activity of thermophilic enzymes at elevated temperatures, polymers capable of increasing electron flux to the enzyme at a cathode would further increase activity compared to current electrochemical enzymatic hydrogen evolution reaction (HER) systems.

Here, we demonstrate the hydrogen evolution activity of *T. kivui* HDCR via direct electron transfer when immobilized at an electrode. We further investigated the capacity of embedding HDCR in a Cc-PAA redox polymer at the cathode and found an enhanced activity of H_2 production by more than 340-fold. Further, Cc-PAA mediated hydrogen production occurred at 90% faradaic efficiency with minimal current loss. Interestingly, HDCR demonstrated reversible inactivation by CO when in the redox polymer, with most activity being regained after CO was removed. Cc-PAA is a promising redox polymer for enzyme-mediated, enhanced hydrogen production even at elevated temperatures, making it an auspicious component of new technology for use in chemical storage for renewable energy.

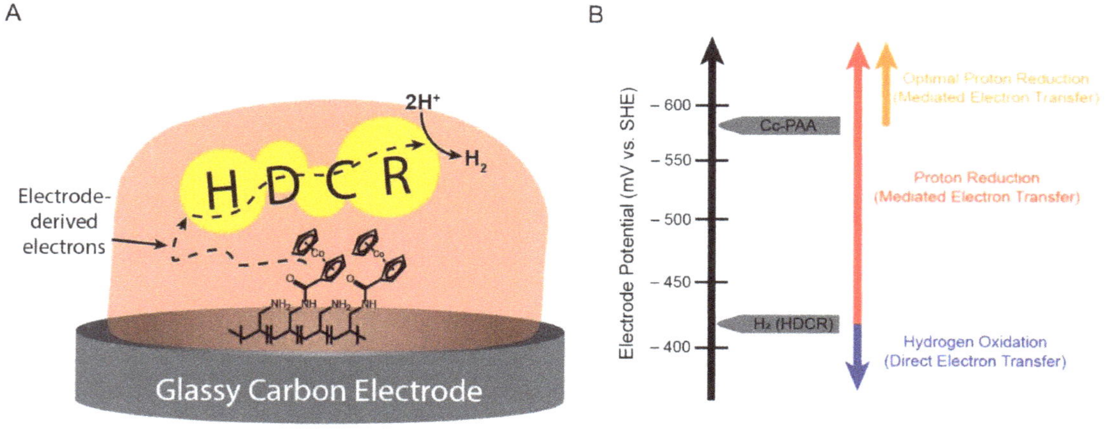

Figure 1. (**A**) In the mediated electron transfer system, electrons are passed from the electrode through the Cc-PAA redox polymer, with self-exchange between cobaltocene functional groups, before entering the HDCR enzyme for use in hydrogen evolution at the hydrogenase subunit's active site. (**B**) An energy diagram shows the regimes of hydrogen oxidation and proton reduction in the mediated Cc-PAA/HDCR system. At potentials above the HDCR hydrogenase midpoint potential, direct electron transfer results in a very small oxidative current from hydrogen oxidation at the enzyme. At potentials below the midpoint potential, proton reduction occurs, mediated by Cc-PAA. When the potential is below the midpoint potential of the Cc-PAA, electron transfer to HDCR occurs at the maximum rate.

2. Results

2.1. Optimization of Electrochemical Buffer Conditions

Electrochemical activity of HDCR was tested first at a 3-electrode, rotating disk electrode (RDE) system. The magnitude of the steady state reductive wave at low potentials of cyclic voltammetry was assumed to be analogous to hydrogen evolution activity as a result of the electrons passing through the electrode for proton reduction. Increasing the temperature of the electrochemical cell from room temperature to 60 °C resulted in a higher rate of H_2 formation but resulted in a higher loss of current density over time (Figure S1). Therefore, an intermediate temperature of 40 °C was selected for further electrochemical studies, as this temperature displayed similar initial current density as a sample at 60 °C at a slower loss of activity over time. At 40 °C, the hydrogenase activity was 190 μmol $H_2 \cdot min^{-1} \cdot [mg\ protein]^{-1}$, and the activity of formate dehydrogenase < 1 μmol formate$\cdot min^{-1} \cdot [mg\ protein]^{-1}$. Thus, only electrochemical hydrogen formation was investigated further. As shown in Figure S2A, pH optimization was carried out using HDCR immobilized in Cc-PAA in two different buffers: 1 M sodium citrate with 200 mM potassium chloride for testing a pH range 4.5–6, and 1.6 M potassium phosphate with 320 mM sodium chloride for a pH range 6–7. The difference in activity observed between both buffers at pH 6 was approximately 4%. The optimum HDCR activity was at pH 5.0, allowing for a high concentration of protons in the electrochemical setup. Subsequent assays were conducted using citrate buffer at pH 5.0, maintaining a 5:1 ratio of sodium citrate to potassium chloride (Figure S2B). An increase in activity with increasing buffer concentration was observed up to the solubility limit of the buffer, resulting in a maximum activity at 2.4 M total solute concentration, with no increase observed in the absence of HDCR. All subsequent electrochemical tests were carried out in the optimal buffer: 2 M sodium citrate, 400 mM potassium phosphate, pH 5.0.

2.2. Hydrogenase Activity Using Methyl Viologen Assays

Hydrogen evolution activity of HDCR in a chemical assay was determined using methyl viologen as electron mediator. Sodium dithionite showed little reducing activity in the citrate buffer, with no color change of methyl viologen upon the addition of dithionite. This is understandable, given the strong dependence of the sodium dithionite redox potential on temperature, pH, and other environmental factors [24]. Titanium citrate (Ti(III) citrate) showed the ability to reduce methyl viologen at low concentrations of sodium citrate in the buffer but not at high sodium citrate concentrations. It, therefore, was used as the electron donor in a 10× diluted electrochemistry buffer (0.2 M sodium citrate, 40 mM sodium chloride), maintaining an excess concentration of Ti(III) citrate. Hydrogen was quantified via GC, and a linear increase in rate was observed (Figure S3). A maximum specific activity of 190 ± 25 μmol·min^{-1}·[mg protein]$^{-1}$ was observed, which was roughly a 4-fold higher specific activity relative to the Fe-Fe hydrogenase CpI from mesophilic *Clostridium pasteurianum* (Figure S4) [18]. Methyl viologen activity assays displayed a similar trend of increasing activity with increasing buffer concentration to that observed in electrochemical buffer optimizations.

2.3. Cobaltocene-Mediated Hydrogen Evolution in Cyclic Voltammetry Experiments

Cobaltocene-mediated hydrogen evolution activity was examined using cyclic voltammetry (CV) (Figure 2). HDCR was immobilized in Cc-PAA, and the reductive current at low potentials was measured. A maximum current density of 3.0 mA·cm^{-2} was achieved. When cobaltocene was absent, a maximum current density of 8.7 μA·cm^{-2} was observed. Therefore, the presence of cobaltocene resulted in an activity increase of 340-fold, indicating significantly enhanced hydrogen evolution activity. This activity was also compared to inactive, O_2-exposed HDCR immobilized in Cc-PAA, and BSA embedded in Cc-PAA as a negative control, both of which displayed significantly lower activities than active HDCR.

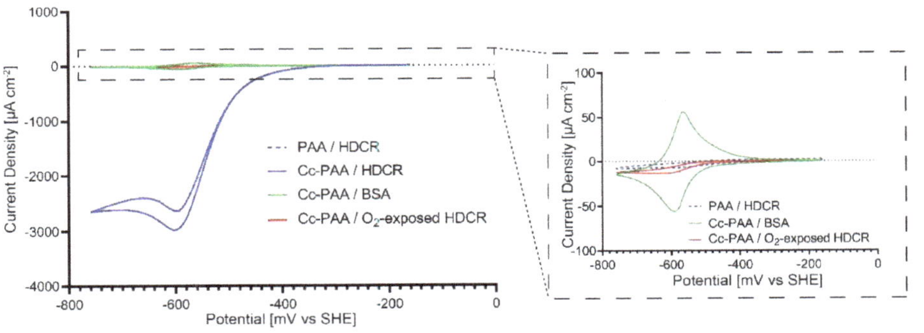

Figure 2. Cyclic voltammetry of HDCR on a rotating disc electrode. A greatly enhanced reductive current at low potentials was observed when cobaltocene-functionalized polyallylamine was present as electron shuttle in the electrode-coating hydrogel. The curve displayed a relatively constant current density with time, at low potentials, which is indicative of mediated enzymatic hydrogen evolution. Inset: Bovine serum albumin (BSA) immobilized in Cc-PAA as a negative activity control showed the expected reduction and oxidation peaks of Cc-PAA. HDCR in non-functionalized polyallylamine and O_2-exposed HDCR in Cc-PAA both displayed minimal activities. Rotation rate: 2.83 μg HDCR, 2000 RPM, scan rate: 10 mV/s. pH 5, 40 °C.

2.4. Stability of Cc-PAA at Elevated Temperatures

In both CV and amperometric i-t experiments, reductive current of hydrogen evolution decreased over time. As the enzyme activity appeared to be stable at those temperatures based on methyl viologen assays (Figure S3), we hypothesized that the loss of electrochemical activity was due to an instability of the redox polymer at elevated temperatures. We tested this directly via cyclic voltammetry by monitoring the peak heights of cobaltocene reduction and oxidation at the electrode surface in the absence of HDCR. As shown in

Figure S5, Cc-PAA stability varied widely with temperature. After 3 h of continuous cycling, electrodes at 25 °C, 40 °C and 60 °C retained 79%, 32% and 25% activity, respectively. Qualitative tests showed that increasing the crosslinker ratio during electrode preparation or increasing the drying time increased stability at high temperatures. However, both experimental variations also resulted in lower initial activity in the presence of HDCR, as seen in Figure S6.

2.5. Electrochemical Hydrogen Evolution Rates by HDCR and Faradaic Efficiency

To determine the relationship between observed reductive current and hydrogen produced, quantitative analysis of faradaic efficiency was conducted in closed-system electrochemical H-cells. Amperometric i-t was conducted at a potential of −650 mV vs. SHE. Hydrogen in the headspace was measured via GC using a thermal conductivity detector, and compared to the theoretical maximum yield, which was calculated from the total current passed through the electrode. Hydrogen production was also measured for a control experiment lacking Cc-PAA (Figure 3). Over a 6.5 h testing period, hydrogen production rate in the cobaltocene-mediated system (6.24 μmol·mg protein^{-1}·min^{-1}) was, on average, 17-fold higher than in the cobaltocene-free control (0.365 μmol·mg protein^{-1}·min^{-1}).

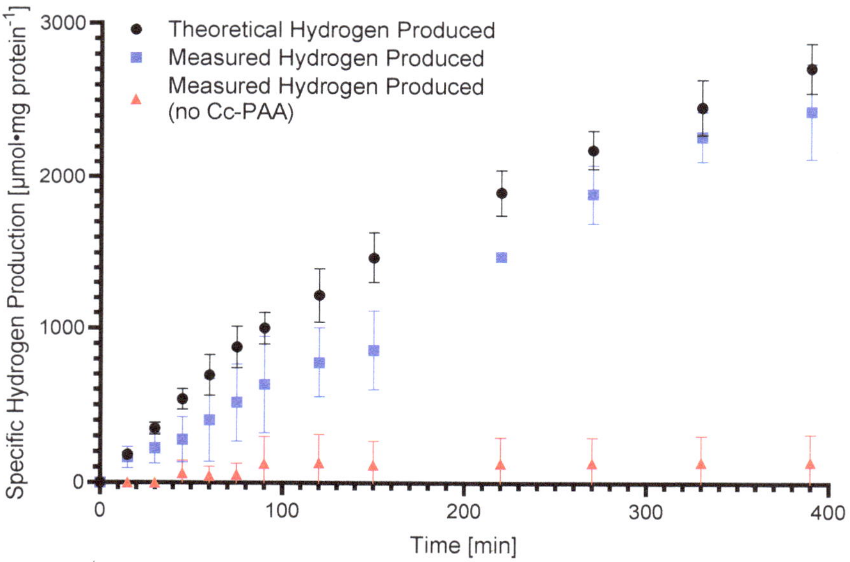

Figure 3. Quantitative comparison of hydrogen produced with and without the presence of Cc-PAA in a closed electrochemical system. Cc-PAA resulted in significantly higher hydrogen production, approaching the maximum theoretical yield based on current passed through the working electrode. Error bars indicate one standard deviation, replicates: N = 2. Potential: −650 mV vs. SHE. pH 5, 40 °C.

The faradaic efficiency was calculated from the ratio of the measured and theoretical hydrogen yields and is shown in Figure S7. Due to the low hydrogen concentrations at early timepoints, variability in measurements is more visible, resulting in a high apparent variance in efficiency. However, over the time span of the experiment, faradaic efficiency of cobaltocene-mediated hydrogen evolution converged at approximately 90%, showing that reductive waves observed during CV and reductive currents observed in amperometric i-t are useful indicators of hydrogen production.

2.6. Reversible HDCR Inactivation by Carbon Monoxide

The effect of carbon monoxide (CO) on H_2-evolution activity of *T. kivui* HDCR in the Cc-PAA-mediated electrochemical system was tested in an RDE setup using amperometric i-t. A CO-saturated buffer solution was injected into the system to reach an approximate concentration of 10 µM CO in the electrochemical cell and hydrogen evolution (Figure S8) or hydrogen oxidation (Figure 4) was tested. Sodium citrate buffer at pH 5 was used for hydrogen evolution, and potassium phosphate buffer at pH 8 was used for hydrogen oxidation, to make oxidation more thermodynamically favorable. Because the RDE system is intrinsically open to the anoxic atmosphere of the glove box, rotation of the electrode resulted in loss of CO over time.

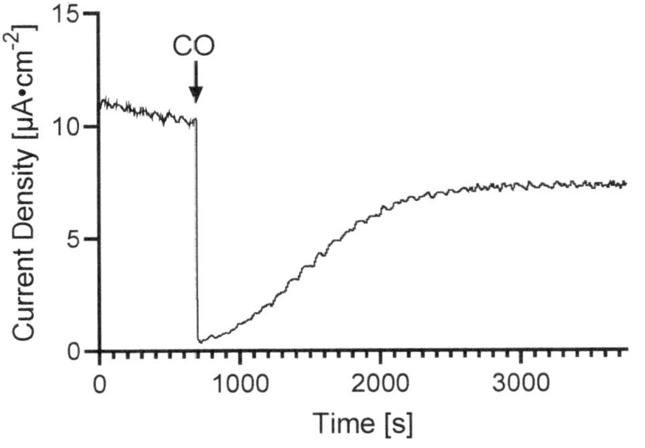

Figure 4. Reversible carbon monoxide inactivation of HDCR H_2 oxidation activity during oxidative amperometric i-t. After initial CO addition, HDCR activity as inferred by current production decreased rapidly, before regaining ca. 75% of activity over a period of 30 min. Rotation rate: 2000 RPM, potential: −200 mV vs. SHE. Phosphate buffer, pH 8, 40 °C.

In both the oxidative and reductive directions, addition of CO resulted in a sharp reduction of activity after stable current consumption or production. Interestingly, activity slowly recovered to 70–75% of the original current density, which is indicative of reversible CO inactivation.

3. Materials and Methods

3.1. Materials and Chemicals

All chemicals were purchased from Sigma-Aldrich (St. Louis, MO, USA) or Fisher Scientific (Waltham, MA, USA) with the following exceptions: 1-(2,5-dioxopyrrolidinylcarboxy)-cobaltocenium hexafluorophosphate was purchased from MCAT (Donaueschingen, Germany). Poly(allylamine) (PAA) (MW 15000, 15% aq. soln.) was purchased from Polysciences, Inc (Warrington, PA, USA). Ethylene glycol diglycidyl ether (EGDGE) was purchased from Tokyo Chemical Industry (Tokyo, Japan). Glassy carbon working electrodes (GCE) (3 mm diameter), platinum counter electrodes, saturated calomel reference electrodes (SCE), and a CHI1230c potentiostat were purchased from CH Instruments, Inc (Austin, TX, USA). A rotating disk electrode workstation (WaveVortex 10) and glassy carbon rotating disk electrodes (5 mm diameter) were purchased from Pine Research Instrumentation, Inc (Durham, NC, USA). Electrode preparation was conducted in an anoxic chamber (COY Laboratory Products, Inc., Grass Lake, MI, USA) with an N_2/H_2 headspace.

3.2. HDCR Expression and Purification

HDCR from *T. kivui* was produced using the homologous production system as recently described [25]. Therefore, the production strain was grown at 66 °C under anoxic conditions in complex medium using 1 L flasks (Müller-Krempel, Bülach, Switzerland) with 500 mL of media with additional 28 mM glucose and 200 $\mu g \cdot mL^{-1}$ kanamycin. The preparation of cell-free extract as well as protein purification via nitrilotriacetic acid (Ni^{2+}-NTA) resin (Qiagen, Hilden, Germany) were carried out as described before [25]. In contrast, 25 mM of Tris was used in all buffers and the equilibration and elution buffer contained 50 mM and 150 mM imidazole, respectively. Buffer exchange of the eluted and pooled proteins was achieved by ultrafiltration in 100-kDa VIASPIN tubes (Sartorius Stedim Biotech GmbH, Germany) using a buffer composition of 25 mM K-phosphate, 10 mM $MgSO_4$, 20 mM KCl, 20% [v/v] glycerol, 2 mM DTE, pH 7.0. Aliquots of the purified HDCR enzyme were frozen in liquid nitrogen and stored at −80 °C [10].

3.3. Synthesis of Cobaltocene-Functionalized Poly(allylamine) (Cc-PAA)

Cc-PAA was synthesized as previously described [18].

3.4. Preparation of HDCR-Embedded Electrodes

The HDCR/Cc-PAA enzymatic electrode was prepared via drop casting 11 µL of a solution containing 7 µL Cc-PAA (5 mg/mL), 2.3 µL EGDGE (10% *v/v*), and 2.3 µL of HDCR (1.3 mg/mL) onto a polished glassy carbon electrode (5 mm diameter). The electrode was then dried for 75 min before testing. The HDCR/PAA enzymatic electrode used the same mixture, substituting poly(allylamine) (PAA) (5 mg/mL) in place of Cc-PAA. The BSA/Cc-PAA electrode was prepared by drop casting 11 µL of a solution containing 7 µL Cc-PAA (5 mg/mL), 2.3 µL EGDGE (10% *v/v*), and 2.3 µL of BSA (2 mg/mL), followed by 75 min of drying before use. O_2-exposed HDCR was prepared by incubating HDCR in a conical centrifuge tube open to air at room temperature for 30 min. Bulk electrolysis enzymatic electrodes were prepared by drop casting 5.5 µL of the solution described above on a polished glassy carbon electrode (3 mm diameter), followed by drying for 75 min. Cobaltocene-free bulk electrolysis controls were prepared using 5.5 µL of a solution containing 2.3 µL HDCR and 9.3 µL BSA, to aid in protein immobilization and minimize side reactions occurring on the exposed glassy carbon surface.

3.5. Electrochemical Methods

Rotating disk studies were performed with a WaveVortex® electrode rotator (Pine Research Instrumentation, Durham, NC, USA). Studies used an SCE reference electrode, GCE working electrode, and platinum mesh counter electrode in a three-electrode configuration. All potentials are reported relative to the standard hydrogen electrode (SHE) ($E^0_{SHE} = E^0_{SCE} + 0.242$ V). Electrode preparation and testing were done in an anoxic glove bag. Preparation occurred at room temperature and tests were conducted at the temperatures indicated.

Bulk electrolysis experiments were completed using an airtight H-cell (Adams and Chittenden, Berkeley, CA, USA) with a Nafion® proton exchange membrane. A silver/silver chloride (Ag/AgCl) reference electrode, glassy carbon disk working electrode, and graphite rod counter electrode were used. Potentials were reported relative to the SHE ($E^0_{SHE} = E^0_{Ag/AgCl} + 0.197$ V). Electrode preparation was done in an anoxic glove bag at room temperature, and testing was conducted at 40 °C, after flushing for 20 min with 100% N_2. Tests were completed using a VMP3 potentiostat (Biologic Science Instruments, Seyssinet-Pariset, France).

3.6. Synthesis of Ti(III) Citrate

Ti(III) citrate was synthesized as previously described [26], with the following modification: pH adjustment to pH 7.0 was completed with 1 M NaOH in place of sodium carbonate.

3.7. Methyl Viologen Assay for Quantification of Chemical Hydrogen Evolution Rates

Chemical hydrogen evolution activity assays were carried out using 5 mL liquid samples with 5 mL headspace in sealed vials. The indicated buffer solutions were combined with 2 mM methyl viologen and 10 mM electron donor. For phosphate buffered samples, sodium dithionite was used as the electron donor. For citrate buffered samples, Ti(III) citrate was used as the electron donor. Vials were flushed with nitrogen for 10 min, followed by addition of HDCR, and incubated at 40 °C. Hydrogen concentration in the headspace was measured by gas chromatography (GC).

4. Discussion and Conclusions

Cc-PAA was previously found to enhance hydrogen electrosynthesis by hydrogenases, with 20- to 40-fold increase in reductive current densities compared to the absence of the redox polymer [18]. In this work, we showed that Cc-PAA, paired with HDCR of thermophilic *T. kivui*, increased current density 340-fold in an RDE system. This resulted in maximum current densities of approximately 3 $mA{\cdot}cm^{-2}$—an unprecedented activity that is significantly higher than previously reported current densities in mediated enzymatic electrochemical hydrogen evolution systems, particularly in systems which contain hydrogen in the headspace [18,22]. In closed-system electrochemical hydrogen experiments, Cc-PAA increased hydrogen production by 17-fold. This difference in activity increase can be attributed to several factors, including primarily mass transfer limitations in the closed-system H-cells which were mixed using stir bars rather than an RDE, and lower apparent specific activity due to loss of enzyme activity over the extended course of the 6.5-h test. Despite the heat inactivation of Cc-PAA, approximately 90% faradaic efficiency was maintained over the 6.5-h testing period. This confirms that at elevated temperatures, Cc-PAA continues to supply electrons to embedded enzymes with high specificity even when partially degraded.

CO inactivation of HDCR was largely reversible, with a fast inactivating step followed by a gradual reactivating step, with 70–75% recovery of activity. This important finding shows promise for the use of this enzyme in CO-containing systems such as syngas utilizing processes. Additional optimization of the system may allow for this recovery to be improved further. While the enzyme preparation used had a bias for hydrogenase activity, given the previously demonstrated activity of the enzyme to catalyze the reaction in either direction in nonelectrochemical settings, it is hypothesized that electron mediation via Cc-PAA to fully active HDCR would also result in CO_2 reduction to formate. If this is the case, it would allow for the production of a highly soluble energy carrier at high rates, and should be investigated in future work [27]. The presented data show that Cc-PAA in conjunction with the *T. kivui* HDCR acts as a highly effective catalyst for hydrogen production, achieving both high current densities and high specificity. This demonstrates an important improvement in catalytic current density of a mediated enzymatic hydrogen evolution system, further optimization of which will allow for industrially relevant rates and volumes to be achieved, aiding in the search for sustainable hydrogen catalysis.

Supplementary Materials: The following are available online at https://www.mdpi.com/article/10.3390/catal11101197/s1, Figure S1. Activity comparison of HDCR activity in Cc-PAA at different temperatures. A: 25 °C, B: 30 °C, C: 40 °C, D: 60 °C. Tests at 40 °C showed high activity, but with less activity loss over time compared to higher temperatures. Phosphate buffer, pH 6, Figure S2. pH and buffer concentration optimizations revealed an optimal pH of 5.0 and higher activities at increasing buffer concentrations. (A) Comparison of the magnitude of the reductive wave for electrochemical hydrogen evolution in buffers of different pH. Citrate buffer was used for pH 6 and below, and phosphate buffer for pH 6 and above, with negligible difference between the buffers at pH 6. (B) Citrate buffer had a molar ratio of 5:1 sodium citrate:KCl, and this ratio was maintained across all solute concentrations. Activity increased with increasing solute concentration up to the solubility limit of the buffer—a similar trend to that observed in previous FeFe hydrogenase, cobaltocene-mediated systems [1]. Rotation rate: 2000 RPM, scan rate: 10 mV/s. Replicates: N = 3, Figure S3. Specific hydrogen production by HDCR in 0.2 M sodium citrate and 40 mM KCl, pH 5.0, when mediated by methyl viologen with Ti(III) citrate as electron donor. Similar trends at different rates were observed in phosphate buffer and HEPES buffer, both at pH 6 with dithionite as the electron donor. Test completed at 40 °C. Replicates: N = 3, Figure S4. HDCR displayed high specific activities for methyl viologen mediated hydrogen evolution, compared to previously reported hydrogenases in their respective optimal electrochemistry buffers [1]. Replicates: N = 3, Figure S5. Cobaltocene redox peak magnitude is lost over time at increasing rates with increasing temperature. Of the tested temperatures, Cc-PAA shows the highest stability at 25 °C (A), intermediate stability at 40 °C (B), and lowest stability at 60 °C (C). Note the different time scale between panels—12 h for A and B, 4 h for C. Cycles are normalized for the maximum magnitude achieved at each temperature, respectively. Rotation rate: 2000 RPM, scan rate: 10 mV/s. pH 5, Figure S6. Increasing crosslinker ratio resulted in less activity loss over time, but with a lower initial activity. Data shown is the magnitude of the reductive wave during cyclic voltammetry of HDCR embedded in Cc-PAA with the usual crosslinker ratio as "low" and a 2× crosslinker concentration as "high". Timepoint 1: 1 min; timepoint 2: 29 min. pH 5, Figure S7. Faradaic efficiency of Cc-PAA-mediated HDCR approached 90% over extended testing periods. Large error bars at early timepoints are due to variation in measurements at very low hydrogen concentrations. Error bars indicate one standard deviation. Replicates: N = 2. pH 5, 40 °C, Figure S8. Carbon monoxide inactivation during reductive amperometric i-t of Cc-PAAembedded HDCR displayed an initial strong decrease in activation before regaining activity to a stable value. Subsequent additional inactivation resulted in another initial loss of activity, followed by ca. 70% of activity being regained. Rotation rate: 2000 RPM, potential: −650 mV vs. SHE. Citrate buffer, pH 5, 40 °C.

Author Contributions: The manuscript was written through contributions of all authors. Conceptualization, A.M.S. and V.M.; methodology, A.M.S., V.M., J.C.R. and F.M.S.; formal analysis, J.C.R.; investigation, J.C.R. and F.M.S.; resources, A.M.S., V.M., J.C.R. and F.M.S.; data curation, J.C.R.; writing—original draft preparation, J.C.R.; writing—review and editing, A.M.S., V.M., J.C.R. and F.M.S.; visualization, A.M.S. and V.M.; supervision, A.M.S. and V.M.; project administration, A.M.S.; funding acquisition, A.M.S. and V.M. All authors have read and agreed to the published version of the manuscript.

Funding: Work in the Spormann lab was funded by the Achievement Rewards for College Scientists (ARCS) Foundation and the Stanford Global Climate and Energy Project (GCEP). Work from the Müller lab was funded by the European Research Council (ERC) under the European Union's Horizon 2020 research and innovation program (grant agreement 741791).

Acknowledgments: We gratefully acknowledge the Spormann Laboratory at Stanford University, the Müller lab at Johann Wolfgang Goethe University in Frankfurt am Main, and Ross D. Milton, University of Geneva, for useful discussions.

Conflicts of Interest: The authors declare no competing financial interests.

Abbreviations

HDCR	Hydrogen-Dependent CO_2 Reductase;
Cc-PAA	cobaltocene-functionalized poly(allylamine);
RDE	rotating disk electrode;
SHE	standard hydrogen electrode;
CV	cyclic voltammetry.

References

1. Lindstrom, P.; Fritsch, D.; Fickling, M.; Chase, N.; Martin, L.; Anti, L.; Hansom, S.; Gross, P.; Palguta, J.; Dyl, K.; et al. *Annual Energy Outlook 2020*; U.S. Energy Information Administration: Washington, DC, USA, 2020.
2. Dresselhaus, M.S.; Thomas, I.L. Alternative energy technologies. *Nature* **2001**, *414*, 332–337. [CrossRef]
3. Ruth, J.C.; Spormann, A.M. Enzyme Electrochemistry for Industrial Energy Applications—A Perspective on Future Areas of Focus. *ACS Catal.* **2021**, *11*, 5951–5967. [CrossRef]
4. Abe, J.O.; Popoola, A.P.I.; Ajenifuja, E.; Popoola, O.M. Hydrogen energy, economy and storage: Review and recommendation. *Int. J. Hydrogen Energy* **2019**, *44*, 15072–15086. [CrossRef]
5. Gotovsky, M.A.; Gotovsky, A.M.; Mikhailov, V.E.; Lychakov, V.D.; Sukhorukov, Y.G.; Sukhorukova, E.A. Formate: The Third Way in Green Energy. *Int. J. Chem. Eng. Appl.* **2019**, *10*, 189–194. [CrossRef]

6. Blanco, H.; Cazzola, P.; Dulac, J.; Fukui, H.; Kim, T.-Y.; Kurban, Z.; Levi, P.; Malischek, R.; McGlade, C.; Petrosyan, K.; et al. *The Future of Hydrogen*; International Energy Agency: Paris, France, 2019.
7. Cadoux, C.M.; Milton, R.D. Recent enzymatic electrochemistry for reductive reactions. *ChemElectroChem* **2020**, *7*, 1974–1986. [CrossRef]
8. Ceccaldi, P.; Schuchmann, K.; Müller, V.; Elliott, S.J. The Hydrogen Dependent CO_2 reductase: The first completely co tolerant fefe-hydrogenase. *Energy Environ. Sci.* **2017**, *10*, 503–508. [CrossRef]
9. Schwarz, F.M.; Ciurus, S.; Jain, S.; Baum, C.; Wiechmann, A.; Basen, M.; Müller, V. Revealing formate production from carbon monoxide in wild type and mutants of Rnf- and Ech-containing acetogens, *Acetobacterium woodii* and *Thermoanaerobacter kivui*. *Microb. Biotechnol.* **2020**, *13*, 2044–2056. [CrossRef]
10. Schwarz, F.M.; Schuchmann, K.; Müller, V. Hydrogenation of CO_2 at ambient pressure catalyzed by a highly active thermostable biocatalyst. *Biotechnol. Biofuels* **2018**, *11*, 237. [CrossRef]
11. Müller, V. New Horizons in Acetogenic Conversion of One-Carbon Substrates and Biological Hydrogen Storage. *Trends Biotechnol.* **2019**, *37*, 1344–1354. [CrossRef]
12. Lubitz, W.; Ogata, H.; Rudiger, O.; Reijerse, E. Hydrogenases. *Chem. Rev.* **2014**, *114*, 4081–4148. [CrossRef]
13. Lienemann, M.; Deutzmann, J.S.; Milton, R.D.; Sahin, M.; Spormann, A.M. Mediator-free enzymatic electrosynthesis of formate by the *Methanococcus maripaludis* heterodisulfide reductase supercomplex. *Bioresour. Technol.* **2018**, *254*, 278–283. [CrossRef]
14. Basso, A.; Serban, S. Industrial applications of immobilized enzymes—A review. *Mol. Catal.* **2019**, *479*, 110607. [CrossRef]
15. Srikanth, S.; Alvarez-Gallego, Y.; Vanbroekhoven, K.; Pant, D. Enzymatic Electrosynthesis of Formic Acid through Carbon Dioxide Reduction in a Bioelectrochemical System: Effect of Immobilization and Carbonic Anhydrase Addition. *ChemPhysChem* **2017**, *18*, 3174–3181. [CrossRef]
16. Sakai, K.; Kitazumi, Y.; Shirai, O.; Takagi, K.; Kano, K. Efficient bioelectrocatalytic CO_2 reduction on gas-diffusion-type biocathode with tungsten-containing formate dehydrogenase. *Electrochem. Commun.* **2016**, *73*, 85–88. [CrossRef]
17. Lee, C.-Y.; Park, H.S.; Fontecilla-Camps, J.C.; Reisner, E. Photoelectrochemical H_2 Evolution with a Hydrogenase Immobilized on a TiO_2-Protected Silicon Electrode. *Angew. Chem. Int. Ed.* **2016**, *55*, 5971–5974. [CrossRef]
18. Ruth, J.C.; Milton, R.D.; Gu, W.; Spormann, A.M. Enhanced Electrosynthetic Hydrogen Evolution by Hydrogenases Embedded in a Redox-Active Hydrogel. *Chem. Eur. J.* **2020**, *26*, 7323–7329. [CrossRef] [PubMed]
19. Morra, S.; Valetti, F.; Sarasso, V.; Castrignanò, S.; Sadeghi, S.J.; Gilardi, G. Hydrogen production at high Faradaic efficiency by a bio-electrode based on TiO_2 adsorption of a new [FeFe]-hydrogenase from *Clostridium perfringens*. *Bioelectrochemistry* **2015**, *106*, 258–262. [CrossRef] [PubMed]
20. Ruff, A. Redox polymers in bioelectrochemistry: Common playgrounds and novel concepts. *Curr. Opin. Electrochem.* **2017**, *5*, 66–73. [CrossRef]
21. Hardt, S.; Stapf, S.; Filmon, D.T.; Birrell, J.A.; Rüdiger, O.; Fourmond, V.; Léger, C.; Plumeré, N. Reversible H2 oxidation and evolution by hydrogenase embedded in a redox polymer film. *Nat. Catal.* **2021**, *4*, 251–258. [CrossRef] [PubMed]
22. Shiraiwa, S.; So, K.; Sugimoto, Y.; Kitazumi, Y.; Shirai, O.; Nishikawa, K.; Higuchi, Y.; Kano, K. Reactivation of standard [NiFe]-hydrogenase and bioelectrochemical catalysis of proton reduction and hydrogen oxidation in a mediated-electron-transfer system. *Bioelectrochemistry* **2018**, *123*, 156–161. [CrossRef] [PubMed]
23. Yuan, M.; Sahin, S.; Cai, R.; Abdellaoui, S.; Hickey, D.P.; Minteer, S.D.; Milton, R.D. Creating a Low-Potential Redox Polymer for Efficient Electroenzymatic CO_2 Reduction. *Angew. Chem. Int. Ed.* **2018**, *57*, 6582–6586. [CrossRef] [PubMed]
24. Mayhew, S.G. The Redox Potential of Dithionite and $SO^{-}{}_2$ from Equilibrium Reactions with Flavodoxins, Methyl Viologen and Hydrogen plus Hydrogenase. *Eur. J. Biochem.* **1978**, *85*, 535–547. [CrossRef] [PubMed]
25. Katsyv, A.; Schoelmerich, M.C.; Basen, M.; Müller, V. The pyruvate:ferredoxin oxidoreductase of the thermophilic acetogen, *Thermoanaerobacter kivui*. *FEBS Open Bio* **2021**, *11*, 1332–1342. [CrossRef] [PubMed]
26. Zehnder, A.J.B.; Wuhrmann, K. Titanium(III) citrate as a nontoxic oxidation-reduction buffering system for the culture of obligate anaerobes. *Science* **1976**, *194*, 1165–1166. [CrossRef]
27. Sidgwick, N.V.; Gentle, J.A.H.R. CCXX1.-The Solubilities of the Alkali Formates and Acetates in Water. *J. Chem. Soc.* **1922**, 1837–1843. [CrossRef]

catalysts

MDPI

Article

From Cell-Free Protein Synthesis to Whole-Cell Biotransformation: Screening and Identification of Novel α-Ketoglutarate-Dependent Dioxygenases for Preparative-Scale Synthesis of Hydroxy-L-Lysine

Jascha Rolf †, Philipp Nerke †, Annette Britner, Sebastian Krick, Stephan Lütz and Katrin Rosenthal *

Chair for Bioprocess Engineering, Department of Biochemical and Chemical Engineering, TU Dortmund University, D-44227 Dortmund, Germany; jascha.rolf@tu-dortmund.de (J.R.); philipp.nerke@tu-dortmund.de (P.N.); annette.britner@tu-dortmund.de (A.B.); sebastian.krick@tu-dortmund.de (S.K.); stephan.luetz@tu-dortmund.de (S.L.)

* Correspondence: katrin.rosenthal@tu-dortmund.de; Tel.: +49-231-755-5115

† These authors have contributed equally to this work.

Abstract: The selective hydroxylation of non-activated C-H bonds is still a challenging reaction in chemistry. Non-heme Fe^{2+}/α-ketoglutarate-dependent dioxygenases are remarkable biocatalysts for the activation of C-H-bonds, catalyzing mainly hydroxylations. The discovery of new Fe^{2+}/α-ketoglutarate-dependent dioxygenases with suitable reactivity for biotechnological applications is therefore highly relevant to expand the limited range of enzymes described so far. In this study, we performed a protein BLAST to identify homologous enzymes to already described lysine dioxygenases (KDOs). Six novel and yet uncharacterized proteins were selected and synthesized by cell-free protein synthesis (CFPS). The subsequent in vitro screening of the selected homologs revealed activity towards the hydroxylation of L-lysine (Lys) into hydroxy-L-lysine (Hyl), which is a versatile chiral building block. With respect to biotechnological application, *Escherichia coli* whole-cell biocatalysts were developed and characterized in small-scale biotransformations. As the whole-cell biocatalyst expressing the gene coding for the KDO from *Photorhabdus luminescens* showed the highest specific activity of 8.6 ± 0.6 U g_{CDW}^{-1}, it was selected for the preparative synthesis of Hyl. Multi-gram scale product concentrations were achieved providing a good starting point for further bioprocess development for Hyl production. A systematic approach was established to screen and identify novel Fe^{2+}/α-ketoglutarate-dependent dioxygenases, covering the entire pathway from gene to product, which contributes to accelerating the development of bioprocesses for the production of value-added chemicals.

Keywords: KDO; Fe^{2+}/α-ketoglutarate-dependent dioxygenase; chaperones; cell-free protein synthesis (CFPS); L-lysine; hydroxy-L-lysine

Citation: Rolf, J.; Nerke, P.; Britner, A.; Krick, S.; Lütz, S.; Rosenthal, K. From Cell-Free Protein Synthesis to Whole-Cell Biotransformation: Screening and Identification of Novel α-Ketoglutarate-Dependent Dioxygenases for Preparative-Scale Synthesis of Hydroxy-L-Lysine. *Catalysts* **2021**, *11*, 1038. https://doi.org/10.3390/catal11091038

Academic Editors: Evangelos Topakas, David D. Boehr and Roland Wohlgemuth

Received: 30 June 2021
Accepted: 25 August 2021
Published: 27 August 2021

1. Introduction

Non-heme Fe^{2+}/α-ketoglutarate-dependent dioxygenases constitute a large superfamily of enzymes. They are capable of catalyzing a plethora of different reactions, such as desaturations, epoxidations, halogenations, oxidations, cyclizations, and predominantly hydroxylations [1,2]. In recent years, Fe^{2+}/α-ketoglutarate-dependent dioxygenases have been discovered, which hydroxylate L-lysine (Lys) to produce hydroxy-L-lysine (Hyl) [3,4]. The enzymes have been termed KDO for lysine dioxygenase [3]. Depending on the respective KDO, different isomers are formed with high regio- and stereospecificity. Hyl is a molecule of industrial interest as it is a building block for the synthesis of a variety of pharmacologically relevant molecules, such as the HIV protease inhibitor palinavir [5,6], or newly identified drug candidates such as tambromycin (anticancerogenic activity), cepafungin I or glidobactin A, both of which are proteasome inhibitors [7–9]. It can also be used for the generation of chiral amino alcohols, which represent relevant chiral building blocks [10,11]. Therefore, the discovery of novel enzymes able to catalyze challenging reactions such as the selective hydroxylation of non-activated C-H bonds in Lys is of high importance.

To screen and identify novel enzymes, in vitro or cell-free protein synthesis (CFPS) has become an established tool for rapid transcription and translation [12–14]. CFPS can complement traditional in vivo protein synthesis to accelerate the discovery of novel enzymes or enzyme variants [15]. However, a common challenge is that proteins, when synthesized with heterologous systems, often do not fold properly and therefore become insoluble. Molecular chaperones prevent protein aggregation and promote protein folding. Exogenous addition of molecular chaperone proteins has effectively facilitated the synthesis of various soluble proteins in CFPS systems [16]. Alternatively, the chaperones can be directly synthesized in the source organisms before CFPS extract preparation [17]. To accomplish this, *Escherichia coli* strains are transformed with plasmids encoding different sets of molecular chaperones. These are synthesized during growth of the strain, which serves as the basis for the cell extract. The extract thus contains not only all components for transcription and translation, but also chaperones that will support the synthesis of soluble protein. This avoids the time-consuming and laborious synthesis and purification of chaperones, or the expensive use of commercially available chaperones. Alternatively, the CFPS reaction mix can be used first to synthesize the chaperones, and only in a second step does the synthesis of the target protein take place [18]. In this case, it is mandatory that the synthesis solution is refreshed after the first synthesis step, as the systems will otherwise have a lower synthesis performance.

In this study, chaperone-enriched CFPS extracts were developed to synthesize KDOs, which are difficult to express in soluble form in *E. coli* [3,8]. Five different chaperones containing CFPS systems were prepared from *E. coli* strains and tested to screen novel and putative KDOs (Figure 1). The CFPS systems allowed the efficient synthesis of soluble enzymes without the need for exogenous addition or co-expression of folding effectors. Subsequent activity assays demonstrated the successful hydroxylation of Lys to Hyl for several KDOs, including six novel and previously biochemically uncharacterized and undescribed enzymes. The novel KDOs were further characterized in whole-cell systems using recombinant *E. coli*. One newly identified homolog was selected and applied in a resting-cell biotransformation on a preparative scale.

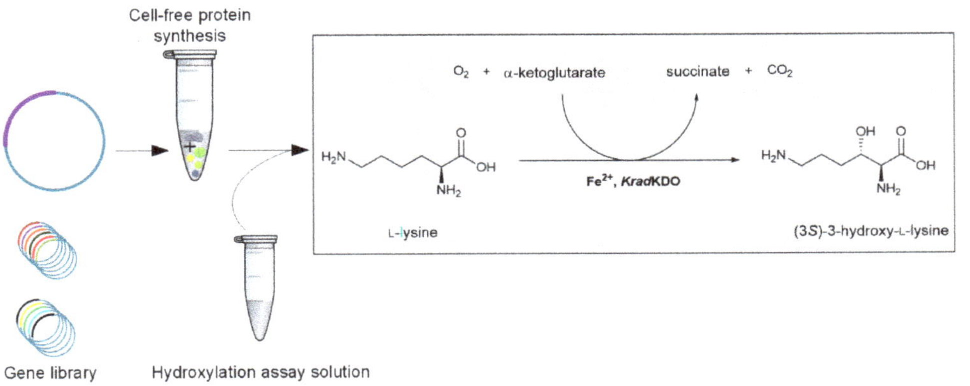

Figure 1. Schematic procedure for screening novel lysine dioxygenases (KDOs) with cell-free protein synthesis (CFPS). Exemplary reaction for the KDO from *Kineococcus radiotolerans* (*Krad*KDO).

2. Results

2.1. Sequence Similarity Search for Novel KDOs in Bacteria

In this study, 13 enzymes from different bacteria (Figure 2) were tested for hydroxylation of Lys to Hyl, of which six were uncharacterized enzymes. To identify novel KDOs belonging to the superfamily of Fe^{2+}/α-ketoglutarate-dependent dioxygenases, we searched for enzyme homologs to the known KDOs, which were discovered by Baud et al. 2014 [3] and Hara et al. 2017 [4], using Protein Basic Local Alignment Search Tool (BLAST) (NCBI, Bethesda, MD, USA). These enzymes catalyze the regio- and stereospecific hydroxylation of Lys to form either (3*S*)-3-hydroxy-L-lysine or (4*R*)-4-hydroxy-L-lysine and were grouped in the clavaminate synthase-like family (IPR014503) [19]. Our BLAST search led to protein sequences of the putative KDOs *Krhi*KDO from *Kineococcus rhizospharae* and *Mint*KDO from *Mycobacterium interjectum* (Figure 3A). They share a protein sequence identity of 47.32% and 42.11%, compared to the sequence of *Caci*KDO from *Catenulispora acidiphila*, respectively.

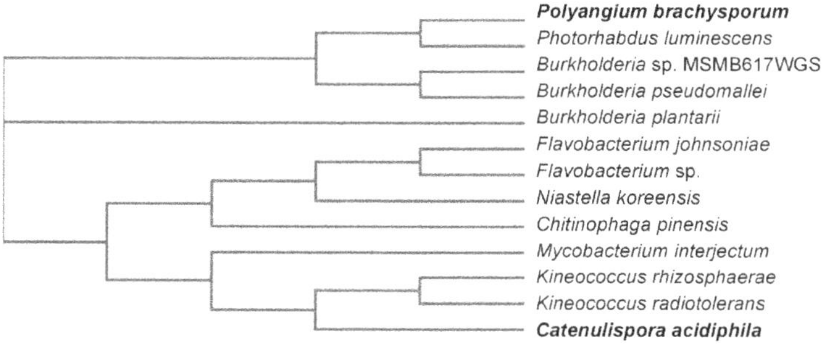

Figure 2. Phylogenetic tree of the organisms of origin of the known and putative KDOs considered in this study. The figure was created using the EMBL-EBI Web Service [20] with the respective protein sequences.

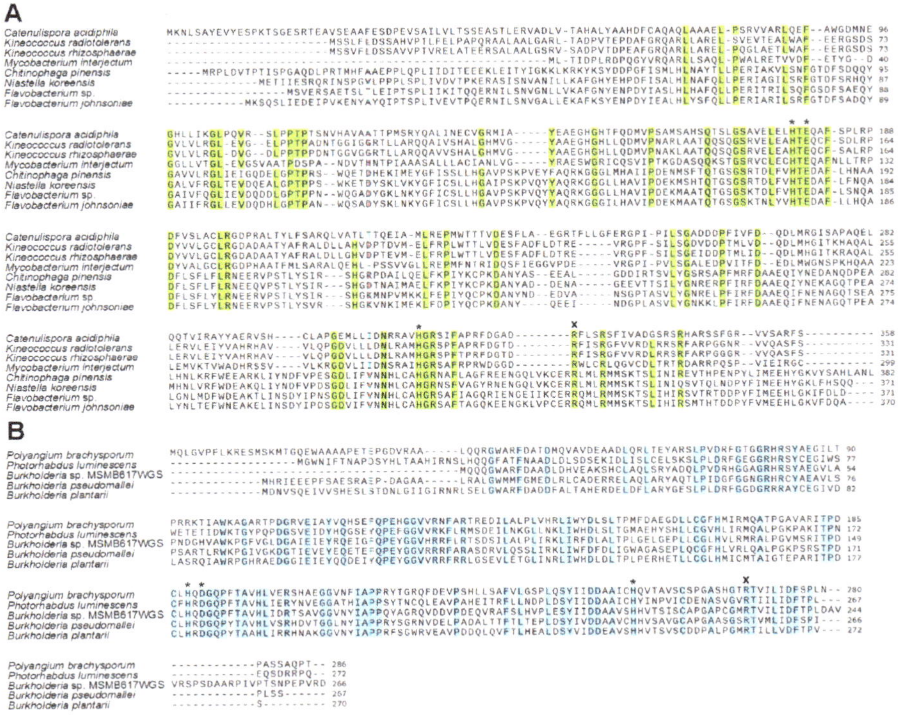

Figure 3. (**A**) Multiple sequence alignment of the KDO from *Catenulispora acidiphila* (*Caci*KDO) with homologs from different species. (**B**) Multiple sequence alignment of the KDO from *Polyangium brachysporum* (*Bpra*KDO) with homologs from different species. Shading indicates a consensus sequence above a threshold of 80%. The alignment was generated using MUSCLE (SnapGene®®, GSL Biotech LLC, Chicago, IL, USA, Version 4.3.11). The asterisk depicts the three amino acids, which form the metal binding triad. The x denotes the arginine, which is involved in α-ketoglutarate (α-KG) binding. Positions were adopted from crystal structures and homology models [9,19].

In 2019, Amatuni et al. discovered that the enzyme GlbB, from the glidobactin synthesis cluster of *Polyangium brachysporum* (now reclassified as *Schlegelella brevitalea* sp. nov. [21]) also belongs to the Fe^{2+}/α-ketoglutarate-dependent dioxygenase family and acts as KDO to produce (4*S*)-4-hydroxy-L-lysine [9]. This enzyme is therefore termed *Pbra*KDO in our study. Interestingly, as the authors mentioned, this enzyme only shares a very low sequence identity with the KDOs discovered previously (e.g., 11.7% to *Caci*KDO). *Pbra*KDO was found to form a novel cluster in the PF10014 (IPR018724) family [9]. The phylogenetic assignment clearly shows the distinction between the two different groups of KDOs belonging to the superfamily of Fe^{2+}/α-ketoglutarate-dependent dioxygenases (Figure 2).

Using the protein sequence of *Pbra*KDO as a query sequence, we found almost 100 protein sequences ranging from a maximum of 59% identity to 31% identity. We selected four sequences with different phylogenetic distance to *Pbra*KDO for our study (Figure 2, Figure 3B); the putative KDOs *Plum*KDO from *Photorhabdus luminescens* (56.73% identity), *Bspe*KDO from *Burkholderia* species MSMB617WGS (57.48% identity), *Bpse*KDO from *Burkholderia pseudomallei* (51.89% identity) and *Bpla*KDO from *Burkholderia plantarii* (50.41% identity). These protein sequences were all annotated as belonging to the Fe^{2+}/α-ketoglutarate-dependent dioxygenase family.

The enzymes from the group of *C. acidiphila* are all larger in size than the KDO from *P. brachysporum* and its homologs (Figure 3). The alignments show the conserved sites for metal ion binding as well as the conserved arginine, which is involved in the binding of α-ketoglutarate (α-KG). Fe^{2+} binding is usually facilitated by a 2-His-1-carboxylate facial triad [22]. While Fe^{2+} binding of the first group is facilitated by His-Glu-His motif (Figure 3A), it is constituted by a His-Asp-His motif in the second group (Figure 3B).

2.2. Cell-Free Protein Synthesis Identifies Novel KDOs

CFPS can be performed within a few hours and thus accelerates the screening of enzyme variants [13,15]. Therefore, we performed the synthesis of known KDOs and yet undescribed homologs in an *E. coli*-based CFPS system with plasmids carrying DNA templates.

The protein syntheses were investigated with SDS-PAGE analysis (Figure 4A, Supplementary Materials Figures S3–S26). Corresponding protein bands could be proven for most protein homologs. For the new variants, no synthesis could be confirmed only for the homolog derived from *B. plantarii*. However, the successful synthesis cannot be ruled out due to the complex protein composition of the CFPS mix in combination with insufficient protein concentrations. In some cases, prominent bands of *E. coli*'s endogenous proteins overlaid the area for the associated sizes. Therefore, no preselection of expressible or non-expressible homologs was carried out and all CFPS reactions were screened for activity in subsequent hydroxylation assays combined with LC-MS-based analysis. The previously described KDOs showed activity towards the hydroxylation of Lys, proven by the formation of Hyl (Figure 4B). Previous studies already identified the products of *Cpin*KDO, *Nkor*KDO, *Fspe*KDO, *Fjoh*KDO to be (4*R*)-4-hydroxy-L-lysine and *Caci*KDO and *Krad*KDO to be (3*S*)-3-hydroxy-L-lysine [3,4,19]. The product of *Pbra*KDO was identified to be (4*S*)-4-hydroxy-L-lysine [8,9]. We were able to distinguish between 5-hydroxy-DL-lysine (analytical standard) and (4*R*)-4-hydroxy-L-lysine with our HPLC-methods. Unfortunately, (3*S*)-3-hydroxy-L-lysine and (4*S*)-4-hydroxy-L-lysine eluted at the same retention time, so no clear discrimination was possible. However, based on the phylogenetic origin, it is likely that *Krhi*KDO and *Mint*KDO form (3*S*)-3-hydroxy-L-lysine and that *Plum*KDO, *Bspe*KDO, *Bpse*KDO and *Bpla*KDO produce (4*S*)-4-hydroxy-L-lysine. The products still need to be investigated further to determine their absolute configuration. For that, synthesis in preparative scale of all products would be required for chemical derivatization and NMR analysis, which was beyond the scope of this work. Nevertheless, these results confirm the applicability of CFPS in combination with a hydroxylation assay for KDOs.

Remarkably, four of six novel KDOs, which have never been characterized biochemically before, catalyzed the synthesis of more than 50 μM Hyl and were thus confirmed as KDOs. Only the enzyme homologs originating from *M. interjectum* and *B.* species MSMB617WGS produced just trace amounts of Hyl. In negative controls with the CFPS mix but without a DNA template, Hyl could not be detected after 20 h. The highest product concentration of 3.66 ± 0.16 mM (~37% yield) was achieved by the KDO of *Flavobacterium* species. As the CFPS reaction solution is a complex mixture, it is possible that enzymes from *E. coli* metabolism, which are present in the extract, degraded Lys or α-KG. Moreover, oxygen limitation, oxidation of Fe^{2+}, or chelation by components of the CFPS reaction solution might be reasons for the incomplete conversion in the hydroxylation assay. It should be noted that identical reaction conditions were tested for all enzymes. However, it has already been described that some KDOs showed higher activities under different conditions, such as increased or decreased temperatures and pH [4]. Nevertheless, higher product concentration may indicate better protein expression, higher enzyme stability or activity, and thus indicate a suitable biocatalyst for the biotransformation of Lys into Hyl.

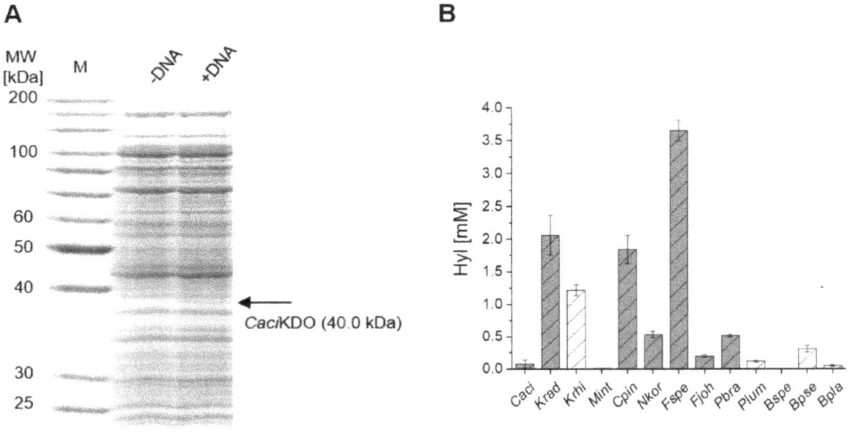

Figure 4. (**A**) SDS-PAGE gel image of cell-free synthesized *Caci*KDO. M, marker; -DNA, CFPS mix without plasmid template; +DNA, CFPS mix containing the plasmid template for *Caci*KDO. (**B**) Hydroxy-L-lysine (Hyl) concentrations achieved in in vitro hydroxylation assays for cell-free synthesized KDOs (*Caci, Catenulispora acidiphila; Krad, Kineococcus radiotolerans, Krhi, Kineococcus rhizosphaerae; Mint, Mycobacterium interjectum; Cpin, Chitinophaga pinensis; Nkor, Niastella koreensis; Fspe, Flavobacterium* sp.; *Fjoh, Flavobacterium johnsoniae; Pbra, Polyangium brachysporum; Plum, Photorhabdus luminescens; Bspe, Burkholderia* sp. MSMB617WGS; *Bpse, Burkholderia pseudomallei; Bpla, Burkholderia plantarii*). Assays were performed on a 100 µL scale with 10 mM L-lysine (Lys) and 20% *v/v* protein solution for 20 h. Protein homologs are sorted by phylogenetic distance. Grey bars indicate described KDOs, while white bars indicate undescribed KDOs. Error bars are a result of biological duplicates.

Hence, the combination of CFPS and the in vitro hydroxylation assay is very well suited for the screening and identification of KDOs. A further simplification and increase in speed would be the use of PCR-based linear DNA templates, thus eliminating any cloning steps [23]. This would allow a much larger number of proteins to be screened for biocatalytic activity in a very short time.

2.3. Chaperone-Assisted Expression Can Improve the Productivity of Cell-Free Synthesized KDOs

In a previous study, suboptimal protein yields of *Caci*KDO and *Pbra*KDO were obtained due to the synthesis of insoluble proteins [7,8]. This issue could be mostly solved by the co-expression of chaperones, in this case, GroEL and GroES. In our study, similar problems were noticed when we analyzed the soluble and total fraction of the cell-free synthesized KDOs (Figure 5). Since correct three-dimensional folding is critical for full enzyme function, these insoluble proteins are usually inactive.

Therefore, we decided to test chaperone-enriched cell extracts for the CFPS of KDOs, which could lead to a higher fraction of soluble enzyme and thereby increased hydroxylation activity. We generated five additional cell extracts, in which the commercially available plasmids pG-KJE8, pKJE7, pGro7, pG-Tf2, and pTf16 were used for the expression of different chaperone sets, consisting of DnaK, DnaJ, GrpE, GroES, GroEL, and tig. These extracts were used for the synthesis of the 13 enzyme variants, which were tested in subsequent hydroxylation assays (Table 1).

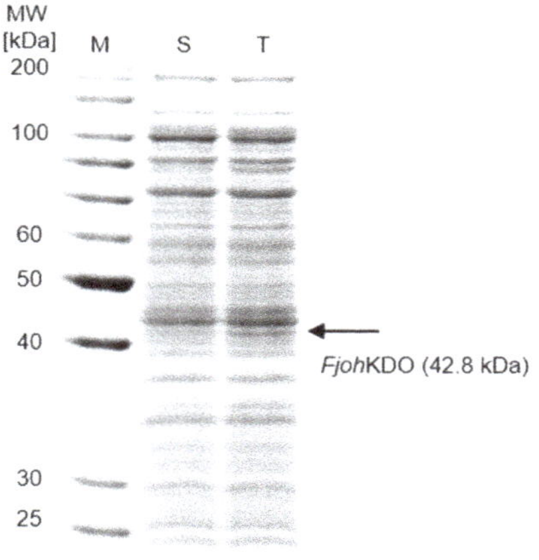

Figure 5. Exemplary SDS-PAGE gel image of cell-free synthesized KDO from *Flavobacterium johnsoniae* (*Fjoh*KDO). M, marker; T, total protein; S, soluble protein fraction (obtained through centrifugation for 10 min at 18,000× *g*).

Table 1. Hyl concentrations in mM achieved in in vitro hydroxylation assays for cell-free synthesized KDOs with chaperone-enriched cell extracts. Assays were performed on a 100 µL scale with 10 mM Lys and 20% *v/v* protein solution for 20 h. Coloring applies to each row individually; dark green is the highest value and red is the lowest value. Values below 0.02 mM are dark red. Standard deviations are a result of biological duplicates.

	Reference	DnaK, DnaJ, GrpE, GroES, GroEL	DnaK, DnaJ, GrpE	GroES, GroEL	GroES, GroEL, tig	tig
Caci	0.08 ± 0.06	0.06 ± 0.00	0.03 ± 0.01	0.05 ± 0.00	0.07 ± 0.01	0.02 ± 0.00
Krad	2.07 ± 0.30	0.83 ± 0.54	0.99 ± 0.34	0.17 ± 0.05	2.02 ± 0.03	1.65 ± 0.09
Krhi	1.22 ± 0.08	0.15 ± 0.02	0.32 ± 0.01	0.06 ± 0.01	0.46 ± 0.01	0.02 ± 0.00
Mint	0.01 ± 0.00	0.02 ± 0.00	0.02 ± 0.00	0.01 ± 0.00	0.02 ± 0.00	0.01 ± 0.00
Cpin	1.85 ± 0.21	0.48 ± 0.24	1.55 ± 0.41	0.34 ± 0.01	0.30 ± 0.02	0.05 ± 0.00
Nkor	0.54 ± 0.05	0.28 ± 0.08	0.44 ± 0.04	0.02 ± 0.00	0.03 ± 0.00	0.01 ± 0.00
Fspe	3.66 ± 0.16	3.13 ± 0.98	2.68 ± 0.20	0.23 ± 0.01	0.27 ± 0.02	0.09 ± 0.00
Fjoh	0.20 ± 0.02	0.08 ± 0.01	0.15 ± 0.01	0.01 ± 0.00	0.02 ± 0.00	0.01 ± 0.00
Pbra	0.52 ± 0.02	0.95 ± 0.20	0.31 ± 0.07	0.03 ± 0.02	0.08 ± 0.04	0.03 ± 0.00
Plum	0.12 ± 0.01	0.12 ± 0.01	0.08 ± 0.02	0.01 ± 0.00	0.01 ± 0.00	0.00 ± 0.00
Bspe	0.00 ± 0.00	0.01 ± 0.00	0.01 ± 0.00	0.01 ± 0.00	0.01 ± 0.00	0.01 ± 0.00
Bpse	0.31 ± 0.06	0.72 ± 0.12	0.33 ± 0.01	0.05 ± 0.01	0.15 ± 0.09	0.03 ± 0.02
Bpla	0.05 ± 0.01	0.11 ± 0.01	0.02 ± 0.00	0.01 ± 0.00	0.01 ± 0.00	0.01 ± 0.00

The concentrations obtained after 20 h incubation with 10 mM Lys varied from 0 to a maximum of 3.66 ± 0.16 mM, which was still obtained with *Fspe*KDO in the cell extract without chaperones. Thus, the maximum obtained yield was 37%, which is significantly lower than the expected yield of up to 100% [4,7,8]. Lys or Hyl degrading enzymes in the complex cell extract or insufficient KDO activity or stability may cause the incomplete conversion of Lys into Hyl. The CFPS reaction volume is too low and complex for a more detailed analysis of the product and substrate progress, so no definite statement can be made at this point. In most cases, lower Hyl concentrations were achieved in the chaperone-containing synthesis mixes compared to the reference. This could be explained by interactions between the enzymes and the chaperones. The chaperones might have a negative influence on the activity or overall stability of the biocatalysts. A previous study showed that an excess of DnaK, DnaJ, and GrpE was inhibitory for protein production, and it was suggested that increased proteolysis could be the explanation [24]. Interestingly, enhanced synthesis in the presence of chaperones cannot be attributed to individual chaperones, but rather occurs in the complex mixture of all chaperones. In the case of the *Pbra*KDO homologs, the combination of DnaK, DnaJ, GrpE, GroES, and GroEL increased the product concentrations for all active variants. Increases of more than 100% were achieved for *Bpse*KDO and *Bpla*KDO. The enzyme variants *Mint*KDO and *Bspe*KDO, which showed only marginal Hyl production in the initial screening, did not show any higher product concentrations in these experiments either. Furthermore, a tendency can be seen that the respective chaperone sets show a similar effect for phylogenetically closely related variants. Thus, chaperone-assisted CFPS is a good tool for the rapid screening of suitable chaperones for the synthesis of difficult-to-synthesize proteins and can lead to more efficient biocatalysts.

2.4. Heterologous Expression of Novel KDOs

*Pbra*KDO is known to catalyze the hydroxylation of Lys to (4*S*)-4-hydroxy-L-lysine. The enzyme and its respective homologs have not yet been characterized for biotechnological application in a whole-cell biocatalyst format. Since all homologs of *Pbra*KDO were shown to synthesize Hyl in the CFPS experiments, we next investigated them as whole-cell biocatalysts with *E. coli*. Although *Bspe*KDO yielded only trace amounts of Hyl in the screening experiments, we decided to include it in the whole-cell experiments to test the transferability of our approach. All proteins were successfully synthesized in vivo in *E. coli* BL21 (DE3), but the percentage of the soluble protein fraction varied significantly among the different homologs (Figure 6A). Especially for *Bspe*KDO, the heterologous expression led almost exclusively to insoluble protein. We then tested the different strains in small-scale resting-cell biotransformations for hydroxylation of Lys and determined the activity of the biocatalysts (Figure 6B). No activity was detected for *E. coli* BL21 (DE3) pET-24a(+)-*Bspe*KDO. This might be a result of misfolded protein and is in accordance with the results obtained from CFPS experiments. *E. coli* BL21 (DE3) pET-24a(+)-*Bpla*KDO and *Plum*KDO exhibited the highest activity of the tested whole-cell biocatalysts of about 2 U g_{CDW}^{-1}. At first glance, these results seem to contradict the results from the in vitro hydroxylation assays, where *Pbra*KDO and *Bpse*KDO gave the highest product concentrations of the five considered enzymes. However, it should be noted that the in vivo and in vitro expression conditions are very different, and the biotransformation conditions also differ from each other. Despite that, the combination of CFPS and the in vitro hydroxylation assay correctly identified the four most active and therefore most promising enzyme homologs.

To test whether the whole-cell biotransformation is limited by mass transfer, we performed resting-cell biotransformations with the addition of 1% *v/v* Triton X-100 as permeabilization agent (Figure 6B). As before, α-KG was added in twofold excess relative to Lys to avoid limitation by insufficient co-substrate concentrations. The permeabilization led to a vast increase in activity for the four active biocatalysts. *E. coli* BL21 (DE3) pET-24a(+)-*Plum*KDO exhibited the highest activity with 8.6 ± 0.6 U g_{CDW}^{-1}, which is more than a threefold improvement compared to the assay without Triton X-100. The activity is on a similar scale as other *E. coli*-based whole-cell biocatalysts from the study of Hara et al. (5–27 U g_{CDW}^{-1}, calculated from the given specific productivities [4]).

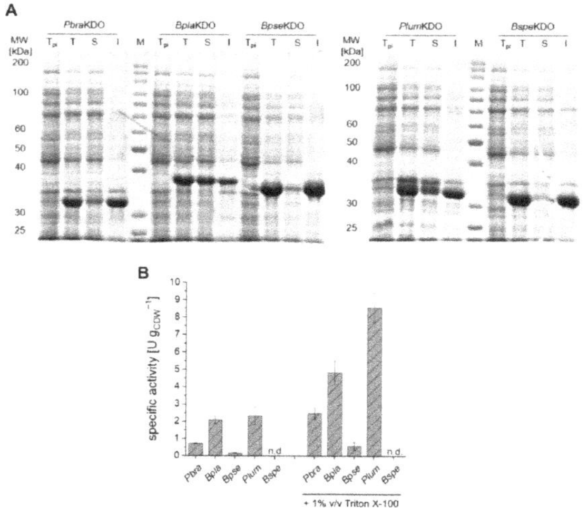

Figure 6. (**A**) SDS-PAGE gel image of the protein content of *E. coli* BL21 (DE3) whole-cell biocatalysts. M, marker; T_{pi}, total protein pre-induction; T, total protein; S, soluble protein fraction; I, insoluble protein fraction. Molecular weights of the proteins of interest are *Pbra*KDO 32.5 kDa, *Bpla*KDO 31.5 kDa, *Bpse*KDO 30.5 kDA, *Plum*KDO 31.8 kDa and *Bspe*KDO 30.2 kDa. (**B**) Specific whole-cell biocatalyst activities, assays were performed in 0.5 mL 50 mM KPi pH 7.4 with 1 g_{CDW} L^{-1}, 10 mM Lys, 20 mM α-KG, 5 mM L(+)-ascorbic acid and 1 mM $FeSO_4$ for 1 h at 30 °C in an Eppendorf® ThermoMixer® C at 500 rpm. n.d., not detected. Standard deviations are a result of two biological replicates.

Limitations due to mass transfer across the bacterial membrane have already been observed with whole-cell biocatalysts in combination with other Fe^{2+}/α-ketoglutarate-dependent dioxygenases [25–27]. In these cases, permeabilization was achieved by Nymeen solution or freezing and thawing of the cells. Alternatively, the overexpression of respective transporters, for example LysP, the lysine permease from *E. coli*, could reduce mass transfer limitations. Such endeavors have already been proven successful for other whole-cell biocatalysts such as the production of 5-aminovalerate [28] or the production of L-pipecolic acid from Lys [29] using recombinant *E. coli*.

2.5. Preparative-Scale Production of Hydroxy-L-Lysine

We employed the most active biocatalyst, *E. coli* BL21 (DE3) pET-24a(+)-*Plum*KDO for the preparative production of Hyl on a 50 mL reaction scale (Figure 7). Using a biocatalyst concentration of 10 g_{CDW} L^{-1} (Figure 7A), 25 mM of Lys were fully converted to Hyl. Motivated by these results, we set up a reaction with 50 mM Lys. From 50 mM initial Lys, 30 mM were converted to Hyl during 12 h of biotransformation, which corresponds to a yield of ca. 60% and a titer of almost 5 g L^{-1} (Figure 7B). After 12 h, no further conversion or degradation of the substrate and the product was observed. Generally, complete conversion of Lys to Hyl is feasible (Figure 7A). This was also demonstrated in studies with other KDOs [4,7,8]. Amatuni et al. fully converted approximately 40 mM Lys to Hyl using a cell lysate from *E. coli* BL21 (DE3) expressing the gene coding for *Pbra*KDO, with a final amount of lysate corresponding to an OD_{600} of 15 [8]. The reaction was carried out overnight at 23 °C in 50 mM KPi pH 8.0. In our study, we employed cells at an OD_{600} of 30 at 30 °C in 50 mM KPi pH 7.4. It is therefore very likely that optimization of the reaction conditions may lead to a higher degree of conversion. Moreover, the use of Triton X-100 may not fully circumvent mass transfer limitation in the whole-cell biotransformation.

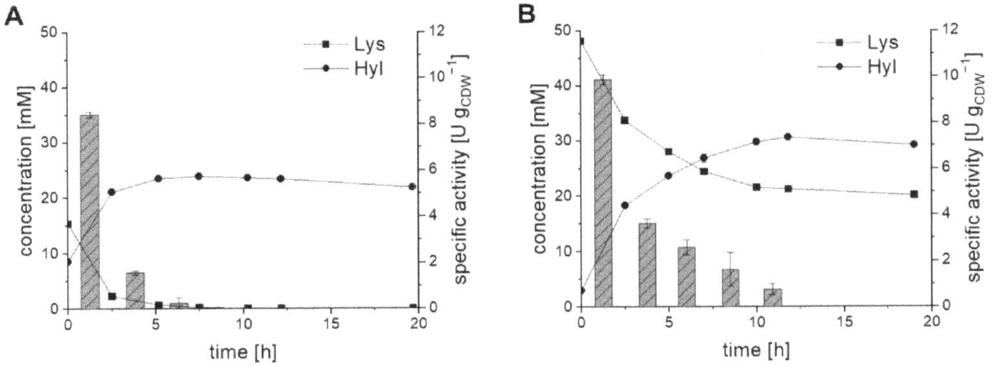

Figure 7. Progress curve of the whole-cell biotransformation on a preparative scale with *E. coli* BL21 (DE) pET-24a(+)-*Plum*KDO; (**A**) 25 mM Lys (initial concentration), (**B**) 50 mM Lys (initial concentration). Specific activities were calculated from the respective product concentrations between the sampling times. Biotransformation was performed in 50 mL at 30 °C and 180 rpm. The reaction mixture contained 10 g_{CDW} L^{-1} cells in 50 mM KPi buffer pH 7.4. Error bars are a result of technical duplicates.

The activity during the first 2.5 h is in the same range as in the small-scale biotransformations, which reflects the principal scalability (Figure 7). While the specific activity during the first 2.5 h of reaction is about 10 U $g_{CDW}{}^{-1}$ for the biotransformation with initially 50 mM Lys, it already drops to 3.6 U $g_{CDW}{}^{-1}$ between 2.5 and 5 h (Figure 7B). As α-KG was added in large excess (twofold concentration of Lys), this is not likely to be the reason for the incomplete conversion. The K_M of *Pbra*KDO is about 34 μM for Lys, so the decreasing substrate concentration is not considered to be the reason for the reduction in the specific activity in the reaction with 50 mM initial Lys [9]. Product inhibition might be an explanation but has not yet been reported for KDOs. The incomplete conversion might also be attributed to enzyme stability (kinetic stability, thermodynamic stability, operational stability). Moreover, Fe^{2+}/α-ketoglutarate-dependent dioxygenases are reported to show uncoupling effects, which might lead to inactivation of the biocatalyst and therefore result in incomplete conversion [30]. Interestingly, Hara et al. were able to completely convert up to 600 mM Lys to Hyl, with *E. coli* whole-cell biocatalysts expressing the gene coding for *Krad*KDO (K4H-1 in their study) using a biocatalyst concentration of OD_{600} = 30 in 52 h [4]. This might indicate higher biocatalyst stability, which is of high importance for efficient scale-up. The reasons for this significantly higher stability are not yet known. Nevertheless, our results are a good basis and they demonstrate the principal applicability of the whole-cell biocatalyst, which already showed suitable productivity and a high titer without a detailed optimization. Through optimization of the reaction parameters and systematic elucidation of the process boundaries, the overall performance of the biotransformation can likely be further increased [31].

3. Materials and Methods

3.1. Chemicals/Strains and Plasmids

Chemicals were purchased from Carl Roth GmbH & Co. KG (Karlsruhe, Germany), VWR International (Radnor, PA, USA), Merck KGaA (Darmstadt, Germany), Sigma-Aldrich Chemie GmbH (Steinheim, Germany), AppliChem GmbH (Darmstadt, Germany) and Thermo Fisher Scientific (Waltham, MA, USA). The kits for DNA extraction (NucleoSpin® Gel and PCR Clean-up), plasmid isolation (NucleoSpin® Plasmid (No Lid)) and genomic DNA extraction (NucleoSpin® Microbial DNA) were purchased from Macherey-Nagel GmbH & CO. KG (Düren, Germany). Q5® High-Fidelity Polymerase (2× Master Mix) and restriction enzymes were purchased from New England Biolabs (Ipswich, MA, USA). Plasmids pG-KJE8, pTf16, pGro7, pG-Tf2, and pKJE7 were obtained from Takara Holdings

Inc. (Kyoto, Japan). The plasmids pET-22b(+)-*Caci*KDO, pET-22b(+)-*Cpin*KDO, and pET-22b(+)-*Fjoh*KDO were a kind gift from Prof. Anne Zaparucha and were described in [3].

A list of strains and plasmids used in this study is shown in Table 2. Nucleotide sequences of the genes are provided in Supplementary Materials (Table S2).

3.2. Cloning

Gene sequences coding for *Krad*KDO and *Krhi*KDO were amplified from genomic DNA of *Kineococcus radiotolerans* and *Kineococcus rhizosphearae* with the primer pairs PPN070/PPN071 and PPN072/PPN073 and cloned into NdeI/NotI digested pET-24a(+) via Gibson cloning. Gene sequences coding for *Pbra*KDO (PPN088/PPN089), *Bpla*KDO (PPN091/PPN092), *Plum*KDO (PPN092/PPN093), *Bpse*KDO (PPN094/PPN095), *Bspe*KDO (PPN096/PPN097) and *Mint*KDO (PPN074/PPN075) were purchased as DNA strings from Thermo Fisher Scientific (Waltham, MA, USA) and amplified by PCR with the indicated primer pairs. The PCR products were purified by gel electrophoresis and cloned into NdeI/NotI digested pET-24a(+) via Gibson cloning [32]. The codon-optimized gene sequence of *Pbra*KDO was adopted from [9], gene sequences of the homologs were codon-optimized for *E. coli* by Thermo Fisher Scientific.

Table 2. Strains and plasmids used in this study.

Strain	Information		
Escherichia coli DH5α	F-, Φ80dlacZΔM15, Δ(lacZYA-argF)U169, deoR, recA1, endA1, hsdR17(rK-mK+), phoA, supE44, λ-, thi 1, gyrA96, relA1		
Escherichia coli BL21 (DE3)	F-, ompT, hsdSB(rB-mB-), gal (c1875, ind1, Sam7, nin5, lacUV5-T7 gene1), dcm (DE3)		
Kineococcus radiotolerans	wild-type, DSM No. 14245		
Kineococcus rhizospharae	wild-type, DSM No. 19711		
Plasmid	Protein	Description	Reference
pET-22b(+)-*Caci*KDO	*Caci*KDO	KDO, L-Lysine 3-hydroxylase	[3]
pET-22b(+)-*Cpin*KDO	*Cpin*KDO	KDO, L-Lysine 4-hydroxylase	[3]
pET-22b(+)-*Fjoh*KDO	*Fjoh*KDO	KDO, L-Lysine 4-hydroxylase	[3]
pET-22b(+)-*Nkor*KDO	*Nkor*KDO	KDO, L-Lysine 4-hydroxylase	[3]
pET-22b(+)-*Fspe*KDO	*Fspe*KDO	KDO, L-Lysine 4-hydroxylase	[3]
pET-24a(+)-*Krad*KDO	*Krad*KDO	KDO, L-Lysine 3-hydroxylase	This study
pET-24a(+)-*Krhi*KDO	*Krhi*KDO	Putative KDO	This study
pET-24a(+)-*Mint*KDO	*Mint*KDO	Putative KDO	This study
pET-24a(+)-*Pbra*KDO	*Pbra*KDO	KDO, L-Lysine 4-hydroxylase	This study
pET-24a(+)-*Bpla*KDO	*Bpla*KDO	Putative KDO	This study
pET-24a(+)-*Bpse*KDO	*Bpse*KDO	Putative KDO	This study
pET-24a(+)-*Plum*KDO	*Plum*KDO	Putative KDO	This study
pET-24a(+)-*Bspe*KDO	*Bspe*KDO	Putative KDO	This study
pAR1219	T7RNAP	T7 RNA-polymerase	[33]
pG-KJE8	DnaK, DnaJ, GrpE, GroES, GroEL	Molecular chaperones	[24]
pKJE7	DnaK, DnaJ, GrpE	Molecular chaperones	[24]
pGro7	GroES, GroEL	Molecular chaperones	[24]
pG-Tf2	GroES, GroEL, tig	Molecular chaperones	[34]
pTf16	tig	Molecular chaperones	[34]

All vector constructs were checked for errors with sanger sequencing (Microsynth Seqlab, Göttingen, Germany). Plasmids used in this study are shown in Table 2 and nucleotide sequences of the genes and primers are provided in the Supplementary Materials (Tables S1 and S2).

3.3. *E. coli* Extract Preparation

The *E. coli* extract was prepared as described by [15] with some modifications, which are stated in the following. *E. coli* BL21 (DE3) was transformed with pAR1219 for over-expression of T7 RNA polymerase (T7RNAP) and with pG-KJE8, pKJE7, pGro7, pG-Tf2, or pTf16 for the overexpression of different chaperone sets, respectively. A preculture of 10 mL lysogeny broth medium (LB, 10 g L^{-1} tryptone, 5 g L^{-1} yeast extract, 5 g L^{-1} NaCl) with 100 μg mL^{-1} ampicillin and 20 μg mL^{-1} chloramphenicol, in case of strains with chaperone-encoding plasmids, was inoculated with a single colony of the source strain for the cell extract. The preculture was grown for 16 h at 200 rpm and 37 °C. The main culture of 125 mL 2xYTPG medium (16 g L^{-1} tryptone, 10 g L^{-1} yeast extract, 5 g L^{-1} NaCl, 7 g L^{-1} K_2HPO_4, 3 g L^{-1} KH_2PO_4, 18 g L^{-1} glucose) in a 1 L baffled shake flask was inoculated to an OD_{600} of 0.1 and grown at 200 rpm at 37 °C. Chaperone expression was induced by addition of 0.5 mg mL^{-1} L-arabinose (Carl Roth, Karlsruhe, Germany) or 5 ng mL^{-1} tetracycline (Thermo Fisher Scientific, Waltham, MA, USA) according to the manual of the chaperone plasmid set. At an OD_{600} of 0.6, 1 mM of isopropyl-β-D-thiogalactopyranoside (IPTG, Carl Roth, Karlsruhe, Germany) was added to induce T7RNAP production. Cells were harvested at an OD_{600} of 3 and pelleted via centrifugation at 5000× *g* for 10 min at 10 °C. The pellets were washed three times with 4 °C cold S30 buffer (10 mM tris acetate, pH 8.2; 14 mM magnesium acetate; 60 mM potassium acetate; and 2 mM dithiothreitol (DTT, Carl Roth, Karlsruhe, Germany), flash-frozen with liquid nitrogen, and stored at −80 °C. For lysis, cells were thawed on ice and resuspended in 1 mL of S30 buffer per gram of wet cells. Three cycles of sonication were performed for 40 s and 2 mM DTT were added. Cellular debris was removed by centrifugation at 18,000× *g* for 10 min at 4 °C. The supernatant was incubated in an Eppendorf® ThermoMixer® C (Eppendorf, Hamburg, Germany) at 450 rpm for 60 min at 37 °C, and then centrifuged at 10,000× *g* for 10 min at 4 °C. The final supernatant was flash-frozen with liquid nitrogen and stored at –80 °C until use. Protein concentrations were determined by the Bradford method using bovine serum albumin (BSA) as a standard [35]. Obtained extracts contained between 40 and 60 mg mL^{-1} total protein.

3.4. *Cell-Free Protein Synthesis (CFPS)*

CFPS reactions with a reaction volume of 20 μL were performed in microtubes containing: *E. coli* cell free extract (see 4.3) amounting to 9.2 to 15.3 mg mL^{-1} protein, 10 mM magnesium glutamate, 130 mM potassium glutamate, 1.5 mM each of 20 amino acids (except leucine), 1.25 mM leucine, 50 mM 4-(2-hydroxyethyl)-1-piperazineethanesulfonic acid (HEPES), 1.5 mM ATP and GTP, 0.9 mM cytidine triphosphate (CTP) and uridine triphosphate (UTP), 0.2 mg mL^{-1} *E. coli* tRNA, 0.26 mM coenzyme A (CoA), 0.33 mM nicotinamide adenine dinucleotide (NAD), 0.75 mM cyclic adenosine monophosphate (cAMP), 0.068 mM folinic acid, 1 mM spermidine, 30 mM 3-phosphoglyceric acid (3-PGA), 2% polyethylene glycol-8000 (PEG), and 5 nM plasmid DNA. Reactions were incubated in an Eppendorf® ThermoMixer® C for 4 h at 450 rpm and 37 °C.

3.5. *In Vitro Biotransformations*

In vitro biotransformation reactions were performed in a total volume of 100 μL, consisting of 20 μL CFPS-reaction solution and 80 μL assay solution (50 mM KPi pH 7.0). Final concentrations were 10 mM Lys, 20 mM α-KG, 5 mM L(+)-ascorbic acid, and 1 mM $FeSO_4$. Reactions were incubated in 1.5 mL reaction tubes at 25 °C and 500 rpm in an Eppendorf® ThermoMixer® C for 20 h and stopped by addition of 100 μL acetonitrile. After centrifugation for 10 min and 17,000× *g*, the supernatant was subjected to LC-MS analysis.

3.6. *Resting-Cell Biotransformations*

For resting-cell biotransformations, LB pre-cultures (10 mL) were inoculated from cryogenic-stocks and incubated at 37 °C and 200 rpm overnight. A 50 mL LB culture with 50 μg mL^{-1} kanamycin was inoculated to an initial OD_{600} of 0.1 and incubated at 37 °C and 200 rpm until an OD_{600} of 0.6–0.8. Cultures were then cooled on ice for 20 min and heterologous gene expression was started by addition of 0.5 mM IPTG. Cultivation was continued at 20 °C for another 20 h.

Cells were harvested by centrifugation (4500× *g*, 10 min, 4 °C) and resuspended in 50 mM KPi buffer pH 7.4 to a biomass concentration of approximately 1 g_{CDW} L^{-1} (an OD_{600} of 1 corresponds to 0.312 g_{CDW} L^{-1}). Resting-cell biotransformations were performed in a total volume of 0.5 mL in 2 mL reaction tubes (1 g_{CDW} L^{-1}, 20 mM α-KG, 5 mM L(+)-ascorbic acid and 1 mM $FeSO_4$). Reactions were performed at 30 °C and 500 rpm in an Eppendorf® ThermoMixer® C. After preconditioning for 5 min, biotransformation was started by addition of 10 mM Lys and the reactions were incubated for 60 min and quenched by addition of 0.5 mL acetonitrile. Following centrifugation for 10 min and 17,000× *g*, the supernatant was analyzed via LC-MS.

3.7. Preparative-Scale Biotransformation

For the preparative-scale biotransformation, *E. coli* BL21 (DE3) pET-24a(+)-*Plum*KDO was cultivated as in 3.6, but the main cultures consisted of 2 × 200 mL LB medium in 2-L baffled shake flasks. After cultivation, the cells were harvested by centrifugation and concentrated to the desired biomass concentration. The biotransformation was performed in a volume of 50 mL at 30 °C and with orbital shaking at 180 rpm (2.5 cm amplitude) in a 500 mL baffled shake flask. The reaction mixture contained 50 or 100 mM α-KG (twofold concentration of Lys), 5 mM $FeSO_4$, 10 mM L(+)-ascorbic acid, 1% *v/v* Triton X-100 and approximately 10 g_{CDW} L^{-1} cells in 50 mM KPi buffer pH 7.4. The cells were preconditioned at the desired temperature for 10 min. Afterwards, the reaction was started by addition of 25 or 50 mM Lys. Aliquots of 0.5 mL were withdrawn after regular time intervals and cells were separated by centrifugation (10 min, 17,000× *g*) and the supernatant was subjected to LC-MS analysis.

3.8. Analytics

Detection and quantification of Lys and Hyl was performed on an Agilent 1260 Infinity II liquid chromatography system coupled to a 6120 single quadrupole ESI mass spectrometer (Agilent Technologies Inc., Santa Clara, CA, USA). Chromatographic separation was realized with an iHILIC®-(P) Classic column (150 × 2.1 mm, 5 µm, 200 Å) in combination with a SeQuant® ZIC®-pHILIC-guard column (20 × 2.1 mm, 5 µm, 200 Å) in isocratic mode (70% acetonitrile, 30% 50 mM ammonium acetate pH 4.0) and a flow rate of 0.2 mL min^{-1} in 20 min. MS-analysis was conducted in positive SIM mode with the following settings: drying gas chamber temperature 350 °C, gas flow 12 L min^{-1}, capillary voltage 3000 V and nebulizer pressure 35 psig. Quantification was conducted by external calibration using standards of Lys (Sigma-Aldrich, St. Louis, MI, USA) and 5-hydroxy-DL-lysine hydrochloride (Carbosynth, Compton, UK) as the latter was the only commercially available Hyl isomer. Lys was retrieved as $[M+H]^+$ (*m/z* 147.1) and Hyl isomers as $[M+H]^+$ (*m/z* 163.1).

4. Conclusions

In this study, CFPS has been combined with subsequent activity assays for the identification of Fe^{2+}/α-ketoglutarate-dependent dioxygenases for the first time. Investigating a set of known and putative KDOs, production of Hyl was confirmed for all published as well as for new, previously undescribed enzymes. This demonstrates that CFPS is a valuable tool to simplify and speed up the identification of new Fe^{2+}/α-ketoglutarate-dependent dioxygenases. In current research, for example, genetic-engineering derived enzyme variants are generated, which possess higher activity or which expand the reaction and substrate scope of Fe^{2+}/α-ketoglutarate-dependent dioxygenases [5,36–38]. It is reasonable that our screening system would also allow for rapid prototyping of enzyme variants, analysis of the substrate scope, or the screening of different reaction conditions. With regard to our study, it will be interesting to investigate the substrate scope of our newly identified enzymes as *Pbra*KDO also accepts L-leucine and L-methionine as substrates [9].

We successfully applied the newly identified KDOs in a whole-cell biocatalyst format. Mass transfer of the substrates and products across the membrane was found to be a major limitation of the biotransformation. Using a permeabilization method with Triton X-100, preparative-scale production of Hyl was accomplished and feasibility was demonstrated. Reaction conditions (pH, temperature) and enzyme stabilities require further investigation and optimization, which would very likely lead to higher product formation. Optimization of gene expression (IPTG concentration and expression temperature) may lead to higher percentages of functionally active protein, which in turn is expected to yield higher reaction rates and conversion. Moreover, our experiments with CFPS and different combinations of chaperones indicate that distinct sets of chaperones (e.g., DnaK, DnaJ, GrpE, GroES, and GroEL in the case of *Pbra*KDO and its homologs) might be beneficial for biocatalyst activity also in a whole-cell format. While some chaperones and combinations thereof had a beneficial influence on product formation, others showed severe detrimental effects. This demonstrates the applicability of CFPS for the screening of chaperones for difficult-to-express proteins.

In summary, a systematic approach from CFPS to screen and identify novel Fe^{2+}/α-ketoglutarate-dependent dioxygenases to a whole-cell biotransformation for the preparative synthesis of Hyl was successfully developed. New homologs have now been identified with suitable reactivity, resulting in a multi-gram scale hydroxylation reaction. These homologs now expand the spectrum of the previously limited number of described KDOs (nine wild-type enzymes) and represent potential new candidates for biotechnological application.

Supplementary Materials: The following are available online at https://www.mdpi.com/article/10.3390/catal11091038/s1, Figure S1. Exemplary plasmid map of pET-22b(+)-*Caci*KDO, Figure S2. Exemplary plasmid map of pET-24a(+)-*Pbra*KDO, Figure S3–S26 SDS-PAGE analysis of cell-free synthesized KDOs (with and without chaperone-enriched extracts), Table S1. Primer sequences, Table S2. Gene sequences, Table S3. Amino acid sequences.

Author Contributions: Conceptualization, J.R., P.N., S.L. and K.R.; methodology, J.R., P.N., S.L. and K.R.; formal analysis, J.R., P.N., A.B. and S.K.; investigation, J.R., P.N., A.B. and S.K.; resources, S.L.; data curation, S.L. and K.R.; writing—original draft preparation, J.R., P.N. and K.R.; writing—review and editing, S.L. and K.R.; visualization, J.R. and P.N.; supervision, S.L. and K.R.; project administration, S.L. and K.R.; funding acquisition, S.L. All authors have read and agreed to the published version of the manuscript.

Funding: This research was supported by the CLIB-Competence Center Biotechnology (CKB) funded by the European Regional Development Fund (EFRE) and the North-Rhine Westphalian Ministry of Economic Affairs, Innovation, Digitalization and Energy (MWIDE). (Grant number: EFRE-0300098).

Data Availability Statement: Data is contained within the article or supplementary material.

Acknowledgments: We thank Markus Nett, Martina Pohl, and Selina Seide for scientific advice and fruitful discussions. The plasmids pET-22b(+)-*Caci*KDO, pET-22b(+)-*Cpin*KDO, and pET-22b(+)-*Fjoh*KDO were a kind gift from Anne Zaparucha. We would like to acknowledge the support with LC-MS analytics by Chantale Zammarelli.

Conflicts of Interest: The authors declare no conflict of interest. The funders had no role in the design of the study; in the collection, analyses, or interpretation of data; in the writing of the manuscript, or in the decision to publish the results.

References

1. Peters, C.; Buller, R. Industrial Application of 2-Oxoglutarate-Dependent Oxygenases. *Catalysts* **2019**, *9*, 221. [CrossRef]
2. Herr, C.Q.; Hausinger, R.P. Amazing Diversity in Biochemical Roles of Fe(II)/2-Oxoglutarate Oxygenases. *Trends Biochem. Sci.* **2018**, *43*, 517–532. [CrossRef]
3. Baud, D.; Saaidi, P.-L.; Monfleur, A.; Harari, M.; Cuccaro, J.; Fossey, A.; Besnard, M.; Debard, A.; Mariage, A.; Pellouin, V.; et al. Synthesis of Mono- and Dihydroxylated Amino Acids with New α-Ketoglutarate-Dependent Dioxygenases: Biocatalytic Oxidation of C-H Bonds. *ChemCatChem* **2014**, *6*, 3012–3017. [CrossRef]
4. Hara, R.; Yamagata, K.; Miyake, R.; Kawabata, H.; Uehara, H.; Kino, K. Discovery of Lysine Hydroxylases in the Clavaminic Acid Synthase-Like Superfamily for Efficient Hydroxylysine Bioproduction. *Appl. Environ. Microbiol.* **2017**, *83*, e00693-17. [CrossRef]
5. Wang, F.; Zhu, M.; Song, Z.; Li, C.; Wang, Y.; Zhu, Z.; Sun, D.; Lu, F.; Qin, H.-M. Reshaping the Binding Pocket of Lysine Hydroxylase for Enhanced Activity. *ACS Catal.* **2020**, *10*, 13946–13956. [CrossRef]

6. Marin, J.; Didierjean, C.; Aubry, A.; Casimir, J.R.; Briand, J.P.; Guichard, G. Synthesis of Enantiopure 4-Hydroxypipecolate and 4-Hydroxylysine Derivatives from a Common 4,6-Dioxopiperidinecarboxylate Precursor. *J. Org. Chem.* **2004**, *69*, 130–141. [CrossRef]
7. Zhang, X.; King-Smith, E.; Renata, H. Total Synthesis of Tambromycin by Combining Chemocatalytic and Biocatalytic C−H Functionalization. *Angew. Chem. Int. Ed.* **2018**, *57*, 5037–5041. [CrossRef]
8. Amatuni, A.; Shuster, A.; Adibekian, A.; Renata, H. Concise Chemoenzymatic Total Synthesis and Identification of Cellular Targets of Cepafungin I. *Cell Chem. Biol.* **2020**, *27*, 1318–1326.e18. [CrossRef] [PubMed]
9. Amatuni, A.; Renata, H. Identification of a Lysine 4-Hydroxylase from the Glidobactin Biosynthesis and Evaluation of Its Biocatalytic Potential. *Org. Biomol. Chem.* **2019**, *17*, 1736–1739. [CrossRef] [PubMed]
10. Baud, D.; Peruch, O.; Saaidi, P.-L.; Fossey, A.; Mariage, A.; Petit, J.-L.; Salanoubat, M.; Vergne-Vaxelaire, C.; de Berardinis, V.; Zaparucha, A. Biocatalytic Approaches towards the Synthesis of Chiral Amino Alcohols from Lysine: Cascade Reactions Combining Alpha-Keto Acid Oxygenase Hydroxylation with Pyridoxal Phosphate- Dependent Decarboxylation. *Adv. Synth. Catal.* **2017**, *359*, 1563–1569. [CrossRef]
11. Fossey-Jouenne, A.; Vergne-Vaxelaire, C.; Zaparucha, A. Enzymatic Cascade Reactions for the Synthesis of Chiral Amino Alcohols from L-Lysine. *J. Vis. Exp.* **2018**, *132*, 56926. [CrossRef] [PubMed]
12. Rolf, J.; Rosenthal, K.; Lütz, S. Application of Cell-Free Protein Synthesis for Faster Biocatalyst Development. *Catalysts* **2019**, *9*, 190. [CrossRef]
13. Quertinmont, L.T.; Orru, R.; Lutz, S. RApid Parallel Protein EvaluatoR (RAPPER), from Gene to Enzyme Function in One Day. *Chem. Commun.* **2015**, *51*, 122–124. [CrossRef] [PubMed]
14. Silverman, A.D.; Karim, A.S.; Jewett, M.C. Cell-Free Gene Expression: An Expanded Repertoire of Applications. *Nat. Rev. Genet.* **2020**, *21*, 151–170. [CrossRef]
15. Rolf, J.; Siedentop, R.; Lütz, S.; Rosenthal, K. Screening and Identification of Novel CGAS Homologues Using a Combination of in Vitro and In Vivo Protein Synthesis. *Int. J. Mol. Sci.* **2019**, *21*, 105. [CrossRef]
16. Stech, M.; Kubick, S. Cell-Free Synthesis Meets Antibody Production: A Review. *Antibodies* **2015**, *4*, 12–33. [CrossRef]
17. Kang, S.-H.; Kim, D.-M.; Kim, H.-J.; Jun, S.-Y.; Lee, K.-Y.; Kim, H.-J. Cell-Free Production of Aggregation-Prone Proteins in Soluble and Active Forms. *Biotechnol. Prog.* **2005**, *21*, 1412–1419. [CrossRef] [PubMed]
18. Yang, H.-J.; Lee, K.-H.; Lim, H.J.; Kim, D.-M. Tandem Cell-Free Protein Synthesis as a Tool for Rapid Screening of Optimal Molecular Chaperones. *Biotechnol. J.* **2019**, *14*, 1800523. [CrossRef] [PubMed]
19. Bastard, K.; Isabet, T.; Stura, E.A.; Legrand, P.; Zaparucha, A. Structural Studies Based on Two Lysine Dioxygenases with Distinct Regioselectivity Brings Insights Into Enzyme Specificity within the Clavaminate Synthase-Like Family. *Sci. Rep.* **2018**, *8*, 8–18. [CrossRef] [PubMed]
20. Madeira, F.; Park, Y.M.; Lee, J.; Buso, N.; Gur, T.; Madhusoodanan, N.; Basutkar, P.; Tivey, A.R.N.; Potter, S.C.; Finn, R.D.; et al. The EMBL-EBI Search and Sequence Analysis Tools APIs in 2019. *Nucleic Acids Res.* **2019**, *47*, W636–W641. [CrossRef]
21. Tang, B.; Yu, Y.; Liang, J.; Zhang, Y.; Bian, X.; Zhi, X.; Ding, X. Reclassification of 'Polyangium Brachysporum' DSM 7029 as Schlegelella Brevitalea Sp. Nov. *Int. J. Syst. Evol. Microbiol.* **2019**, *69*, 2877–2883. [CrossRef] [PubMed]
22. Koehntop, K.D.; Emerson, J.P.; Que, L. The 2-His-1-Carboxylate Facial Triad: A Versatile Platform for Dioxygen Activation by Mononuclear Non-Heme Iron(II) Enzymes. *J. Biol. Inorg. Chem.* **2005**, *10*, 87–93. [CrossRef] [PubMed]
23. Schinn, S.M.; Broadbent, A.; Bradley, W.T.; Bundy, B.C. Protein Synthesis Directly from PCR: Progress and Applications of Cell-Free Protein Synthesis with Linear DNA. *N. Biotechnol.* **2016**, *33*, 480–487. [CrossRef] [PubMed]
24. Nishihara, K.; Kanemori, M.; Kitagawa, M.; Yanagi, H.; Yura, T. Chaperone Coexpression Plasmids: Differential and Synergistic Roles of DnaK-DnaJ-GrpE and GroEL-GroES in Assisting Folding of an Allergen of Japanese Cedar Pollen, Cryj2, in Escherichia Coli. *Appl. Environ. Microbiol.* **1998**, *64*, 1694–1699. [CrossRef]
25. Falcioni, F.; Blank, L.M.; Frick, O.; Karau, A.; Bühler, B.; Schmida, A. Proline Availability Regulates Proline-4-Hydroxylase Synthesis and Substrate Uptake in Proline-Hydroxylating Recombinant Escherichia Coli. *Appl. Environ. Microbiol.* **2013**, *79*, 3091–3100. [CrossRef]
26. Hara, R.; Nishikawa, T.; Okuhara, T.; Koketsu, K.; Kino, K. Ectoine Hydroxylase Displays Selective Trans-3-Hydroxylation Activity towards l-Proline. *Appl. Microbiol. Biotechnol.* **2019**, *103*, 5689–5698. [CrossRef]
27. Zhang, C.; Ma, J.; Li, Z.; Liang, Y.; Xu, Q.; Xie, X.; Chen, N. A Strategy for L-Isoleucine Dioxygenase Screening and 4-Hydroxyisoleucine Production by Resting Cells. *Bioengineered* **2017**, *9*, 72–79. [CrossRef]
28. Li, Z.; Xu, J.; Jiang, T.; Ge, Y.; Liu, P.; Zhang, M.; Su, Z.; Gao, C.; Ma, C.; Xu, P. Overexpression of Transport Proteins Improves the Production of 5-Aminovalerate from l-Lysine in Escherichia Coli. *Sci. Rep.* **2016**, *6*, 30884. [CrossRef]
29. Cheng, J.; Huang, Y.; Mi, L.; Chen, W.; Wang, D.; Wang, Q. An Economically and Environmentally Acceptable Synthesis of Chiral Drug Intermediate L-Pipecolic Acid from Biomass-Derived Lysine via Artificially Engineered Microbes. *J. Ind. Microbiol. Biotechnol.* **2018**, *45*, 405–415. [CrossRef]
30. Mantri, M.; Zhang, Z.; McDonough, M.A.; Schofield, C.J. Autocatalysed Oxidative Modifications to 2-Oxoglutarate Dependent Oxygenases. *FEBS J.* **2012**, *279*, 1563–1575. [CrossRef]
31. Kadisch, M.; Willrodt, C.; Hillen, M.; Bühler, B.; Schmid, A. Maximizing the Stability of Metabolic Engineering-Derived Whole-Cell Biocatalysts. *Biotechnol. J.* **2017**, *12*, 1600170. [CrossRef]

32. Gibson, D.G.; Young, L.; Chuang, R.-Y.; Venter, J.C.; Hutchison, C.A.; Smith, H.O. Enzymatic Assembly of DNA Molecules up to Several Hundred Kilobases. *Nat. Methods* **2009**, *6*, 343–345. [CrossRef]
33. Davanloo, P.; Rosenberg, A.H.; Dunn, J.J.; Studier, F.W. Cloning and Expression of the Gene for Bacteriophage T7 RNA Polymerase. *Proc. Natl. Acad. Sci. USA* **1984**, *81*, 2035–2039. [CrossRef]
34. Nishihara, K.; Kanemori, M.; Yanagi, H.; Yura, T. Overexpression of Trigger Factor Prevents Aggregation of Recombinant Proteins in Escherichia Coli. *Appl. Environ. Microbiol.* **2000**, *66*, 884–889. [CrossRef] [PubMed]
35. Bradford, M.M. A Rapid and Sensitive Method for the Quantitation of Microgram Quantities of Protein Utilizing the Principle of Protein-Dye Binding. *Anal. Biochem.* **1976**, *72*, 248–254. [CrossRef]
36. Meyer, F.; Frey, R.; Ligibel, M.; Sager, E.; Schroer, K.; Snajdrova, R.; Buller, R. Modulating Chemoselectivity in a Fe(II)/α-Ketoglutarate-Dependent Dioxygenase for the Oxidative Modification of a Nonproteinogenic Amino Acid. *ACS Catal.* **2021**, *11*, 6261–6269. [CrossRef]
37. Papadopoulou, A.; Meierhofer, J.; Meyer, F.; Hayashi, T.; Schneider, S.; Sager, E.; Buller, R.M.U. Re-programming and Optimization of a L-proline Cis-4-hydroxylase for the Cis-3-halogenation of Its Native Substrate. *ChemCatChem* **2021**, 1–7. [CrossRef]
38. Hüttel, W. Biocatalytic Production of Chemical Building Blocks in Technical Scale with α-Ketoglutarate-Dependent Dioxygenases. *Chemie-Ingenieur-Technik* **2013**, *85*, 809–817. [CrossRef]

catalysts

MDPI

Article

Novel Transaminase and Laccase from *Streptomyces* spp. Using Combined Identification Approaches

Erica E. Ferrandi [1], Jelena Spasic [2], Lidija Djokic [2], Yevhen Vainshtein [3], Ramsankar Senthamaraikannan [4], Sandra Vojnovic [2], Christian Grumaz [3], Daniela Monti [1,*] and Jasmina Nikodinovic-Runic [2,*]

1 Istituto di Scienze e Tecnologie Chimiche (SCITEC), Consiglio Nazionale delle Ricerche, via Mario Bianco 9, 20131 Milan, Italy; erica.ferrandi@scitec.cnr.it

2 Institute of Molecular Genetics and Genetic Engineering, University of Belgrade, Vojvode Stepe 444a, 11221 Belgrade, Serbia; spasa91@gmail.com (J.S.); lidijadjokic@imgge.bg.ac.rs (L.D.); sandravojnovic@imgge.bg.ac.rs (S.V.)

3 Fraunhofer Institute for Interfacial Engineering (IGB), In-Vitro Diagnostics, Nobelstrasse 12, 70569 Stuttgart, Germany; yevhen.vainshtein@igb.fraunhofer.de (Y.V.); chris.grumaz@gmail.com (C.G.)

4 AMBER Centre, Trinity College Dublin, College Green, Dublin 2, Ireland; Ramsankar.S@tcd.ie

* Correspondence: daniela.monti@scitec.cnr.it (D.M.); jasmina.nikodinovic@gmail.com or jasmina.nikodinovic@imgge.bg.ac.rs (J.N.-R.); Tel.: +39-02-28500025 (D.M.); +381-11-397-60-34 (J.N.-R.)

Abstract: Three *Streptomyces* sp. strains with a multitude of target enzymatic activities confirmed by functional screening, namely BV129, BV286 and BV333, were subjected to genome sequencing aiming at the annotation of genes of interest, in-depth bioinformatics characterization and functional expression of the biocatalysts. A whole-genome shotgun sequencing followed by *de novo* genome assembly and annotation was performed revealing genomes of 6.4, 9.4 and 7.3 Mbp, respectively. Functional annotation of the proteins of interest resulted in between 2047 and 2763 putative targets. Among the various enzymatic activities that the three *Streptomyces* strains demonstrated to produce by functional screening, we focused our attention on transaminases (TAs) and laccases due to their high biocatalytic potential. Bioinformatics search allowed the identification of a putative TA from *Streptomyces* sp. BV333 as a potentially novel broad substrate scope TA and a putative laccase from *Streptomyces* sp. BV286 as potentially novel blue multicopper oxidase. The two sequences were cloned and overexpressed in *Escherichia coli* and the two novel enzymes, transaminase Sbv333-TA and laccase Sbv286-LAC, were characterized. Interestingly, both enzymes resulted to be exceptionally thermostable, Sbv333-TA showing a melting temperature (T_M = 85 °C) only slightly lower compared to the T_M of the most thermostable transaminases described to date (87–88 °C) and Sbv286-LAC being even thermoactivated at temperature >60 °C. Moreover, Sbv333-TA showed a broad substrate scope and remarkably demonstrated to be active in the transamination of β-ketoesters, which are rarely accepted by currently known TAs. On the other hand, Sbv286-LAC showed an improved activity in the presence of the cosolvent acetonitrile. Overall, it was shown that a combination of approaches from standard microbiological and biochemical screens to genome sequencing and analysis is required to afford novel and functional biocatalysts.

Keywords: *Streptomyces* spp.; transaminase; laccase; whole-genome; biocatalysis

Citation: Ferrandi, E.E.; Spasic, J.; Djokic, L.; Vainshtein, Y.; Senthamaraikannan, R.; Vojnovic, S.; Grumaz, C.; Monti, D.; Nikodinovic-Runic, J. Novel Transaminase and Laccase from *Streptomyces* spp. Using Combined Identification Approaches. *Catalysts* **2021**, *11*, 919. https://doi.org/10.3390/catal11080919

Academic Editor: David D. Boehr

Received: 29 June 2021
Accepted: 26 July 2021
Published: 29 July 2021

1. Introduction

Actinomycetes, especially Streptomycetes, have been in the focus of systems biology and genome metabolic modelling approaches to further intensify their explorations in terms of antibiotic discovery. However, their particular development and specificity of the niches they occupy led our research towards exploring them as a potential source of novel biocatalysts [1,2].

The enzyme market is fast-growing and was valued at USD 7082 million in 2017, and is projected to reach USD 10,519 million in 2024, while the cost of enzymes for biofuel applications alone should total USD 1.0 billion in 2020 [3,4]. More stringent environmental norms coupled with health and environmental awareness are contributing towards intensified research and development efforts in widening the diversity of enzyme applications beyond the laboratory scale. Therefore, carbohydrate hydrolases and proteases are expected to dominate the enzyme market due to their applications in the production of biofuels and detergents, however, other enzyme classes, including laccases and transaminases, are gaining importance due to their increasing usage in food and beverage, textile, pharmaceutical/cosmetic and recycling industries [5].

Streptomyces spp. are usually saprophytic soil-dwelling aerobic filamentous bacteria playing a crucial role in soil ecology by effectively hydrolyzing a wide range of polysaccharides (cellulose, chitin, xylan, and agar) and other natural macromolecules [6]. This biodegradation efficiency is based on the secretion of potent hydrolases (enzymes classified as EC 3 in the EC number classification) such as cellulases, lipases, proteases, xylanases, and cutinases, while other enzymes of environmental and industrial importance have also been identified from this genus [7–9]. Of special interest are laccases and aminotransferases.

Aminotransferases (ATAs) or transaminases (TAs) are pyridoxal-5′-phosphate (PLP)-dependent enzymes capable of performing the transfer of an amino group between an amino donor and a prochiral ketone generating optically pure chiral amines. The numerous examples of ATA applications in the synthesis of chiral amines in the pharmaceutical and fine chemical industries demonstrate the high biocatalytic potential of these enzymes, the most remarkable example being the industrial application of ATAs in the production of the anti-hyperglycaemic drug sitagliptin [10,11]. To date, despite ATAs being quite ubiquitous in nature, ATAs exploited for biocatalytic application are mainly from bacteria and fungi. Curiously, to the best of our knowledge only a few ATAs have been identified and characterized in Streptomycetes. For example, two ATAs from *S. griseus* were reported as capable to catalyse the transamination of ω-aminoacids [12,13], while an ATA from *S. venezuelae* was shown to be involved in the biosynthesis of the rare sugar desosamine [14].

Laccases are blue multicopper oxidoreductases able to catalyze the oxidation of a wide range of substrates [15]. These enzymes find a broad range of applications in the paper and pulp, textile, petrochemical and pharmaceutical industries [16]. Moreover, they can be used for the treatment of industrial wastes by performing detoxifications and bioremediation processes or surface modifications of plastic materials [17]. Bacterial laccases, showing a thermophilic character and working in a broad pH range, result to be particularly attractive for industrial applications [17]. *Streptomyces* laccases are among the best-characterized bacterial laccases and have been identified for example in *S. cyaneus* [18], *S. coelicolor* [19], and *S. sviceus* [20,21].

We previously conducted an extensive activity screen of our in-house library of *Streptomyces* spp. rhizosphere isolates, and showed their biocatalytic potential [2]. The goal of this study was to identify and evaluate potentially novel biocatalysts from selected *Streptomyces* spp. Thus, three of the strains with target activities confirmed by functional screening, namely *Streptomyces* sp. BV129, BV286, and BV333 were subjected to genome sequencing aiming at the annotation of genes of interest, and in-depth bioinformatic characterization that led to cloning and heterologous expression in *E. coli* of one aminotransferase (Sbv333-TA) and one laccase (Sbv286-LAC) sequence.

2. Results

Three *Actinobacteria* strains, namely *Streptomyces* sp. BV129, *Streptomyces* sp. BV286 and *Streptomyces* sp. BV333, were selected in this study to identify and evaluate potentially novel biocatalysts. These strains were selected due to the fact that they showed excellent activities during the phenotypic screen [2]. BV129, BV286 and BV333 strains grew well on the mannitol–soy flower medium and sporulated well after 7 days at 30 °C (Figure 1a). BV286 extruded deep brown pigment on this solid medium even after 24 h growth. Further morphological differences were revealed by SEM (Figure 1b) which revealed chains of smooth and oval spores for BV129 and BV333, while BV286 only sporadically sporulated within this time frame.

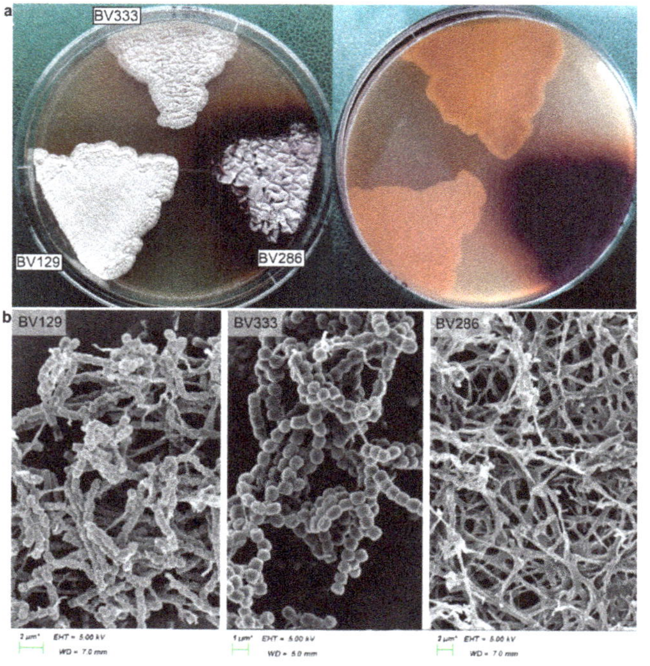

Figure 1. *Streptomyces* sp. BV129, BV333 and BV286 morphology upon growth for 7 days on mannitol-soy flower medium (**a**) agar plates; (**b**) electron micrographs of material collected from the plates.

2.1. Genome Sequencing and Analysis

A whole-genome shotgun sequencing (WGS) followed by *de novo* genome assembly and annotation was performed on these three strains. Genomes were assembled using ABySS v.1.5.2 and annotated with Prokka 1.12 and analyzed by a variety of bioinformatics tools (Table S1). Corresponding assembly statistics are represented in Table 1. The resulting whole-genome shotgun project, including assembly and raw sequence reads, was deposited at NCBI (PRJNA739376; https://www.ncbi.nlm.nih.gov/bioproject/739376).

Table 1. Genome assembly statistics of *Streptomyces* sp. BV129, BV333 and BV286.

Sample ID	*Streptomyces* sp. BV129	*Streptomyces* sp. BV286	*Streptomyces* sp. BV333
Genome size	6,475,242	9,426,047	7,323,588
Genome coverage	83.50	59.02	72.73
Number of contigs	82	212	699
Longest contig	704,247	515,226	151,113
N50 *	144,188	114,510	28,611
L50 **	15	26	80
GC content (%)	71.86	70.77	73.2
Predicted genes	5933	8326	6373
Predicted tRNAs	84	92	83
Predicted CDS	5848	8233	6289
Predicted tmRNA	1	1	1

*—N50—the minimum contig length among contigs required to cover 50% of the whole genome sequence length; can be used as a measure of a quality of assembled genome. Genomes of comparable size with the higher N50 number will be less fragmented and therefore should have higher assembly quality. **—L50 is a minimum number of contigs to reach half of the assembly size. Together with N50 serves as an assembly quality measure.

To reliably classify the *Streptomyces* strains, we performed a phylogenetic analysis based on whole genome-based sequence comparison using a TYGS (The Type (Strain) Genome Server) [22]. TYGS is a web-based bioinformatics pipeline utilizing well-established tools to perform sequence-based taxonomical classification (the full list of tools used for the analysis with corresponding references is given in Table S1). The phylogenetic tree inferred with the FastME 2.1.6.1 [23] from the type-based species clustering from the whole-genome comparison is depicted in Figure 2. Based on the phylogenetic analysis, all three strains belong to different species clusters and do not share common characteristics, such as GC content or the number of predicted genes. Nevertheless, the phylogenetic analysis based on *de novo* assembled sequences reveals a higher similarity between BV129 and BV286 strain (Figure 2).

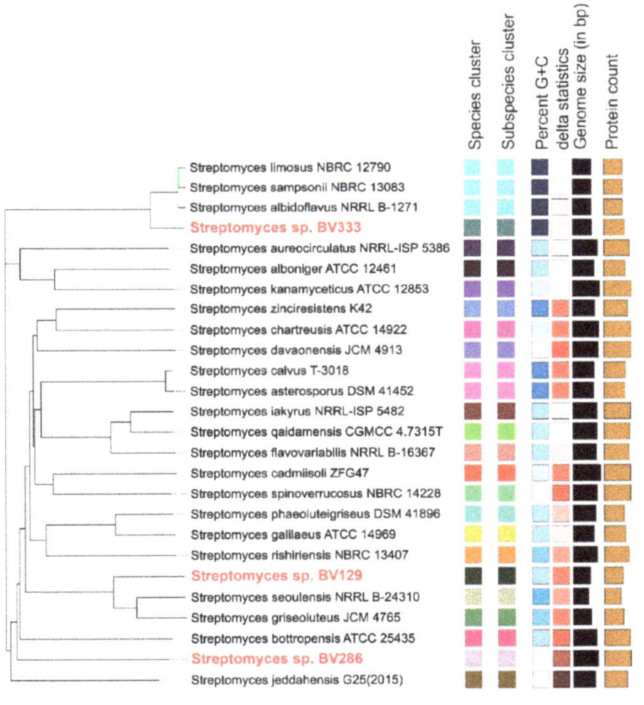

Figure 2. Phylogenetic tree inferred with the FastME 2.1.6.1. using whole-genome comparison with *Streptomyces* sp. BV129, BV333 and BV286 highlighted in red.

A gene-annotation-based search was carried out by using Prokka [24] to find putative gene coding for biotechnologically important enzymatic activities (Table 2). In particular, the search was focused on (i) hydrolases, e.g., lipases, cutinases, cellulases, gelatinase, PHA-depolymerase, and proteases, (ii) oxidoreductases, e.g., lignin peroxidase, laccases and tyrosinases, and (iii) aminotransferases. According to gene annotation, the most abundant enzymes among the selected ones are hydrolases belonging to a rather broad EC category "proteases" (Table 2), followed by "lipases" and "cellulases", while other hydrolases are much less abundant or not present. One or two laccase-coding genes are predicted in all strains, while aminotransferases are largely represented with >30 annotated genes in each strain, this fact being not surprising due to their functional role in the cell metabolism. Nevertheless, many enzymes which are not predicted by the algorithm underlying Prokka were previously identified by alternative methods [2]. Indeed, activities such as gelatinase, PHA-depolymerase and lignin peroxidase were confirmed in the functional screen but revealed non-present by sequence analysis. One of the possible explanations is the way Prokka processes the data. It is a computational pipeline, utilizing different tools for every step of its analysis. For example, for the genes/CDS prediction, Prokka uses the Prodigal [25]. At the later stages of the Prokka pipeline, the prediction results first are refined with the BLAST and then made compatible with an NCBI genome submission policy using the tbl2asn tool. Such conversion sometimes removes additional information from the protein annotation as not compliant with the submission standards.

Table 2. Prokka-based prediction of selected genes in the *de novo* assembled genomes of *Streptomyces* sp. BV129, BV333 and BV286. Enzymes potentially acting in the degradation of different substrates are shown together with the corresponding EC category, other categories for which no genes were predicted by Prokka are not shown.

	EC no.	*Streptomyces* sp. BV129	*Streptomyces* sp. BV286	*Streptomyces* sp. BV333
lipase	3.1.1.3	7	8	4
cutinase	3.1.1.74	0	1	0
cellulase	3.2.1.4	3	11	3
protease	3.4.-.-	124	160	136
laccase	1.10.3.1	2	1	1
tyrosinase	1.14.18.1	1	1	0
aminotransferase (transaminase)	2.6.1.-	32	35	38
Predicted genes with EC number		2049	2763	2114

2.2. Bioinformatics Analysis of Transaminases and Laccases

A deeper bioinformatic analysis of the three genome sequences of *Streptomyces* sp. BV129, BV286 and BV333 was carried out in the search of genes coding for transaminases (ATAs) and laccases. Known ATA sequences, either (*S*)- or (*R*)-selective (Table S2), or laccase sequences (Table S3) were used to perform a multiple sequence alignment with the predicted ORFs of the above mentioned *Streptomyces* genomes by using the LAST program (http://last.cbrc.jp/).

Concerning the search for (*S*)-selective ATAs, 11 potential (*S*)-selective transaminases were identified in the genome of *Streptomyces* strain BV129, 6 sequences in the genome of strain BV286 and 10 sequences in the genome of strain BV333, respectively. Based on the LAST alignment score, 8 of these candidates (BV129_00199, BV129_05035, BV286_01057, BV286_02269, BV286_05820, BV333_00884, BV333_03408, BV333_03485) were selected for further bioinformatics analysis. Five of the eight selected sequences were discarded since they lacked the high-(*S*)-selective-ATAs conserved residues suggested by Steffen-Munsberg et al. [26]. The three remaining sequences shared high sequence identity (88% similarity of BV129_05035 sequence and 91% BV286_05820 sequence compared to BV333_03408, respectively). Therefore, only one of these was selected for further characterization, i.e., the BV333_03408 sequence, encoding for a protein of 459 aa (named Sbv333-TA). Remarkably, BLAST analysis revealed that Sbv333-TA is identical (100% query cover) to an uncharacterized protein annotated as an aspartate aminotransferase from *Streptomyces* sp. M10 (WP_047470642.1). It is also noteworthy that the alignment of Sbv333-TA with available ATA genes from other sources revealed a high sequence similarity with ATA sequences from extremophiles. In particular, Sbv333-TA showed high sequence similarity with the TA from *Thermomicrobium roseii* (Tr-TA, WP_015922033) (43% identity, 89% query cover) [27] and a TA from hot spring metagenomes (B3-TA, KX505389) (41% identity, 94% query cover) [28] (Figure 3). Concerning the search for (*R*)-selective transaminases, only two sequences showing similarity with known (*R*)-selective ATAs were found, but the alignment scores were low and they lacked the important conserved residues suggested by Hohne et al. [29].

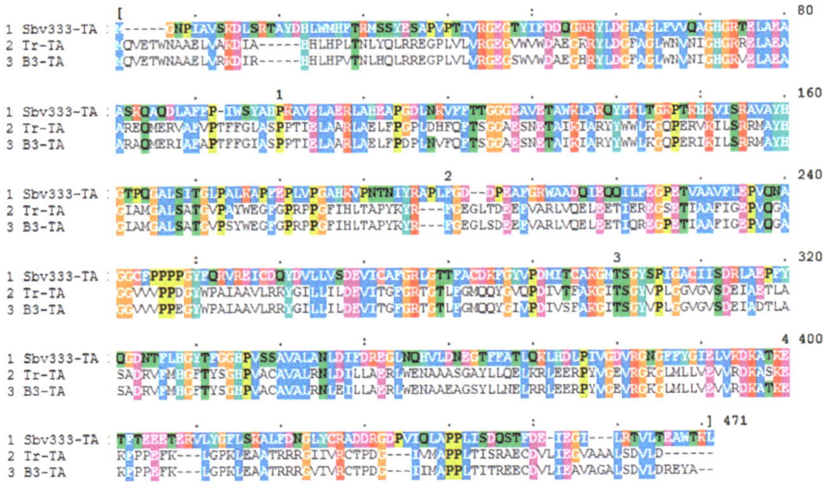

Figure 3. Multiple sequence alignment of Sbv333-TA, Tr-TA and B3-TA.

Concerning the search for laccase homologs, five sequences (BV129_02175, BV129_04916, BV286_03089, BV333_01214 and BV333_01561) showing similarity with known laccases were found in the genomes of the selected *Streptomyces* isolates and in particular three of them showed high LAST alignment score (BV129_04916, BV286_03089, and BV333_01561). These three sequences were quite similar to each other, sharing around 72–74% identity, therefore, among them, the 975 bp-long BV286_03089 sequence, corresponding to a protein of 325 amino acids, was selected for cloning ad expression in *E. coli*. By BLAST analysis, it was shown that BV286_03089 has similarities with a putative copper oxidase from *Streptomyces umbrinus* (Genbank GHH35467.1) (92% identity) while a search in the PDB database revealed that the closest homologue to this sequence is the laccase Ssl1 from *Streptomyces sviceus* (82% identity) (pdb 6YZY). In addition, a Clustal Omega alignment was carried out to compare BV286_03089 with other already characterized laccase sequences (Table S3). This analysis confirmed the high similarity of BV286_03089 with different *Streptomyces* laccases (79–82% identity) and low identity with other bacterial laccases (18%–25% identity).

2.3. Cloning, Expression and Functional Analysis of Novel Transaminase and Laccase

The two selected sequences, BV333_03408 and BV286_03089, potentially coding for an (*S*)-ATA and a laccase, respectively, were cloned into the pETite vector in frame with a C-term His-tag sequence and the resulting plasmids (pETite-Sbv333-TA and pETite-Sbv286-LAC; Table S4) were subsequently transformed and overexpressed in *E. coli* Rosetta cells.

As shown by SDS-PAGE analysis, Sbv333-TA was successfully produced by this host, but mostly accumulated in the cells as inclusion bodies (Figure S1), thus leading to poor recovery yields (about 4 mg L^{-1} after Ni-NTA purification). To overcome these solubility issues, plasmid pETite-Sbv333-TA was expressed in *E. coli* BL21(DE3) containing the plasmid pGro7 (Takara Bio Inc.; Table S4) which allows the co-expression of the target protein with the chaperon proteins GroES and GroEL. This expression system significantly improved the solubility of Sbv333-TA yielding 70 mg L^{-1} of pure enzyme. On the contrary, the laccase from *Streptomyces* sp. BV286 (Sbv286-LAC) was obtained in soluble form in *E. coli* Rosetta cells. Expression conditions were further optimized, and the expression levels were higher when cells were cultivated in terrific broth (TB) media and at 17 °C for 72 h. After successful expression in *E. coli* Rosetta, Sbv286-LAC was purified using QIAGEN Ni-NTA column and 12 mg of pure protein were obtained from 1 L culture

(Figure S2). Cell lysate from cells maintained at 17 °C for 72 h after induction contained the highest concentration of Sbv286-LAC and a lesser concentration of other proteins in comparison to lysates obtained from other induction conditions.

2.3.1. Functional Analysis of Sbv333-TA

The functional characterization of Sbv333-TA was carried out by investigating the influence of different reaction conditions on enzyme activity (Figure 4). The transaminase activity of Sbv333-TA was evaluated by spectrophotometric assays at 245 nm following the formation of the product acetophenone from the benchmark substrates (*S*)-methyl benzyl amine ((*S*)-MBA) and pyruvate. The influence of pH on Sbv333-TA activity was evaluated at pH values ranging from 7.0 to 9.5. As shown in Figure 4a, Sbv333-TA, similarly to most of the known (*S*)-ATAs, shows its maximum activity at alkaline pHs, specifically at pH 9.0 (0.43 U mg^{-1} pure protein at 30 °C).

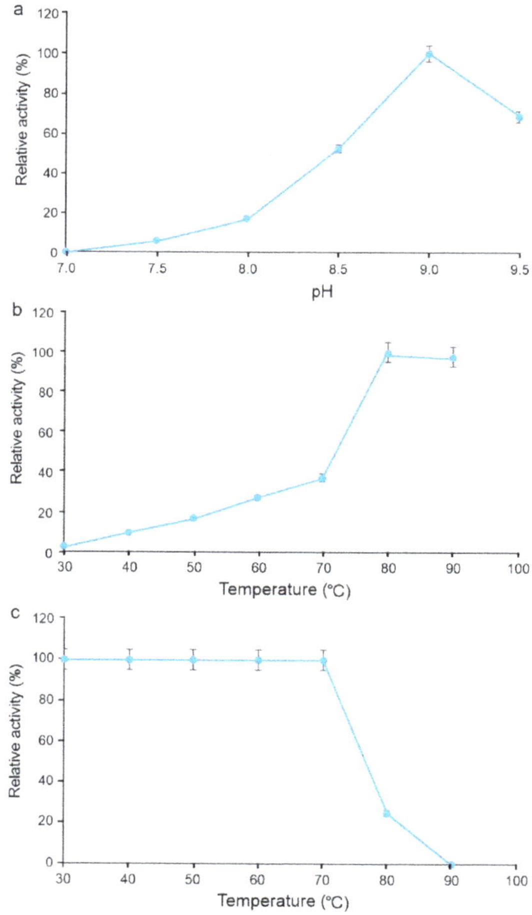

Figure 4. The influence of pH (**a**) and temperature (**b**) on Sbv333-TA activity and influence of temperature (**c**) on Sbv333-TA stability. Values are means of two independent experiments ± standard deviations (SD).

The influence of temperature on the enzyme activity was evaluated in the range of temperature between 20 and 90 °C at the optimal pH (9.0) (Figure 4b). Surprisingly, Sbv333-TA showed a thermophilic character with a constant activity increase with temperature up to 90 °C. Accordingly, Sbv333-TA showed remarkable thermostability. In fact, it retains 100% of starting activity after 3 h incubation at temperatures ranging from 30 °C to 70 °C and it maintains 25% of initial activity even when incubated at 80 °C for 3 h (Figure 4c). To support these results, the melting temperature (T_M) of Sbv333-TA was evaluated by CD analysis, by monitoring structural changes at 220 nm. CD analysis clearly defined a T_M of 85 °C (Figure S3), a value very close to those estimated for the hyperthermophilic enzymes B3-TA (T_M 88 °C) and Tr-TA (T_M 87 °C) [27,28].

To evaluate Sbv333-TA activity toward a series of (*R*)- or (*S*)-aromatic amines, bearing a methyl, ethyl, or propyl side-chain adjacent to the amine function, a spectrophotometric assay was carried out using pyruvate as amino acceptor. Sbv333-TA resulted to be active only toward (*S*)-MBA, indicating that more sterically hindered aromatic amines are not the substrate of this enzyme and, on the other hand, confirming that the enzyme is strictly (*S*)-selective, as inferred by sequence analysis. Subsequently, several keto acids, ketones, esters and aldehydes were evaluated as potential amino acceptors in transamination reactions using (*S*)-MBA as an amino donor (Table 3). The reaction's outcome was estimated after 24 h by GC-MS analysis and Sbv333-TA demonstrated a broad substrate scope in this regard (see Discussion for details).

Table 3. Amino acceptor spectrum of Sbv333-TA.

Amino Acceptor		Conversion (%)
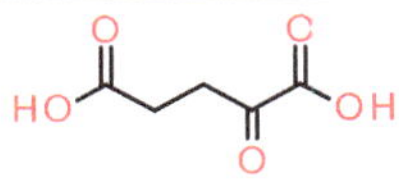	α-ketoglutarate	-
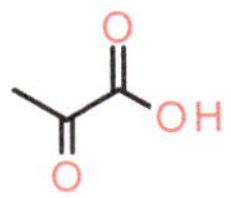	Pyruvate	54
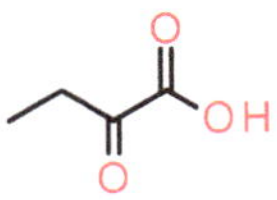	2-Oxobutyrate	46
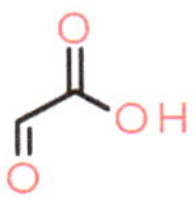	Glioxylate	88
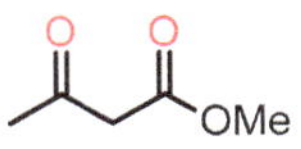	Methylacetoacetate	52
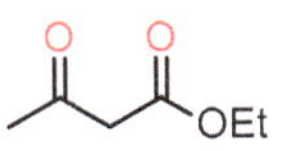	Ethylacetoacetate	35
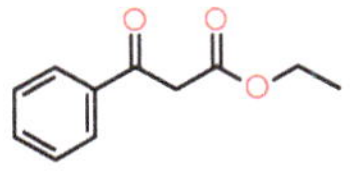	Ethylbenzoylacetate	47

Table 3. *Cont.*

Amino Acceptor		Conversion (%)
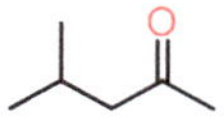	Methyl isobutyl ketone	7
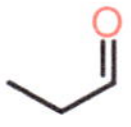	Propionaldehyde	100
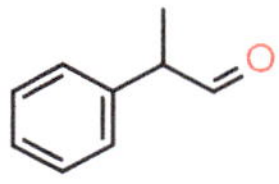	Phenylpropionaldehyde	64

2.3.2. Functional Analysis of Sbv286-LAC

Laccase activity profile at different pHs was determined using ABTS and syringaldazine as the substrate with buffers of different pH (pH 3.0–9.0) and results are presented as relative enzyme activity where maximal activity was set as 100% (Figure 5). Sbv286-LAC reached its maximum activity towards ABTS at pH 4.0 (0.3 U mg^{-1} pure protein at 30 °C, Figure 5a), and was active from pH 3.0 to pH 8.0, while its activity peaked at alkaline pH using syringaldazione as substrate (0.05 U mg^{-1} pure protein at 30 °C). The temperature optimum (60 °C) was determined by enzyme assays with ABTS at temperatures ranging from 20 to 80 °C at pH 4.0 (Figure 5b).

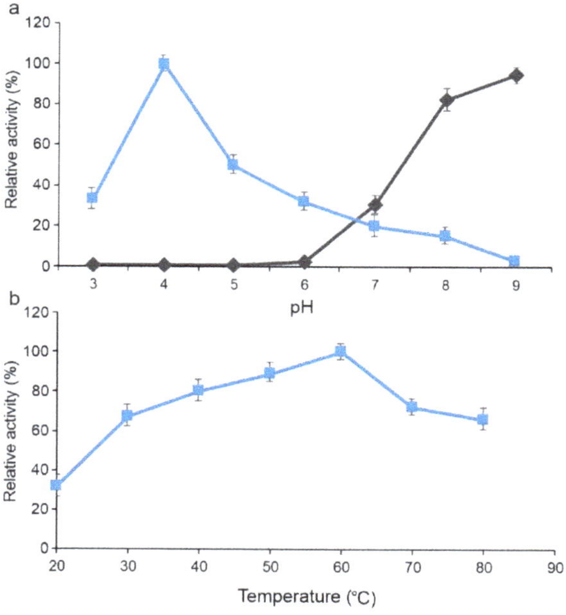

Figure 5. pH and temperature optimum of the recombinant Sbv286-LAC. Relative laccase activity at (**a**) different pH using ABTS (□) and syringaldazione (◇) as substrates and (**b**) different temperatures. Values are means of two independent experiments ± standard deviations (SD).

Sbv286-LAC exhibited not only a remarkable thermophilicity, but also a high thermostability at 60 °C and 80 °C, as well as a 1.5-fold increase in enzyme activity even after 2 h of incubation at high temperatures in comparison to control (enzyme activity measured before incubation at 60 °C and 80 °C) (Figure 6a). Enzyme activity was affected by the presence of 10% (v/v) of solvents in the reaction buffer (Figure 6b). While DMF, DMSO and methanol reduced the enzyme activity between 20–70% in comparison to no-solvents control, acetonitrile enhanced the Sbv286-LAC by 1.5-fold upon incubation for 15 and 30 min (Figure 6b).

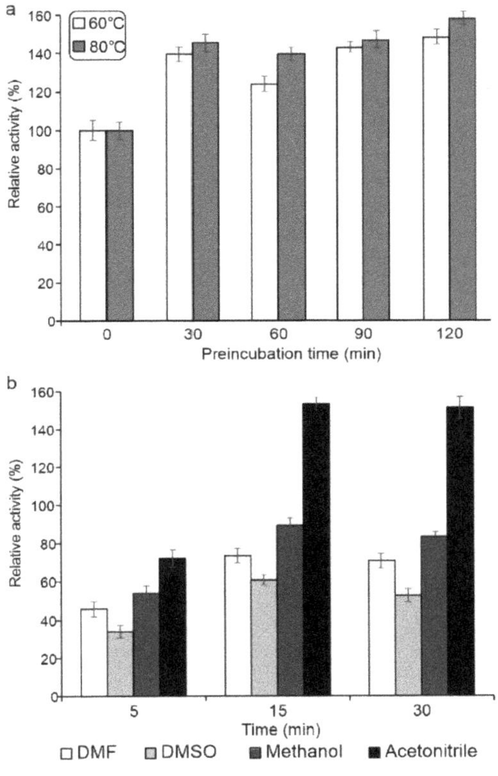

Figure 6. Thermal and solvent stability of the recombinant Sbv286-LAC. Relative laccase activity after incubation (**a**) at 60 °C and 80 °C for up to 2 h and (**b**) in different solvents. Untreated control reactions were set to 100%. Values are means of two independent experiments ± standard deviations (SD).

3. Discussion

Following our previous work of screening of a library of *Streptomyces* spp. rhizosphere isolates to explore their biocatalytic potential [2], in this work, we selected three strains from this library, i.e., *Streptomyces* sp. BV129, BV286 and BV333, which, according to functional screening, showed the capability to produce interesting enzymatic activities. The genome of the three strains was sequenced, *de novo* assembled and submitted for genome annotation. It is noteworthy that we found some apparent discrepancies between the results obtained by functional screening [2] and gene annotation by Prokka. For example, no gene was annotated as gelatinases, but strain BV129 and strain BV333 resulted to produce these enzymes. Or, on the other hand, various genes were annotated as lipases in all three strains, but no lipase activity was detected in strain BV286 by functional screening. However, these differences could be due to inaccuracies of the annotation algorithm or to the substrates used during the functional screening that could not be suitable for the enzymes expressed by the selected *Streptomyces* strains.

With the three new genomes in our hands, we focused our search for novel enzymes on transaminases and laccases, enzymes that are of particular interest for biocatalytic applications. Transaminase and laccase candidates were found in the genomes of all three strains and the most promising candidates (Sbv333-TA and Sbv286-LAC) were selected based on sequence alignment score and presence of key active site residues. It must be pointed out that only homologs to (*S*)-selective ATAs were identified. Considering that (*S*)-selective ATAs are mainly produced by bacteria whereas (*R*)-selective ATAs belong to a completely different fold type class (PLP fold type IV) from that of (*S*)-ATA (PLP fold type I) and are usually produced by fungi [29,30], the identification of only (*S*)-ATA homologs in *Streptomyces* is not surprising. Moreover, it is worth noting that the Prokka algorithm predicted around 30 aminotransferases in each genome, but much fewer ATA candidates have been found by a more specific bioinformatics search. It is important to highlight that our search was restricted to homologues of experimentally confirmed ATAs showing activity toward a wide range of substrates (the so-called high-activity ATAs) [26], and these enzymes are relatively rare when compared to transaminases accepting as amino donors only α-amino acids.

Bioinformatic analysis showed that neither Sbv333-TA nor Sbv286-LAC corresponded to already characterized enzymes and, interestingly, Sbv333-TA shared high sequence identity with ATA sequences from extremophiles, whereas Sbv286-LAC showed high similarity to the so-called small laccases, typical two-domains bacterial laccases [17].

Sbv333-TA and Sbv286-LAC were cloned and overexpressed in *E. coli* Rosetta, and while Sbv286-LAC resulted to be successfully expressed in this host, Sbv333-TA formed inclusion bodies and its successful production was obtained by co-expression with co-chaperons GroES and GroEL. Activity assays using (*S*)-MBA and ABTS as substrates, respectively, revealed that, as expected, the two proteins were functionally active transaminase and laccase, thus both proteins were submitted for functional characterization.

Although *Streptomyces* BV333 and BV286 are mesophilic strains, both enzymes showed a thermophilic character. Sbv333-TA activity increased constantly with temperature up to 90 °C and remarkably, the same behavior is described in literature only for B3-TA [10], the most thermostable wild-type TA reported to date. Indeed, these data are in correlation with the relatively high sequence identity between these two proteins (41% identity). Bacterial laccases and laccase-like multi-copper oxidases are usually described as thermophilic enzymes and Sbv286-LAC, with an optimum temperature of 60 °C, is in line with these findings [19,31]. Furthermore, the two enzymes showed remarkable thermal stability, Sbv333-TA retaining 100% of starting activity after 3 h incubation at temperatures up to 70 °C and Sbv286-LAC resulting even thermoactivated at 60 °C and 80 °C after 2 h incubation. Sbv333-TA thermostability is comparable to that of other thermophilic TAs, such as the TA from *Thermomicrobium roseum*, which retains 100% of starting activity after 3 h incubation at 70 °C, as well. The exceptional thermostability of Sbv333-TA was confirmed also by determining its melting temperature (T_M) which resulted to be of 85 °C, a value only slightly lower than the T_M reported for the thermostable B3-TA (88 °C) and the ATA from *T. roseum* (87 °C) [27]. On the other side, the thermoactivation effect observed with Sbv286-LAC at high temperature was previously shown by Ece and coworkers with a laccase from *S. cyaneus* [18]. It is worth noting that the thermostability of these enzymes makes these biocatalysts quite attractive for industrial applications. In fact, (thermo)stable enzymes usually result suitable to be used under industrial process harsh conditions, such as at high temperatures and/or in the presence of organic solvents.

The relative activities of the two enzymes were also tested at different pHs and Sbv333-TA showed the best activity at pH 9.0, which is in agreement with the pH optimum of most of the transaminases described to date, whereas Sbv286-LAC reached maximum activity at pH 4.0 or 8.5 according to the tested substrate (ABTS or syringaldazine respectively) like other laccases, such as the one from *S. sviceus* [20].

In addition, Sbv286-LAC tolerance to organic co-solvents (10% (v/v)) was evaluated and this enzyme resulted to be quite stable in MeOH and DMF, while, surprisingly, acetonitrile enhanced its activity by 1.5 fold. Some bacterial laccases have exhibited high tolerance to different solvents as well, including ethanol, methanol, DMF, acetonitrile, acetone, and DMSO. For example *Bacillus pumilus* W3 laccase retains > 50% of its activity in solvents such as ethanol, methanol, DMF and DMSO 10% (v/v) [32] while the laccase from *Bacillus licheniformis* retained 55% of its activity in the presence of 40% DMSO [33]. Similar to our study, Wu and co-workers tested different fungal laccases and they also observed an activity enhancement by 1.5- to 4.0-fold by enzyme pre-incubation in acetone, methanol, ethanol, DMSO, and DMF [34].

Finally, the activity of Sbv333-TA was tested toward three differently substituted amino donors and toward a wide array of amino acceptors. Concerning the amino donors, only (*S*)-MBA, bearing a small methyl group, was accepted, this result suggesting that the substrate-binding pocket is not large enough to accept bulkier substituents. Moreover, the enzyme did not convert (*R*)-MBA, demonstrating to be strictly (S)-selective, as expected given the similarity to class I (S)-selective TAs. As far as amino acceptor concerns, this transaminase accepted a broad range of substrates (Table 3). Comparing the amino donor spectra of the new enzyme with the spectra of the closest relative B3-TA and the spectra of the most characterized (*S*)-ATA from *Vibrio fluvialis* [10], it is possible to appreciate some analogies and differences. α-Ketoacids, such as pyruvate and 2-oxo butyrate are accepted by all three enzymes, while the di-ketoacid α-ketoglutarate is not accepted by any of them. In the respect of aldehydes, glyoxylic acid was one of the best substrates in all three cases, while propionaldehyde and phenyl propionaldehyde are mostly preferred by Sbv333-TA. Very low conversions were obtained with aliphatic ketones such as methyl isobutyl-ketone (7% conv.). Remarkably, Sbv333-TA showed activity toward β-ketoesters, and in particular, it was active toward methyl acetoacetate (52% conv.) that gave low conversions with B3-TA and Vf-TA and toward ethyl benzoylacetate (47% conv.), a precursor of β-phenylalanine, which is not accepted by most of the described transaminases [35,36]. This result is quite interesting considering that amination of β-keto acids is quite challenging, as they decompose easily in an aqueous solution and amination of more stable β-keto esters would be a valuable alternative.

4. Materials and Methods

4.1. Reagents and Chemicals

Microbiological media and components such as yeast extract, tryptone, casamino acids were from Oxoid (Hampshire, UK). Amino donors, amino acceptors, ABTS, PLP, IPTG, inorganic salts were purchased from Sigma-Aldrich (Munich, Germany). All other reagents were of analytical grade and commercially available. All strains and plasmids used in this study are presented in Table S4.

4.2. Analytical Methods

GC-MS analyses were carried out on an HP-5MS column (30 m × 0.25 mm × 0.25 μm, Agilent) on a Finnigan TRACE DSQ GC/MS instrument (ThermoQuest, San Jose, CA, USA). Acetylation of (S)-MBA before injection was done as described in Ferrandi et al. 2017 [10]. GC-MS analysis was carried out by keeping column temperature at 60 °C for 1 minute, then raising the temperature to 150 °C at 6 °C/min, hold 1 min and finally raising the temperature from 150 °C to 210°C at 12 °C/min. Under these conditions retention times were: acetophenone, 7.6 min; (S)-α-methylbenzylamine ((S)-MBA), 16.3 min.

CD analysis was performed on nitrogen-flushed Jasco J-1100 spectropolarimeter (Easton, MD, USA) interfaced with a thermostatically controlled cell holder. For the determination of apparent T_M, spectral scans at increasing temperatures were performed at 210 nm, varying the temperatures as follows: 20 up to 65 °C at 5 °C/min data pitch each 2 °C, hold 30 s; 65 up to 90 °C at 2.5 °C/min, data pitch each 0.5 °C, hold 30 s; 90 up to 95 °C at 5 °C/min, pitch data each 2 °C, hold 30 s. This analysis was carried out using a purified Sbv333-TA sample diluted in degassed water (0.15 mg mL^{-1} final concentration) in quartz cuvettes with 0.1 cm path length.

Scanning electron micrographs of three Streptomyces strains grown on MSF agar were obtained by a high-resolution field emission Zeiss Ultra Plus-SEM (Carl Zeiss AG, Oberkochen, Germany) using InLens detector with an accelerating voltage of 5 kV at a working distance of 5 mm. Prior to imaging, strains were fixed onto the SEM stubs using carbon tape and sputtered with gold/palladium (80/20 ratio) for 10 s.

4.3. *Streptomyces sp. BV129, BV286, BV333 Genome Sequencing, Annotation, and Analysis*

Streptomyces spp. genomic DNA (gDNA) was isolated by the method of Nikodinovic et al. [37].

The sequencing library for an Illumina HiSeq2500 was prepared from extracted DNA with the Nextera DNA kit (Illumina) using a standard protocol. Sequencing was performed in a paired-end mode with 2 × 150 cycles for every species. After de-multiplexing with Illumina's software bcl2fastq-1.8.4 with default settings for adapter trimming (at least 90% match of bases should match) and no mismatches allowed in sequencing barcode, all reads undergo quality-based trimming to remove potential contaminants, low quality reads and sequencing adapters with the help of BBDuk from the BBMap package version 34.41 (*https://sourceforge.net/projects/bbmap/*). To pass the quality filter, read quality needed to be higher than a Phred score of 20 and achieve a minimal length of 50 bp after quality-based and adapter trimming. Additionally, every sample was quality-controlled before and after trimming with the FastQC (*http://www.bioinformatics.babraham.ac.uk/projects/fastqc/*). FastQC evaluates per base sequence quality, average base composition, GC content, sequence length distribution and adapter contaminations after trimming. Samples statistics are given in Table S5.

The assembly was carried out on a computational cluster running under CentOS 6.9 by ABySS version 1.5.2 [38] after iterative k-mer length optimization with the k = 85 (BV129, BV333) and k = 92 (BV286). Gene annotation was performed with the Prokka 1.12 [24].

The phylogenetic tree was inferred with the FastME 2.1.6.1 [23] and the tree was estimated from the distance matrix and rooted at the midpoint [39].

4.4. *In Silico Screening for Novel Transaminases and Laccases*

Bioinformatic search for novel biocatalysts was carried out by performing alignment analysis with known sequences retrieved from the NCBI database as queries. The program LAST (http://lastweb.cbrc.jp/) was used with default settings [40]. Bioinformatic analysis was done for three novel *Streptomyces* genomes in order to identify novel sequences coding for transaminases (*S*- and *R*-selective) and laccases. Sequence alignment was done using ClustalΩ [41,42].

4.5. *Gene Cloning and Recombinant Strains*

Selected enzyme-coding sequences (BV333_03408 and BV286_03089) were amplified from *Streptomyces* gDNA under standard PCR conditions using XtraTaq polymerase (Genespin, Milan, Italy) and primers (Table S6). Standard PCR amplifications were carried out in 50 μL reaction mixtures containing 100 ng gDNA, primers (1 μM each), dNTPs (0.2 mM each), 2 U of Xtra.Taq polymerase and 5 μL of buffer containing $MgCl_2$. All PCR reagents were from Genespin. PCR conditions were as follows: 95 °C for 3 min, followed by 35 cycles at 95 °C for 30 s, 65 °C for 30 s, 72 °C for 100s, and then 72 °C for 10 min. Amplified sequences were purified from agarose gel (1.0%, w/v) using the Wizard®SV Gel and PCR Clean-Up System (Promega, Wisconsin, USA) before cloning. Sequences BV333_03408 and BV286_03089 were cloned in the pETite C-His Kan plasmid using the Expresso T7 Cloning and Expression kit from Lucigen (Wisconsin, USA). According to the Expresso T7 Cloning and Expression kit manual, amplified BV333_03408 or BV286_03089 were mixed with the pETite linear plasmid and transformed in chemically competent *HI-Control 10 G* cells. The resulting plasmids pETite-Sbv333-TA and pETite-Sbv286-LAC were purified using the E.Z.N.A. Plasmid Mini kit II (Omega/VWR). Correct insertion of amplified sequences was confirmed by sequencing on both strands (Biofab Research

(Rome, Italy)) using primers T7 promoter and pETite reverse (Table S7). Purified pETite-Sbv333-TA and pETite-Sbv286-LAC were subsequently transformed in *E. coli* Rosetta for protein expression.

4.5.1. Sbv333-TA Expression and Purification

E. coli Rosetta (DE3)-Sbv333-TA cells were grown overnight at 37 °C, 220 rpm, in LB medium supplemented with 30 μg mL^{-1} kanamycin and 34 μg mL^{-1} chloramphenicol ($LB_{kan30cam34}$, 100 mL). Amounts of 20 mL of precultures were subsequently inoculated in 500 mL $LB_{kan30cam34}$ medium and maintained at 37 °C, 220 rpm, till the OD_{600} cell density reached 0.5–1. Recombinant protein expression was induced by the addition of IPTG (1 mL of 1 M solution in water) and the culture was transferred to 30 °C with shaking 220 rpm and grown overnight. Cells were thus recovered by centrifugation (5000 rpm, 30 min, 4 °C), resuspended in 20 mL wash buffer (20 mM KP buffer, pH 7.0, 500 mM NaCl, 20 mM imidazole) and lysed by sonication. Cell extracts were separated from insoluble debris by centrifugation (11,000 rpm, 30 min) and the presence of soluble recombinant protein in clear lysates was assessed by SDS-PAGE (10% T, 2.6% C).

To perform protein purification, clear cell extracts containing the overexpressed protein were incubated with the Ni Sepharose 6 Fast Flow agarose resin (Ni-NTA) (GE Healthcare, Italy) for 90 min at 4 °C under mild shaking. The mixture was then loaded onto a glass column (10 × 110 mm), the resin was washed with 20 mL of wash buffer (20 mM imidazole, 500 mM NaCl, 20 mM potassium phosphate buffer pH 8.0). His-tagged Sbv333-TA was thus eluted using a 3-step gradient (10 mL washing buffer with 100, 200, and 300 mM imidazole, respectively), dialyzed against 20 mM KP buffer, pH 9.0, at 4 °C for 16 h and stored at −80 °C. Protein content was estimated according to the method of Bradford with the Bio-Rad Protein Assay and protein purity was assessed by SDS-PAGE analysis (10% T, 2.6% C). The molecular weight protein standard mixture from Bio-Rad (Karlsruhe, Germany) was used as a reference. Gels were stained for protein detection with Coomassie Brilliant Blue.

Due to unsatisfactory results obtained using *E. coli* Rosetta (DE3)-Sbv333-TA, pETite-Sbv333-TA was transformed into *E. coli* BL21(DE3) containing the plasmid pGro7 (Takara Bio Inc., Kyoto, Japan), obtaining the *E. coli* BL21(DE3)-GroES, GroEL-Sbv333-TA strain. An overnight preculture of *E. coli* BL21(DE3)-GroES, GroEL-Sbv333-TA grown in LB_{kan30} at 37 °C was then inoculated in 1 L of LB $_{kan30cam20}$ and 2 mg mL^{-1} arabinose was added for induction of chaperone proteins. When cell density reached OD_{600} 0.4–0.6, recombinant expression of Sbv333-TA was induced by the addition of IPTG (1 mL of 1 M solution in water) and the culture was kept at 30 °C overnight. Finally, cells were harvested by centrifugation, lysed by sonication and Sbv333-TA was purified as described above.

4.5.2. Sbv286-LAC Expression and Purification

E. coli Rosetta (DE3)-Sbv286-LAC cells were grown at 37 °C and 180 rpm for 16 h in $LB_{kan30cam34}$ (100 mL). The following day, 1% overnight inoculum was added to fresh LB, M9 or TB media supplemented with antibiotics (30 μg mL^{-1} kanamycin and 34 μg mL^{-1} chloramphenicol) and cultures were induced with 1 mM IPTG at OD_{600} between 0.4–0.6. Enzyme expression was carried out at 37 °C for 3 h, 28 °C for 16 h and at 17 °C for 72 h with shaking at 180 rpm. Sbv286-LAC was purified using QIAGEN Ni-NTA spin kit following the manufacturer's instruction (QIAGEN, Hilden, Germany). Purified Sbv286-LAC, as well as whole cells and cell extracts samples collected during recombinant expression optimization, were analyzed by SDS-PAGE (10% T, 2.6% C), as previously described for Sbv333-TA.

4.6. Functional Characterization of Sbv333-TA and Sbv286-LAC

Sbv333-TA activity was determined by spectrophotometric assays in quartz cuvettes (assay volume 0.5 mL) containing transaminase assay solution (2.5 mM pyruvate, 2.5 mM (*S*)-MBA in 0.1 M KP buffer, pH 9.0 and 0.25% (v/v) dimethyl sulfoxide (DMSO)). Formation of acetophenone upon enzyme addition (10–50 μL of purified Sbv333-TA, 70–350 μg) was followed at 245 nm (ε_{245} = 12 mM^{-1} cm^{-1}) on a Jasco V-530 UV/VIS spectrophotometer. One unit of activity is defined as the enzyme activity that produces 1 μmol of acetophenone per minute under the assay conditions described above. Sbv286-LAC activity was detected spectrophotometrically at 420 nm using ABTS (2,2′-azino-bis(3-ethylbenzothiazoline-6-sulphonic acid) or syringaldazine at 530 nm as substrates (ε_{420} = 36.0 mM^{-1}cm^{-1}, ε_{530} = 65.0 mM^{-1}cm^{-1} respectively). The assay was performed in acrylic cuvettes (total volume 1 mL) in 20 mM Na-acetate buffer pH 4.0 0.2 mM $CuSO_4$, containing 1 mM ABTS or 20 mM Tris-HCl buffer pH 8.5 0.2 mM $CuSO_4$ containing 20 μM syringaldazine and adding 10–20 μL (28–56 μg) of purified Sbv286-LAC. One unit of enzymatic activity was defined as the amount of enzyme that oxidizes 1 mmol of ABTS or 1 mmol of syringaldazine per minute under the described assay conditions.

The optimum pH of Sbv333-TA was determined within a pH range of 7.0 to 9.5 in 20 mM potassium phosphate buffer (pH 7.0–9.0) or 20 mM Tris-HCl (pH 9.5) at 30 °C, while temperature optimum was determined by heating the assay solutions (20 mM potassium phosphate buffer pH 9.0) in cuvettes in a water bath in the range of 30–90 °C for 15 min before adding the purified enzyme (10–50 μL of purified Sbv333-TA, 70–350 μg). The thermal stability of Sbv333-TA was evaluated by incubating enzyme samples for 3 h at temperatures ranging from 30 to 90 °C and then measuring spectrophotometrically the residual activity following the formation of acetophenone. Experiments were done in duplicate on two independent occasions.

The influence of pH on Sbv286-LAC activity was determined within a pH range of 3.0 to 10.0 using 20 mM acetate buffer (pH 3.0–5.0), 20 mM potassium phosphate buffer (pH 6.0–7.0), 20 mM Tris-HCl (pH 8.0–9.0), respectively. Sbv286-LAC temperature optimum was determined by heating the assay solutions in cuvettes in a water bath in the range of 20–80 °C for 15 minutes before adding the purified enzyme. The thermal stability of purified laccase was assessed by measuring the residual activity after incubation of purified Sbv286-LAC solution at 60 °C and 80 °C. Enzymes samples were taken at 30, 60, 90 and 120 min, quickly cooled on ice and activity was determined by the ABTS assay described above in 20 mM acetate buffer, pH 4.0. Solvent stability of Sbv286-LAC was assessed in 10% (v/v) dimethyl formamide (DMF), dimethyl sulfoxide (DMSO), methanol and acetonitrile by measuring residual enzyme activity using ABTS assay after incubation in solvents at room temperature for 5, 15 and 30 min. Experiments were done at least in duplicate.

The enantioselectivity of Sbv333-TA was evaluated by performing the acetophenone assay in the presence of (*R*)-α-methylbenzylamine as substrate under the same conditions described for (*S*)-MBA. The formation of propiophenone and butyrophenone was determined under the same conditions at 242 nm using either (*S*)- or (*R*)-ethylbenzylamine and (*S*)- or (*R*)-phenylbutylamine, respectively, as substrates. Sbv333-TA amino acceptor spectrum was evaluated at 30 °C in 0.5 mL reaction mixture containing 0.1 M KP buffer, pH 9.0, 10 mM (*S*)-MBA, 10 mM acceptor, 1 mM PLP, and 0.5 mg of purified enzyme. Conversions of (*S*)-MBA (after derivatization) into acetophenone were evaluated after 24 h by GC-MS analysis as described in the analytical methods section.

Supplementary Materials: The following are available online at https://www.mdpi.com/article/10.3390/catal11080919/s1, Figure S1: SDS-PAGE *E. coli* Rosetta expressing aminotransferase from pETite_BV333 TA. 1) before the induction 2) after the induction with IPTG 3) cell lysate 4) purified enzyme M) Marker – SDS-PAGE Standard Broad Range (Bio-Rad, USA), Figure S2: SDS-PAGE gel of purified laccase Sbv286-LAC. 1) before the induction 2) after the induction with IPTG 3) purified enzyme M) Marker – SDS-PAGE Standard Broad Range (Bio-Rad, USA), Figure S3: Sbv333-TA melting curve obtained by circular dichroism spectroscopic analysis at 210 nm Table S1: Bioinformatics tools used to perform phylogenetic classification, Table S2: GenBank accession number of query sequences used in the bioinformatics search for novel ATA. Entry 1-11: (*S*)-selective ATA, Entry 12–18: (*R*)-selective ATA, Table S3: GenBank accession number of query sequences used in the bioinformatics search for novel laccases, Table S4: Bacterial strains and plasmids used in this study, Table S5: Sequencing statistics, Table S6: Oligonucleotide primers used in this study.

Author Contributions: Conceptualization, D.M. and J.N.-R. methodology, E.E.F., Y.V., C.G. and R.S.; investigation, J.S., L.D., S.V., and Y.V.; resources, D.M. and J.N.-R.; data curation, E.E.F., J.S., Y.V.; writing—original draft preparation, J.S., L.D., Y.V., R.S. and E.E.F.; writing—review and editing, S.V., E.E.F., D.M. and J.N.-R.; funding acquisition, D.M. and J.N.-R. All authors have read and agreed to the published version of the manuscript.

Funding: This research was funded by Ministry of Education, Science and Technological Development of the Republic of Serbia, 451-03-9/2021-14/ 200042 and S.V. and J.NR. are partially funded from the European Union's Horizon 2020 research and innovation program under grant agreement No 870292 (BioICEP).

Data Availability Statement: Publicly available datasets were analyzed in this study (https://www.ncbi.nlm.nih.gov/bioproject/739376). The additional data presented in this study are available on request from the corresponding author.

Conflicts of Interest: The authors declare no conflict of interest.

References

1. Spasic, J.; Mandic, M.; Djokic, L.; Nikodinovic-Runic, J. *Streptomyces* spp. in the biocatalysis toolbox. *Appl. Microbiol. Biotechnol.* **2018**, *102*, 3513–3536. [CrossRef]
2. Spasic, J.; Mandic, M.; Radivojevic, J.; Jeremic, S.; Vasiljevic, B.; Nikodinovic-Runic, J.; Djokic, L. Biocatalytic potential of *Streptomyces* spp. isolates from rhizosphere of plants and mycorrhizosphere of fungi. *Biotechnol. Appl. Biochem.* **2018**, *65*, 822–833. [CrossRef]
3. Lopes, A.M.; Ferreira Filho, E.X.; Moreira, L.R.S. An update on enzymatic cocktails for lignocellulose breakdown. *J. Appl. Microbiol.* **2018**, *125*, 632–645. [CrossRef]
4. Pellis, A.; Cantone, S.; Ebert, C.; Gardossi, L. Evolving biocatalysis to meet bioeconomy challenges and opportunities. *New Biotechnol.* **2018**, *40*, 154–169. [CrossRef]
5. Markets&Markets. Biocatalysis & Biocatalysts Market by Type, Application, and Geography-Forecast to 2019. 2015. Available online: https://www.marketsandmarkets.com/Market-Reports/biocatalysis-biocatalyst-market-71612080.html (accessed on 12 May 2021).
6. Kelly, S.A.; Magill, D.J.; Megaw, J.; Skvortsov, T.; Allers, T.; McGrath, J.W.; Allen, C.C.R.; Moody, T.S.; Gilmore, B.F. Characterisation of a solvent-tolerant haloarchaeal (*R*)-selective transaminase isolated from a Triassic period salt mine. *Appl. Microbiol. Biotechnol.* **2019**, *103*, 5727–5737. [CrossRef] [PubMed]
7. Sun, X.; Li, Y.; Tian, Z.; Qian, Y.; Zhang, H.; Wang, L. A novel thermostable chitinolytic machinery of *Streptomyces* sp. F-3 consisting of chitinases with different action modes. *Biotechnol. Biofuels* **2019**, *12*, 136. [CrossRef] [PubMed]
8. Prakash, D.; Nawani, N.; Prakash, M.; Bodas, M.; Mandal, A.; Khetmalas, M.; Kapadnis, B. Actinomycetes: A repertory of green catalysts with a potential revenue resource. *BioMed Res. Int.* **2013**, *2013*, 264020. [CrossRef] [PubMed]
9. Cho, S.S.; Park, D.J.; Simkhada, J.R.; Hong, J.H.; Sohng, J.K.; Lee, O.H.; Yoo, J.C. A neutral lipase applicable in biodiesel production from a newly isolated *Streptomyces* sp. CS326. *Bioprocess Biosyst. Eng.* **2012**, *35*, 227–234. [CrossRef] [PubMed]
10. Ferrandi, E.E.; Monti, D. Amine transaminases in chiral amines synthesis: Recent advances and challenges. *World J. Microbiol. Biotechnol.* **2017**, *34*, 13. [CrossRef]
11. Savile, C.K.; Janey, J.M.; Mundorff, E.C.; Moore, J.C.; Tam, S.; Jarvis, W.R.; Colbeck, J.C.; Krebber, A.; Fleitz, F.J.; Brands, J.; et al. Biocatalytic asymmetric synthesis of chiral amines from ketones applied to sitagliptin manufacture. *Science* **2010**, *329*, 305–309. [CrossRef]
12. Yonaha, K.; Suzuki, K.; Toyama, S. *Streptomyces* beta-alanine: Alpha-ketoglutarate aminotransferase, a novel omega-amino acid transaminase. Purification, crystallization, and enzymologic properties. *J. Biol. Chem.* **1985**, *260*, 3265–3268. [CrossRef]
13. Yonaha, K.; Suzuki, K.; Toyama, S. 4-Aminobutyrate:2-oxoglutarate aminotransferase of *Streptomyces griseus*: Purification and properties. *Eur. J. Biochem.* **1985**, *146*, 101–106. [CrossRef]
14. Burgie, E.S.; Thoden, J.B.; Holden, H.M. Molecular architecture of DesV from *Streptomyces venezuelae*: A PLP-dependent transaminase involved in the biosynthesis of the unusual sugar desosamine. *Protein Sci.* **2007**, *16*, 887–896. [CrossRef] [PubMed]
15. Bassanini, I.; Ferrandi, E.E.; Riva, S.; Monti, D. Biocatalysis with laccases: An updated overview. *Catalysts* **2021**, *11*, 26. [CrossRef]
16. Zerva, A.; Simić, S.; Topakas, E.; Nikodinovic-Runic, J. Applications of microbial laccases: Patent review of the past decade (2009–2019). *Catalysts* **2019**, *9*, 1023. [CrossRef]
17. Guan, Z.-B.; Luo, Q.; Wang, H.-R.; Chen, Y.; Liao, X.-R. Bacterial laccases: Promising biological green tools for industrial applications. *Cell. Molec. Life Sci.* **2018**, *75*, 3569–3592. [CrossRef]
18. Ece, S.; Lambertz, C.; Fischer, R.; Commandeur, U. Heterologous expression of a *Streptomyces cyaneus* laccase for biomass modification applications. *AMB Express* **2017**, *7*, 86. [CrossRef]
19. Sherif, M.; Waung, D.; Korbeci, B.; Mavisakalyan, V.; Flick, R.; Brown, G.; Abou-Zaid, M.; Yakunin, A.F.; Master, E.R. Biochemical studies of the multicopper oxidase (small laccase) from *Streptomyces coelicolor* using bioactive phytochemicals and site-directed mutagenesis. *Microbial Biotechnol.* **2013**, *6*, 588–597. [CrossRef]

20. Gunne, M.; Urlacher, V.B. Characterization of the alkaline laccase Ssl1 from *Streptomyces sviceus* with unusual properties discovered by genome mining. *PLoS ONE* **2012**, *7*, e52360. [CrossRef] [PubMed]
21. Gunne, M.; Höppner, A.; Hagedoorn, P.L.; Urlacher, V.B. Structural and redox properties of the small laccase Ssl1 from *Streptomyces sviceus*. *FEBS J.* **2014**, *281*, 4307–4318. [CrossRef]
22. Meier-Kolthoff, J.P.; Göker, M. TYGS is an automated high-throughput platform for state-of-the-art genome-based taxonomy. *Nat. Comm.* **2019**, *10*, 2182. [CrossRef]
23. Lefort, V.; Desper, R.; Gascuel, O. FastME 2.0: A comprehensive, accurate, and fast distance-based phylogeny inference program. *Mol. Biol. Evol.* **2015**, *32*, 2798–2800. [CrossRef]
24. Seemann, T. Prokka: Rapid prokaryotic genome annotation. *Bioinformatics* **2014**, *30*, 2068. [CrossRef] [PubMed]
25. Hyatt, D.; Chen, G.-L.; LoCascio, P.F.; Land, M.L.; Larimer, F.W.; Hauser, L.J. Prodigal: Prokaryotic gene recognition and translation initiation site identification. *BMC Bioinform.* **2010**, *11*, 119. [CrossRef]
26. Steffen-Munsberg, F.; Vickers, C.; Kohls, H.; Land, H.; Mallin, H.; Nobili, A.; Skalden, L.; van den Bergh, T.; Joosten, H.J.; Berglund, P.; et al. Bioinformatic analysis of a PLP-dependent enzyme superfamily suitable for biocatalytic applications. *Biotechnol. Advan.* **2015**, *33*, 566–604. [CrossRef] [PubMed]
27. Mathew, S.; Deepankumar, K.; Shin, G.; Hong, E.Y.; Kim, B.-G.; Chung, T.; Yun, H. Identification of novel thermostable ω-transaminase and its application for enzymatic synthesis of chiral amines at high temperature. *RSC Adv.* **2016**, *6*, 69257–69260. [CrossRef]
28. Ferrandi, E.E.; Previdi, A.; Bassanini, I.; Riva, S.; Peng, X.; Monti, D. Novel thermostable amine transferases from hot spring metagenomes. *Appl. Microbiol. Biotechnol.* **2017**, *101*, 4963–4979. [CrossRef] [PubMed]
29. Höhne, M.; Schätzle, S.; Jochens, H.; Robins, K.; Bornscheuer, U.T. Rational assignment of key motifs for function guides in silico enzyme identification. *Nat. Chem. Biol.* **2010**, *6*, 807. [CrossRef] [PubMed]
30. Sayer, C.; Martinez-Torres, R.J.; Richter, N.; Isupov, M.N.; Hailes, H.C.; Littlechild, J.A.; Ward, J.M. The substrate specificity, enantioselectivity and structure of the (*R*)-selective amine: Pyruvate transaminase from *Nectria haematococca*. *FEBS J.* **2014**, *281*, 2240–2253. [CrossRef] [PubMed]
31. Ihssen, J.; Reiss, R.; Luchsinger, R.; Thöny-Meyer, L.; Richter, M. Biochemical properties and yields of diverse bacterial laccase-like multicopper oxidases expressed in *Escherichia coli*. *Sci. Rep.* **2015**, *5*, 10465. [CrossRef]
32. Guan, Z.-B.; Song, C.-M.; Zhang, N.; Zhou, W.; Xu, C.-W.; Zhou, L.-X.; Zhao, H.; Cai, Y.-J.; Liao, X.-R. Overexpression, characterization, and dye-decolorizing ability of a thermostable, pH-stable, and organic solvent-tolerant laccase from *Bacillus pumilus* W3. *J. Mol. Catal. B Enzym.* **2014**, *101*, 1–6. [CrossRef]
33. Tonin, F.; Melis, R.; Cordes, A.; Sanchez-Amat, A.; Pollegioni, L.; Rosini, E. Comparison of different microbial laccases as tools for industrial uses. *New Biotechnol.* **2016**, *33*, 387–398. [CrossRef] [PubMed]
34. Wu, M.-H.; Lin, M.-C.; Lee, C.-C.; Yu, S.-M.; Wang, A.H.J.; Ho, T.-H.D. Enhancement of laccase activity by pre-incubation with organic solvents. *Sci. Rep.* **2019**, *9*, 9754. [CrossRef] [PubMed]
35. Buß, O.; Voss, M.; Delavault, A.; Gorenflo, P.; Syldatk, C.; Bornscheuer, U.; Rudat, J. β-Phenylalanine ester synthesis from stable β-keto ester substrate using engineered ω-transaminases. *Molecules* **2018**, *23*, 1211. [CrossRef]
36. Midelfort, K.S.; Kumar, R.; Han, S.; Karmilowicz, M.J.; McConnell, K.; Gehlhaar, D.K.; Mistry, A.; Chang, J.S.; Anderson, M.; Villalobos, A.; et al. Redesigning and characterizing the substrate specificity and activity of *Vibrio fluvialis* aminotransferase for the synthesis of imagabalin. *Protein Eng. Des. Sel.* **2012**, *26*, 25–33. [CrossRef]
37. Nikodinovic, J.; Barrow, K.D.; Chuck, J.A. High yield preparation of genomic DNA from Streptomyces. *BioTechniques* **2003**, *35*, 932–934. [CrossRef]
38. Simpson, J.T.; Wong, K.; Jackman, S.D.; Schein, J.E.; Jones, S.J.; Birol, I. ABySS: A parallel assembler for short read sequence data. *Genome Res.* **2009**, *19*, 1117–1123. [CrossRef]
39. Farris, J.S. Estimating phylogenetic trees from distance matrices. *Am. Nat.* **1972**, *106*, 645–668. [CrossRef]
40. Kielbasa, S.M.; Wan, R.; Sato, K.; Horton, P.; Frith, M.C. Adaptive seeds tame genomic sequence comparison. *Genome Res.* **2011**, *21*, 487–493. [CrossRef]
41. Sievers, F.; Wilm, A.; Dineen, D.; Gibson, T.J.; Karplus, K.; Li, W.; Lopez, R.; McWilliam, H.; Remmert, M.; Soding, J.; et al. Fast, scalable generation of high-quality protein multiple sequence alignments using Clustal Omega. *Mol. Syst. Biol.* **2011**, *7*, 539. [CrossRef]
42. Thompson, J.D.; Higgins, D.G.; Gibson, T.J. CLUSTAL W: Improving the sensitivity of progressive multiple sequence alignment through sequence weighting, position-specific gap penalties and weight matrix choice. *Nucleic Acids Res.* **1994**, *22*, 4673–4680. [CrossRef] [PubMed]

catalysts

MDPI

Article

Biocatalytic Silylation: The Condensation of Phenols and Alcohols with Triethylsilanol

Emily I. Sparkes [1,2], Chisom S. Egedeuzu [1,2], Billie Lias [1,2], Rehana Sung [1], Stephanie A. Caslin [1,2], S. Yasin Tabatabaei Dakhili [1,2], Peter G. Taylor [3], Peter Quayle [2] and Lu Shin Wong [1,2,*]

[1] Manchester Institute of Biotechnology, University of Manchester, 131 Princess Street, Manchester M1 7DN, UK; emilysparkes11@gmail.com (E.I.S.); chisom.egedeuzu@manchester.ac.uk (C.S.E.); billie.lias@student.manchester.ac.uk (B.L.); rehana.sung@manchester.ac.uk (R.S.); stephanie.caslin@sky.com (S.A.C.); s.yasin.tabatabaei.d@gmail.com (S.Y.T.D.)

[2] Department of Chemistry, University of Manchester, Oxford Road, Manchester M13 9PL, UK; Peter.Quayle@manchester.ac.uk

[3] Faculty of Science, Technology, Engineering and Mathematics, Open University, Walton Hall, Milton Keynes MK7 6AA, UK; peter.taylor@open.ac.uk

* Correspondence: l.s.wong@manchester.ac.uk

Abstract: Silicatein-α (Silα), a hydrolytic enzyme derived from siliceous marine sponges, is one of the few enzymes in nature capable of catalysing the metathesis of silicon–oxygen bonds. It is therefore of interest as a possible biocatalyst for the synthesis of organosiloxanes. To further investigate the substrate scope of this enzyme, a series of condensation reactions with a variety of phenols and aliphatic alcohols were carried out. In general, it was observed that Silα demonstrated a preference for phenols, though the conversions were relatively modest in most cases. In the two pairs of chiral alcohols that were investigated, it was found that the enzyme displayed a preference for the silylation of the *S*-enantiomers. Additionally, the enzyme's tolerance to a range of solvents was tested. Silα had the highest level of substrate conversion in the nonpolar solvents *n*-octane and toluene, although the inclusion of up to 20% of 1,4-dioxane was tolerated. These results suggest that Silα is a potential candidate for directed evolution toward future application as a robust and selective biocatalyst for organosiloxane chemistry.

Keywords: silicatein; condensation; silyl ether; organosiloxanes; biocatalysis

Citation: Sparkes, E.I.; Egedeuzu, C.S.; Lias, B.; Sung, R.; Caslin, S.A.; Tabatabaei Dakhili, S.Y.; Taylor, P.G.; Quayle, P.; Wong, L.S. Biocatalytic Silylation: The Condensation of Phenols and Alcohols with Triethylsilanol. *Catalysts* **2021**, *11*, 879. https://doi.org/10.3390/catal11080879

Academic Editors: Evangelos Topakas, Roland Wohlgemuth and David D. Boehr

Received: 26 June 2021
Accepted: 20 July 2021
Published: 22 July 2021

1. Introduction

During the multi-step chemical synthesis of complex molecules, silyl ethers are often employed for the protection of hydroxyl groups [1–7], where their utility arises from orthogonality to other commonly used acid- and base-labile protecting groups. Typically, the introduction of these silyl groups involves the use of an electrophilic silylating reagent such as a silyl chloride (i.e., chlorosilane) or triflate, the latter of which is itself produced from the corresponding silyl chloride [8]. The trimethylsilylation of alcohols has also been demonstrated with *N,O*-bis-silyl trifluoroacetamide [9], *N,N′*-bis-silyl urea [10] and hexamethydisilazane [11]. It has also been known for many years that the silylation of alcohols can be affected by hydrosilanes (silyl hydrides), with the dehydrogenative condensation catalysed by transition metals or strong Lewis acids [12–14]. Alternatively, silylation can also be affected by activation of the silanol to attack by an alcohol (or phenol) under Mitsunobu conditions [15].

However, in all cases, the necessary reagents are energy intensive to produce, and their use results in the generation of stoichiometric amounts of by-products that are hazardous or environmentally undesirable (e.g., triflic acid, hydrogen chloride, hydrogen). In this regard, the capability of silylate hydroxy groups through the condensation of the corresponding silanol and alcohol would circumvent the need for harsh reagents and only release water as the by-product. However, only one example of this dehydrative condensation has been reported, catalysed by rare-earth Lewis acids such as $Yb(OTf)_3$ and $Sc(OTf)_3$ [16]. It would therefore be preferable to identify a reaction pathway that better conforms to the principles of green chemistry—in particular, the harnessing of biological catalysts that can be sustainably sourced and avoid any requirement for rare metals [17].

Silicatein-α (Silα), an enzyme responsible for the condensation of inorganic Si–O bonds in marine demosponges [18,19], has also previously been shown to catalyse the condensation of a variety of organosilanols with aliphatic alcohols to produce the corresponding silyl ether [20]. This enzyme is monomeric and does not require any co-factors for activity, so may offer a relatively benign alternative biocatalytic approach for silylation. However, the general substrate scope of this enzyme is currently not well established and requires further investigation before synthetically useful methods can be subsequently developed.

In this study, the substrate scope from the perspective of the alcohol component was investigated by surveying the enzyme's ability to form triethylsilyl ethers with a range of phenols and aliphatic alcohols. In addition, Silα's tolerance for a variety of polar and nonpolar solvents was tested to examine the optimum reaction conditions for Silα in organic media. In the original report of Silα-catalysed condensations, a preference for phenolic hydroxy groups was found [20]. However, more recent studies found that the hexahistidine affinity tag used for isolation of the enzyme was also contributing to nonspecific catalysis [21]. Thus, in order to more accurately determine the substrate scope and catalysis mediated by the enzyme active site itself, in this study we used an enzyme construct consisting of the ribosomal chaperone protein trigger factor fused to the *N*-terminus and a Strep II tag fused to the *C*-terminus (henceforth referred to as TF-Silα-Strep).

2. Results and Discussion

2.1. *Triethylsilylation of Phenols*

The triethylsilylation of various substituted phenols was first examined to investigate the effect of the substituents on the enzyme-catalysed reaction. Chloro, methyl and methoxy substituents were chosen to represent electron-donating and -withdrawing groups, and all three positions (*ortho*, *meta* and *para*) were investigated. The condensation reactions were carried out using the procedure previously reported in [20], whereby the enzyme was used in the form of a lyophilised solid (in a matrix of potassium salts and 18-crown-6) in organic solvent at 75 °C (Scheme 1). The product conversions were measured after 72 h by GC-MS.

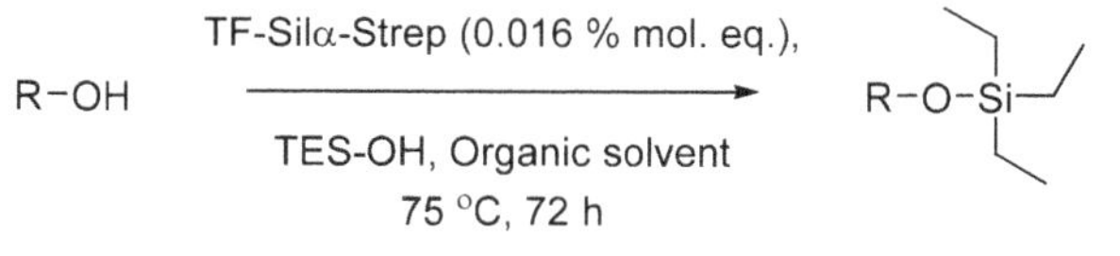

Scheme 1. General reaction scheme for the condensation of the alcohol with triethylsilanol (TES-OH) to form the corresponding silyl ether. R: aliphatic or substituted phenyl groups.

For all phenols tested, some degree of product formation was observed, even in control reactions where the enzyme was omitted. However, in all cases, the inclusion of Silα resulted in enhanced conversions, though some were deemed not to be statistically significant (Figure 1). Unsubstituted phenol was well accepted by the enzyme, giving a gross conversion of 88% after 72 h and a net conversion of 61% after subtraction of the percentage conversion from the nonenzymatic reaction (Table S1 in the Supplementary Information). High net conversions were also observed with *p*-methoxyphenol at 62% (75% gross). This result equated to a nearly 6-fold higher conversion for the enzyme-catalysed reaction vs. the nonenzymatic reaction (i.e., 75% vs. 13%). *o*-Methoxyphenol was the least reactive substrate under these conditions, giving gross and net conversions of 15% and 11%, respectively. Even so, this result still represented a 3.7-fold increase in conversion, which is attributable to the enzyme because the control reaction gave very little conversion.

In all cases, the remaining material was the unconverted alcohol, silanol and the disiloxane self-condensation product.

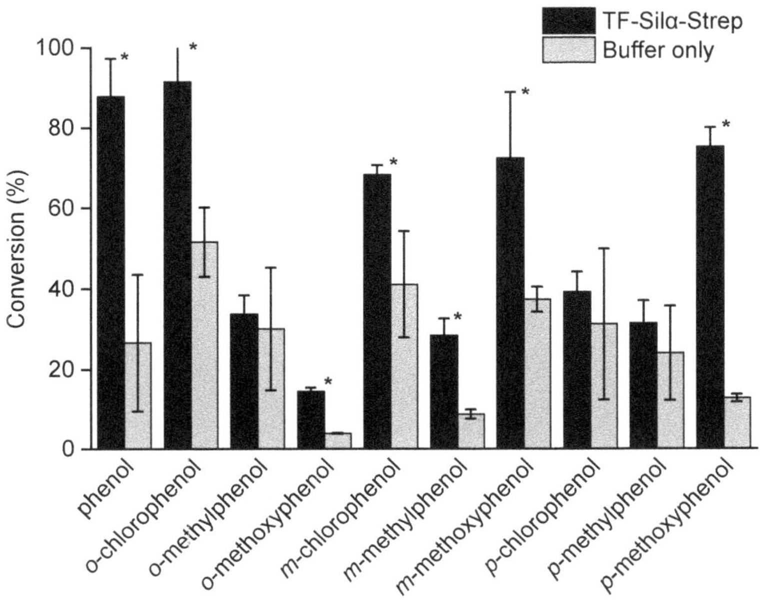

Figure 1. Graph of percentage conversions of phenols to the corresponding silyl ethers after 72 h. The error bars indicate standard deviations. A one-tailed Student's t-test assuming unequal variance was performed, comparing each enzyme to its control. Comparisons resulting in a $p < 0.05$ were deemed to be significant and are marked with *.

In comparing the *ortho*-substituted substrates, 3.7- and 1.8-fold enhancements in the conversion were observed for the methoxy- and chlorophenols, respectively. There was no significant difference in conversion for *o*-methylphenol compared to its control. Since the methoxy-bearing substrate is the most sterically demanding (i.e., volume occupied as quantified by their ligand repulsive energies [22]), its preference could be due to favourable electronic interactions with the enzyme such as hydrogen bonding or dipole–dipole interactions. Similar interactions could also be responsible for the enhancement observed with *o*-chlorophenol.

The phenols with the *meta*-substituents all exhibited enzymatic enhancements to their conversion that were statistically significant, but no trends could be identified. The greatest degree of enzymatic enhancement was found with *m*-methylphenol with a 3.2-fold improvement over the control, though the net percentage conversion was low at just 19%. The methoxy- and chloro-substituted phenols gave higher absolute conversions, but lower levels of enzymatic enhancement at 1.9- and 1.7-fold respectively, since the control reactions gave relatively high conversions even in the absence of the enzyme.

Of the three *para*-substituted phenols, only *p*-methoxyphenol showed a statistically significant conversion enhancement. Indeed, the enzyme gave the highest conversion improvement amongst all the tested phenols (Figure 1). Since this substrate presents the bulkiest group of the three *para*-substituted phenols, it is unlikely that this preference is due to steric factors and suggests that the molecule is forming specific interactions that are favourable to either binding or catalysis.

From the perspective of the type of substituent (Figure S1 and Table S1 in Supplementary Information), the chlorophenols generally gave higher conversions for both enzyme-catalysed and uncatalysed reactions, and thus the lowest fold improvements (<1.8 in all cases). Conversely, the methoxyphenols gave low conversions in the control reactions, but were the most improved by the addition of the enzyme with fold increases ranging from 1.9 to 5.8. The cresols (methylphenols) showed a mixed picture in terms of fold increase and exhibited generally low gross and net conversions.

Overall, no clear trends could be identified either in terms of the type or position of the substituent. An analysis of gross and net conversions as well as fold increase also showed no clear trends with respect to substrate pK_a or Hammett substituent constants [23]. Thus, the results must arise from a complex interplay of stereoelectronic interactions in the active site of the enzyme. However, as the crystallographic structure of the active enzyme is currently unknown, it is not possible to infer specific interactions.

2.2. Triethylsilylation of Aliphatic Alcohols

The condensation of a range of alcohols was then investigated under identical reaction conditions to the phenols. Initially, 1-octanol was tested but gave no conversion above the control reaction even after 192 h (Figure 2, Table S1 in Supplementary Information). This result contrasts with the results from the same protein bearing a hexahistidine affinity tag that gave a net conversion of 86% after just 72 h [20], but is consistent with the observations from a more recent report showing that the exchange of this tag for a streptavidin affinity tag resulted in diminished activity [21]. Similar results are reported with *E*-3-penten-2-ol, with essentially no activity above baseline.

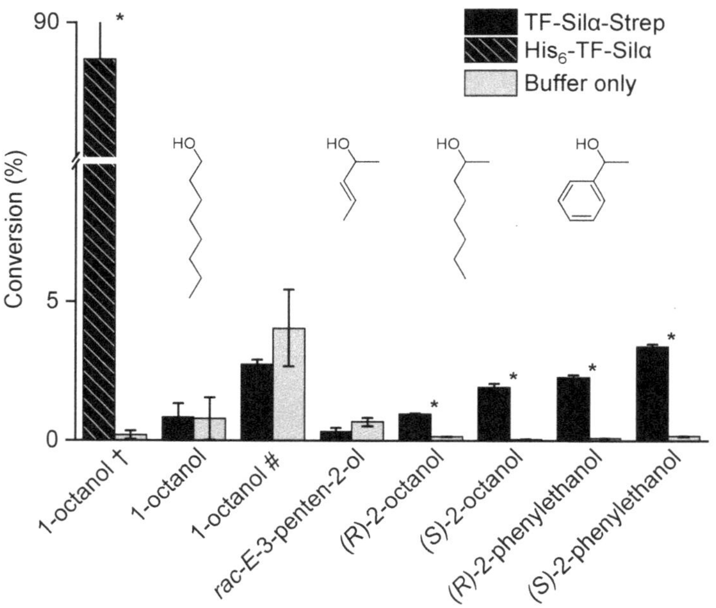

Figure 2. Graph of percentage conversions of alcohols to the corresponding silyl ethers after 72 h. The error bars indicate standard deviations. A one-tailed Student's *t*-test assuming unequal variance was performed, comparing each enzyme to its control. Comparisons resulting in a $p < 0.05$ were deemed to be significant and are marked with *. The † indicates data for the hexahistidine-tagged enzyme taken from [20]. The # indicates a reaction time of 192 h. The structures of the alcohols are also shown inset.

Two pairs of chiral alcohols, 2-octanol and 1-phenylethanol, were also investigated in this survey to assess any potential enantioselectivity displayed by the enzyme. Notably, despite the lack of significant conversions with 1-octanol or (racemic) 3-penten-2-ol, both enantiomers of 2-octanol did result in significant conversions above that of the control reactions. These conversions were extremely low at 0.8% for the *R*-enantiomers and 1.9% for the *S*-enantiomers (net), but were statistically significant. Furthermore, these values represented 6.3- and 38.8-fold increases in conversion compared to the controls. This result demonstrates that the enzyme preferentially catalysed the condensation of the *S*-enantiomer. For the phenylethanols, slightly higher gross conversions of 2.2% and 3.2% (net) were respectively observed for the *R*- and *S*-enantiomers. However, the 25.5- and 20-fold increases compared to the control indicated a lower level of enantioselectivity.

From these results, only very general trends regarding the reactivity could be drawn. The aliphatic alcohols were overall less reactive than the phenols in both enzymatic and control reactions, which may be related to their Brønsted acidity. No clear insights were gained when comparing the individual aliphatic alcohols. Indeed, the results superficially appear to be contradictory. Large hydrophobic alkyl chains such as those presented by the 2-octanols were accepted (albeit with low conversion), but 1-octanol was not. Likewise, both 1-phenylethanols were accepted, but *rac-E*-3-penten-2-ol was not. Substrates containing an aromatic ring appeared to have higher conversions (the phenols and phenylethanols), which suggests preferential binding with the enzyme binding site. This observation appears consistent with earlier computational modelling that indicated a possible cation–π interaction between the ring and a nearby arginine residue [20]. Conversely, the very low level of activity of Silα with the aliphatic substrates could simply be due to the lack of any attractive interactions with these otherwise largely unfunctionalized molecules.

2.3. Screening of Reaction Media

The current procedure for silyl ether condensation with Silα is based on previous work that utilises octane as the reaction medium [20]. However, this solvent greatly limits the range of substrates that can be applied, due to their low solubility. Thus, we tested a selection of solvents in an attempt to address this shortcoming. Several polar solvents were chosen (ethyl acetate, tetrahydrofuran, 1,4-dioxane and diisopropyl ether) and toluene was used as a model aromatic solvent. *m*-Methoxyphenol was chosen as the substrate, as the above results showed a moderate conversion, allowing for leeway to demonstrate both increases and decreases in conversion. As before, the reactions were carried out using lyophilised enzyme at 75 °C and analysed by GC-MS after 72 h. In line with earlier results, some silyl ether product was formed in all cases, even when the enzyme was omitted (Figure 3, Table S2 in Supplementary Information). Increased conversions above that of the control reactions were observed with the nonpolar solvents *n*-octane and toluene, with both giving similar net and gross conversions. In contrast, no net conversion was observed in any of the polar solvents.

To gain further insight, an analysis was performed to correlate net conversion with solvent polarity. For this purpose, the solvent polarity was quantified using empirical "normalised electronic transition energies" (E_T^N) as described by Reichardt [24,25]. This measure is based on the energy of the π–π* transition of a solvatochromic dye in a solvent and adjusted to a range between 0.0 (for tetramethylsilane) and 1.0 (water). Notably, this analysis showed that the silanol condensations were sharply demarcated, whereby solvents with $E_T^N > 0.1$ gave essentially no conversions over the control experiments (Figure 3, Table S2 in Supplementary Information).

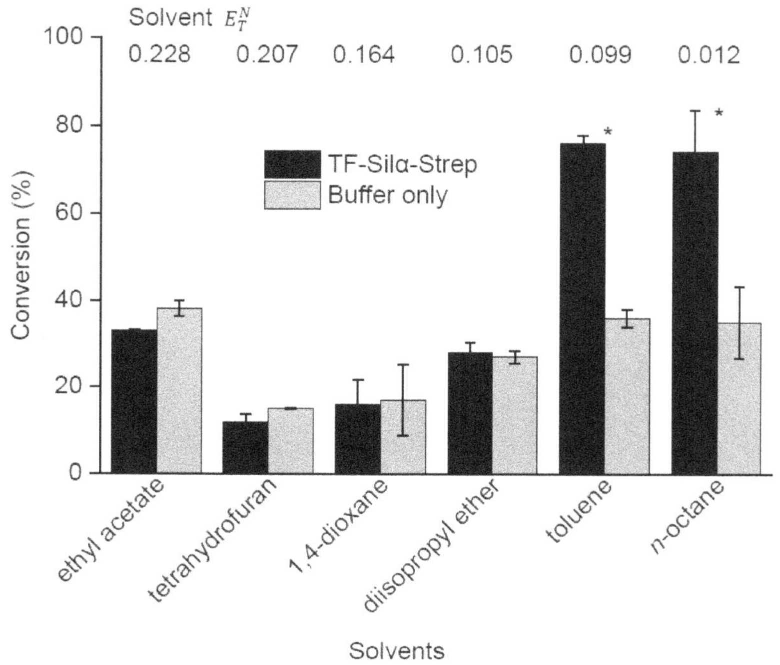

Figure 3. Graph showing percentage conversions of *m*-methoxyphenol and triethylsilanol condensations in a range of solvents after 72 h. A one-tailed Student's *t*-test assuming unequal variance was performed, comparing each enzyme to its control. Comparisons resulting in a $p < 0.05$ were deemed to be significant and are marked with *. The normalised electronic transition energies for the corresponding solvents are shown inset.

It has long been known that when using lyophilised enzymes in organic solvents, nonpolar solvents generally give a higher level of activity because the more polar (or hydrophilic) solvents remove the essential aqueous monolayer that surrounds the lyophilised enzyme and causes protein denaturation [26,27]. The results presented here conform to this postulation and therefore suggest that catalysis requires a correctly folded (not denatured) enzyme; that is, the results are not solely due to simple acid–base catalysis afforded by the presence of acidic or basic amino acid residues. Indeed, acid–base catalysis via a classical S_N2 mechanism would have been expected to produce higher conversions in polar aprotic solvents, which is not the case here.

As attempts to increase the polarity of the reaction media with a single solvent were unsatisfactory, mixtures containing increasing proportions of 1,4-dioxane in *n*-octane were assessed. 1,4-Dioxane was selected due to its lower polarity (and is therefore less likely to remove the aqueous monolayer) and boiling point (101 °C) that more closely matches that of *n*-octane. In general, the net conversion fell with increasing proportions of polar solvent (Figure 4, Table S3 in Supplementary Information). Reactions with 30% 1,4-dioxane gave no statistically significant difference between the enzyme and control conversions, presumably due to the removal of structural water crucial to enzyme function, as mentioned above.

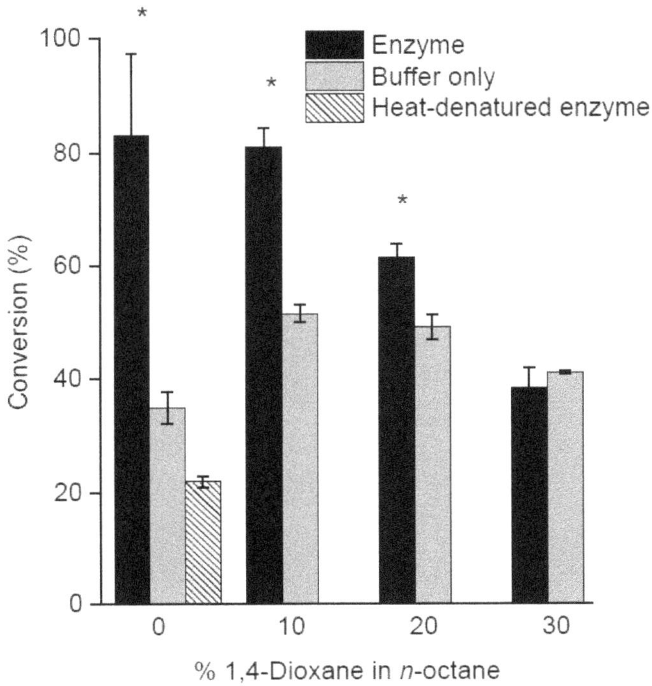

Figure 4. Graph showing percentage conversions of *m*-methoxyphenol and triethylsilanol in varying percentages of 1,4-dioxane in *n*-octane (solvent mixtures) after 72 h. A one-tailed Student's *t*-test assuming unequal variance was performed, comparing each enzyme to its control. Comparisons resulting in a $p < 0.05$ were deemed to be significant and are marked with *.

As a final control experiment, the enzymatic condensation of the *m*-methoxyphenol in neat *n*-octane was carried out using a sample of enzyme that was denatured by heating prior to lyophilisation, to confirm that the enzymatic condensation was indeed the result of specific catalysis and not simply the presence of the polypeptide chain. It was found that the heat-denatured enzyme gave a conversion that was no better than the negative control where no enzyme was added (Figure 4).

3. Materials and Methods

3.1. Materials and Equipment

All solvents and reagents were purchased from either Sigma-Aldrich (now Merck), VWR or Fisher Scientific. All solvents used were supplied as anhydrous, except *n*-octane and isopropyl ether, and used without further purification. Authentic samples of all product silyl ethers were prepared by conventional synthetic silylation of the corresponding alcohol or phenol with chlorotriethylsilane under basic conditions [8] (see Supplementary Information), and were used as standards for the GC-MS analysis. The enzyme was heterologously produced in *E. coli* as previously described [21].

The condensation reactions were carried out in crimp-sealable 8 mm vials (Chromacol C4008-741) that were heated and shaken on an Eppendorf Thermomixer 5350. GC-MS analyses were carried out using an Agilent 5975 Series MSD with the experimental parameters stated below (Table 1).

Table 1. Summary of GC-MS experimental parameters.

Parameter	Setting
Instrument	Agilent 5975 Series MSD
Carrier Gas	9.9995% ultra-high-purity helium
GC Inlet, Split	240 °C, split flow 100 mL min^{-1}, split ratio 50
MS Ionisation	Electron ionisation
GC Temperature Program	50 °C (2 min)→240 °C (8.5 min) at 30 °C min^{-1}, 10.7 min total run time
GC Column	VF-5ht
Volume Injected	1 µL

3.2. Preparation of Lyophilised Enzyme and Matrix

The purified enzyme was buffer exchanged into the lyophilising buffer (100 mM KH_2PO_4, 100 mM K_2HPO_4, 20 mM KCl, pH 7) via PD-10 gravity-fed desalting columns in 2.5 mL batches (with multiple columns used in series for large batches). The protein concentration was adjusted to 5 mg mL^{-1} and 18-crown-6 was added to a 0.04 mM concentration in the final solution. Aliquots of 100 µL of the enzyme solution were placed in the glass vials, flash frozen by plunging them into liquid nitrogen, and then lyophilised. For the negative control where the enzyme was omitted, 100 µL aliquots containing only lyophilising buffer with 18-crown-6 were flash frozen and lyophilised. For the control experiment using heat-denatured enzyme, the 100 µL aliquots in the glass vials were heated at 95 °C for 30 min and allowed to cool to ambient temperature before being flash frozen and lyophilised.

3.3. Enzymatic Condensation Reactions

A stock solution was first prepared by mixing the desired alcohol (1.26 mmol, 420 mM) and triethylsilanol (6.33 mmol) in the desired solvent (3 mL). 100 µL of this mixture was added into each vial containing the lyophilised enzyme (see above) and the vial crimp sealed. Reaction vessels were then heated at 75 °C while shaking at 650 rpm. At the desired time point, hexane (1 mL) was added, and the mixture centrifuged (17,000× g, 10 min) to separate the solid matter. 1 mL of the supernatant was transferred to a clean vial and subjected to GCMS analysis. Each reaction was performed in triplicate, and error bars presented in the figures refer to the standard deviation of these three independent data point measurements. For quantification of conversion rates, the GC-MS was first calibrated (Figures S2–S13 in Supplementary Information) using the synthetically prepared standards.

4. Conclusions

In summary, a survey into the reactivity of TF-Silα-Strep with a selection of aromatic and aliphatic alcohols has been conducted. In general, these enzymatic silyl condensations show a preference for aromatic alcohols (i.e., phenols) over aliphatic alcohols, as evidenced by much higher net conversions. However, greater improvements in conversions were achieved with the chiral aliphatic alcohols, as quantified by the fold increase, since the uncatalysed reactions with these alcohols showed proportionally much lower levels of product formation. In addition, Silα has a preference for the *S* enantiomers of the substrates tested, albeit with only low levels of conversion. A subsequent survey of solvents for this reaction showed that condensation could be effected only in nonpolar solvents ($E_T^N > 0.1$), though the addition of a small amount of polar solvent (up to 20% 1,4-dioxane) was tolerated.

As the main aim of this study was to investigate the substrate scope of Silα, these results are not currently synthetically useful except perhaps in a few cases (e.g., phenol, *p*-methoxyphenol). Nevertheless, this intrinsic activity offers a good starting point for directed evolution [28] into expanding the substrate scope of Silα. These results thus lay the foundation for future exploitation of Silα's chemo- and enantioselectivity toward a "silyl etherase" for practical biocatalysis. Additional work in this direction would benefit from screening a wider variety of aromatic hydroxyl groups, competition experiments with two alcohols and further elucidation of Silα's enantioselective capabilities through screening racemic alcohol mixtures.

Supplementary Materials: The following are available online at https://www.mdpi.com/article/10.3390/catal11080879/s1, Figure S1: Graph of percentage conversions of phenols to the corresponding silyl ethers after 72 h. Table S1: Percentage conversion, net enzymatic conversion and conversion enhancement for the condensation of aromatic alcohols and triethylsilanol after 72 h. Table S2: Percentage conversion, net enzymatic conversion and conversion enhancement for the condensation of *m*-methoxyphenol and triethylsilanol in various solvents after 72 h. Table S3: Percentage conversion, net enzymatic conversion and conversion enhancement for the condensation of *m*-methoxyphenol and triethylsilanol in various mixtures of 1,4-dioxane and *n*-octane. Figures S2–S13: Calibration graphs of area under the peak corresponding to silyl ethers in the GC-MS trace against concentration, compound characterisation data for the authentic silyl ether products.

Author Contributions: Conceptualisation, S.Y.T.D. and L.S.W.; methodology, E.I.S., B.L., R.S., S.A.C. and S.Y.T.D.; validation, E.I.S. and C.S.E.; formal analysis, E.I.S., B.L., P.Q. and L.S.W.; investigation, E.I.S., C.S.E., B.L. and S.A.C.; resources, S.A.C. and S.Y.T.D.; data curation, E.I.S., C.S.E. and L.S.W.; writing—original draft preparation, E.I.S.; writing—review and editing, C.S.E., P.G.T., P.Q. and L.S.W.; visualisation, E.I.S. and C.S.E.; supervision, L.S.W.; project administration, L.S.W.; funding acquisition, P.G.T., P.Q. and L.S.W. All authors have read and agreed to the published version of the manuscript.

Funding: This research was funded by the UK Engineering and Physical Sciences Research Council (research grant EP/S013539/1 and graduate studentship EP/M506436/1 to E.I.S.), the UK Biotechnology and Biological Sciences Research Council (research grant BB/L013649/1 and graduate studentship BB/J014478/1 to S.A.C.) and the Tertiary Education Trust Fund of Nigeria (graduate scholarship to C.S.E.). The GC-MS analysis was performed in the analytical facilities of the Manchester Synthetic Biology Research Centre for Fine and Specialty Chemicals (SYNBIOCHEM), funded by grant BB/M017702/1.

Data Availability Statement: The datasets generated during during the current study are available from the corresponding author on reasonable request.

Conflicts of Interest: The authors declare no conflict of interest. The funders had no role in the design of the study; in the collection, analysis, or interpretation of data; in the writing of the manuscript or in the decision to publish the results.

References

1. Smith, A.B.; Tomioka, T.; Risatti, C.A.; Sperry, J.B.; Sfouggatakis, C. Gram-scale synthesis of (+)-spongistatin 1: Development of an improved, scalable synthesis of the f-ring subunit, fragment union, and final elaboration. *Org. Lett.* **2008**, *10*, 4359–4362. [CrossRef]
2. Wang, B.; Hansen, T.M.; Wang, T.; Wu, D.; Weyer, L.; Ying, L.; Engler, M.M.; Sanville, M.; Leitheiser, C.; Christmann, M.; et al. Total synthesis of phorboxazole a via de novo oxazole formation: Strategy and component assembly. *J. Am. Chem. Soc.* **2011**, *133*, 1484–1505. [CrossRef]
3. Fürstner, A.; Bouchez, L.C.; Funel, J.-A.; Liepins, V.; Porée, F.-H.; Gilmour, R.; Beaufils, F.; Laurich, D.; Tamiya, M. Total syntheses of amphidinolide h and g. *Angew. Chem. Int. Ed.* **2007**, *46*, 9265–9270. [CrossRef]
4. Nicolaou, K.C.; Jiang, X.; Lindsay-Scott, P.J.; Corbu, A.; Yamashiro, S.; Bacconi, A.; Fowler, V.M. Total synthesis and biological evaluation of monorhizopodin and 16-epi-monorhizopodin. *Angew. Chem. Int. Ed.* **2011**, *50*, 1139–1144. [CrossRef]
5. Fortner, K.C.; Kato, D.; Tanaka, Y.; Shair, M.D. Enantioselective synthesis of (+)-cephalostatin 1. *J. Am. Chem. Soc.* **2010**, *132*, 275–280. [CrossRef] [PubMed]
6. Takamura, H.; Kikuchi, S.; Nakamura, Y.; Yamagami, Y.; Kishi, T.; Kadota, I.; Yamamoto, Y. Total synthesis of brevenal. *Org. Lett.* **2009**, *11*, 2531–2534. [CrossRef]
7. Zhang, Y.; Rohanna, J.; Zhou, J.; Iyer, K.; Rainier, J.D. Total synthesis of brevenal. *J. Am. Chem. Soc.* **2011**, *133*, 3208–3216. [CrossRef] [PubMed]
8. Crouch, R.D. Recent advances in silyl protection of alcohols. *Synth. Commun.* **2013**, *43*, 2265–2279. [CrossRef]
9. Stalling, D.L.; Gehrke, C.W.; Zumwalt, R.W. A new silylation reagent for amino acids bis (trimethylsilyl)trifluoroacetamide (bstfa). *Biochem. Biophys. Res. Commun.* **1968**, *31*, 616–622. [CrossRef]
10. Verboom, W.; Visser, G.W.; Reinhoudt, D.N. *N,N′*-bis[trimethylsilyl]-urea: A useful silylating agent for alcohols and carboxylic acids. *Synthesis* **1981**, *1981*, 807–809. [CrossRef]

11. Smith, E.D.; Sheppard, H. Quantitative gas chromatography of amino-acids as trimethylsilyl derivatives. *Nature* **1965**, *208*, 878–880. [CrossRef]
12. Kim, B.-h.; Woo, H.-G. Dehydrocoupling, redistributive coupling, and addition of main group 4 hydrides. *Adv. Organomet. Chem.* **2004**, *52*, 143–174. [CrossRef]
13. Lukevics, E.; Dzintara, M. The alcoholysis of hydrosilanes. *J. Organomet. Chem.* **1985**, *295*, 265–315. [CrossRef]
14. Blackwell, J.M.; Foster, K.L.; Beck, V.H.; Piers, W.E. B(c_6f_5)$_3$-catalyzed silation of alcohols: A mild, general method for synthesis of silyl ethers. *J. Org. Chem.* **1999**, *64*, 4887–4892. [CrossRef]
15. Clive, D.L.J.; Kellner, D. A new method for silylation of hydroxylic compounds: Reaction of phenols and alcohols with silanols mediated by diethyl azodicarboxylate and triphenylphosphine. *Tetrahedron Lett.* **1991**, *32*, 7159–7160. [CrossRef]
16. Nishiwaki, H.; Kiyomori, A.; Kubota, T. Silylation of Hydroxyl Groups. U.S. Patent US20030139619A1, 24 July 2003.
17. Ran, N.; Zhao, L.; Chen, Z.; Tao, J. Recent applications of biocatalysis in developing green chemistry for chemical synthesis at the industrial scale. *Green Chem.* **2008**, *10*, 361–372. [CrossRef]
18. Cha, J.N.; Shimizu, K.; Zhou, Y.; Christiansen, S.C.; Chmelka, B.F.; Stucky, G.D.; Morse, D.E. Silicatein filaments and subunits from a marine sponge direct the polymerization of silica and silicones in vitro. *Proc. Natl. Acad. Sci. USA* **1999**, *96*, 361–365. [CrossRef]
19. Shimizu, K.; Cha, J.; Stucky, G.D.; Morse, D.E. Silicatein α: Cathepsin l-like protein in sponge biosilica. *Proc. Natl. Acad. Sci. USA* **1998**, *95*, 6234–6238. [CrossRef]
20. Tabatabaei Dakhili, S.Y.; Caslin, S.A.; Faponle, A.S.; Quayle, P.; de Visser, S.P.; Wong, L.S. Recombinant silicateins as model biocatalysts in organosiloxane chemistry. *Proc. Natl. Acad. Sci. USA.* **2017**, *114*, E5285–E5291. [CrossRef] [PubMed]
21. Sparkes, E.I.; Kettles, R.A.; Egedeuzu, C.S.; Stephenson, N.L.; Caslin, S.A.; Tabatabaei Dakhili, S.Y.; Wong, L.S. Improved production and biophysical analysis of recombinant silicatein-α. *Biomolecules* **2020**, *10*, 1209. [CrossRef] [PubMed]
22. White, D.P.; Anthony, J.C.; Oyefeso, A.O. Computational measurement of steric effects: The size of organic substituents computed by ligand repulsive energies. *J. Org. Chem.* **1999**, *64*, 7707–7716. [CrossRef]
23. Hansch, C.; Leo, A.; Taft, R.W. A survey of hammett substituent constants and resonance and field parameters. *Chem. Rev.* **1991**, *91*, 165–195. [CrossRef]
24. Reichardt, C. Solvatochromic dyes as solvent polarity indicators. *Chem. Rev.* **1994**, *94*, 2319–2358. [CrossRef]
25. Reichardt, C.; Harbusch-Görnert, E. Über pyridinium-*n*-phenolat-betaine und ihre verwendung zur charakterisierung der polarität von lösungsmitteln, x. Erweiterung, korrektur und neudefinition der e_t-lösungsmittelpolaritätsskala mit hilfe eines lipophilen penta-*tert*-butyl-substituierten pyridinium-*n*-phenolat-betainfarbstoffes. *Liebigs Ann. Der Chem.* **1983**, *1983*, 721–743. [CrossRef]
26. Klibanov, A.M. Enzymatic catalysis in anhydrous organic solvents. *Trends Biochem. Sci.* **1989**, *14*, 141–144. [CrossRef]
27. Stepankova, V.; Bidmanova, S.; Koudelakova, T.; Prokop, Z.; Chaloupkova, R.; Damborsky, J. Strategies for stabilization of enzymes in organic solvents. *Acs Catal.* **2013**, *3*, 2823–2836. [CrossRef]
28. Arnold, F.H. Design by directed evolution. *Acc. Chem. Res.* **1998**, *31*, 125–131. [CrossRef]

catalysts

MDPI

Review

The Power of Biocatalysts for Highly Selective and Efficient Phosphorylation Reactions

Roland Wohlgemuth [1,2]

[1] Institute of Molecular and Industrial Biotechnology, Lodz University of Technology, 90-537 Lodz, Poland; roland.wohlgemuth.1@p.lodz.pl
[2] Swiss Coordination Committee Biotechnology (SKB), 8021 Zurich, Switzerland

Abstract: Reactions involving the transfer of phosphorus-containing groups are of key importance for maintaining life, from biological cells, tissues and organs to plants, animals, humans, ecosystems and the whole planet earth. The sustainable utilization of the nonrenewable element phosphorus is of key importance for a balanced phosphorus cycle. Significant advances have been achieved in highly selective and efficient biocatalytic phosphorylation reactions, fundamental and applied aspects of phosphorylation biocatalysts, novel phosphorylation biocatalysts, discovery methodologies and tools, analytical and synthetic applications, useful phosphoryl donors and systems for their regeneration, reaction engineering, product recovery and purification. Biocatalytic phosphorylation reactions with complete conversion therefore provide an excellent reaction platform for valuable analytical and synthetic applications.

Keywords: phosphorus chemistry; enzymatic phosphorylation; phosphorylation biocatalysts; biocatalysis; phosphoryl donor

Citation: Wohlgemuth, R. The Power of Biocatalysts for Highly Selective and Efficient Phosphorylation Reactions. *Catalysts* **2022**, *12*, 1436. https://doi.org/10.3390/catal12111436

Academic Editor: Jie Fan

Received: 16 August 2022
Accepted: 11 November 2022
Published: 15 November 2022

1. Introduction

From the discovery of the element phosphorus and its chemistry to the current status of knowledge, many light as well as dark aspects have been emerging, connected to value and waste, to health and disease and to life and death. The element phosphorus is of central importance for all living organisms as well as for various human activities, from mining to agriculture, industry, science, the environment and society. Therefore, adequate attention must be paid to the global phosphorus cycle, which involves biochemical as well as geochemical reactions and pathways [1]. The resource-efficient use and recycling of phosphorus is essential at different scales on our planet as the biochemical flow of phosphorus has been identified as a planetary boundary at high risk [2]. As the phosphorus cycle is perturbed by human activities such as the increased utilization of phosphorus mineral resources and increasing abundance of phosphate in aqueous environments, work on closing the phosphorus loop by phosphorus recovery and recycling is important [3]. As phosphoric acid is produced as a key industrial intermediate for phosphate fertilizers on a very large scale by a sulfuric acid treatment of phosphate rock raw materials in a wet process, phosphoric acid is also of much interest for replacing white phosphorus as a precursor in manufacturing phosphorus-containing nonfertilizer chemicals [4]. Thereby, the amount of phosphogypsum waste generated in the wet process is four times larger than the phosphorus fertilizer and also contains toxic metals and radionuclides, which limit the resource efficiency and sustainability of this phosphorus value chain [5]. It is however not only at the global scale but also at the regional and local scale where a sustainable utilization of the nonrenewable element phosphorus is of key importance for a balanced phosphorus cycle, for example for the life and death of biological organisms in nature, for the maintenance of a resilient agriculture and for minimizing environmental pollution.

The long-standing scientific interest in many fundamental aspects of reactions involving the transfer of phosphorus-containing groups is therefore of key importance in obtaining new knowledge about reactions in phosphorus chemistry and their mechanisms, thermodynamics and kinetics. The unique kinetic and thermodynamic characteristics of phosphorus-containing molecules and ionized phosphate groups are of much interest in prebiotic phosphorylation [6], in the highly important role of phosphorus for the biosphere, life and nature on planet earth [7–9], as well as for present and future resource-efficient and sustainable phosphorylation reactions [10]. Spontaneous sugar phosphorylation has been discovered to occur in microdroplets at ambient pressure and temperature, and the ΔG of the phosphorylation reaction in microdroplets has been demonstrated to be negative and much lower than for the reaction in a bulk solution, thus making the phosphorylation reaction more favorable in microdroplets than in bulk solutions [11].

The thermodynamic challenge of a positive Gibbs free energy change (ΔG) for phosphorylation in bulk solutions can be overcome by different approaches depending on the type of phosphorylation reaction. The two main chemical approaches in stoichiometric phosphorylations, for example the conversion of alcohols to phosphate monoesters, require three reaction steps, including an oxidation step and protecting group removal when using trivalent P(III) or two reaction steps including a hydrolysis step when pentavalent P(V) is involved. The direct esterification of alcohols under mild conditions has been pioneered by Cramer using an activated phosphoric acid [12,13]. In the condensation of phosphoric acid with alcohols, the reaction can be favorably shifted towards the product by removing the water formed in the esterification azeotropically from the reaction [14]. The mixed anhydride acetyl phosphate, which was prepared by the activation of phosphoric acid by acetic anhydride, used a high energy phosphoryl donor in the monophosphorylation of alcohols [15]. Good isolated yields have been obtained by the use of tetrabutylammonium dihydrogenphosphate as a phosphate donor and trichloroacetonitrile as an esterification agent [16]. The need for robust and selective one-step phosphorylation reactions, which are sustainable, protecting-group free and versatile, has made the development of highly selective and efficient catalytic phosphorylation reactions an important goal. In the case of the direct phosphorylation of alcohols catalyzed by tetrabutylammonium hydrogensulfate, a favorable Gibbs free energy change can be achieved by the use of the high-energy phosphoryl donor potassium phosphoenolpyruvate [17].

Due to the importance of highly selective and efficient catalytic phosphorylation reactions in the biosphere, phosphorylation biocatalysts, phosphoryl donors and biocatalytic systems are omnipresent in nature. Biocatalytic phosphorylations have therefore been of much interest for both in vitro applications such as biocatalytic syntheses as well for in vivo biotransformations, for example, biocatalytic prodrug activation to biologically active pharmaceuticals in the human body or biocatalytic antibiotic deactivation in drug-resistant microbes by the phosphorylation of antibiotics [10]. The discovery and characterization of phosphoryl-group transferring enzymes from nature (see Figure 1 for a schematic representation of key classes of phosphorylation biocatalysts and the involved reactions) continues therefore to be of much interest. In addition, computational and experimental enzyme engineering methodologies and tools have enabled the development of novel biocatalysts for specific reaction conditions, for catalyzing the phosphorylation of non-natural substrates, or for new-to-nature reactions.

The main goal of this review is to highlight the current state of the art and advances in highly selective biocatalytic phosphorylation reactions, fundamental and applied aspects of phosphorylation biocatalysts, novel phosphorylation biocatalysts, discovery methodologies and tools, analytical as well as synthetic applications, useful phosphoryl donors and systems for their regeneration, reaction engineering, product recovery and purification. The significant advances in all these areas towards improving the resource-efficiency of the overall bioprocess have made selective and efficient biocatalytic phosphorylation reactions an excellent reaction platform for valuable industrial applications.

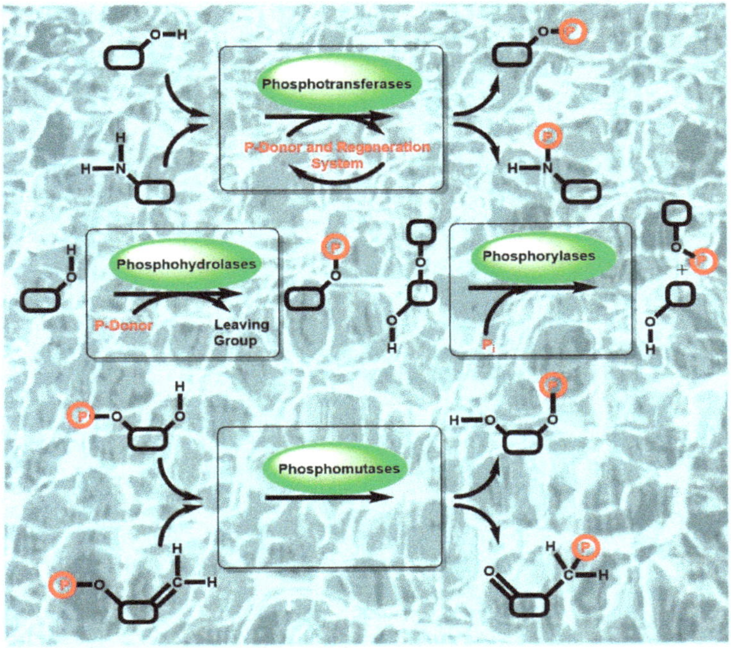

Figure 1. Schematic representation of key classes of phosphorylation biocatalysts and the involved reactions.

2. Structures, Functions and Mechanisms of Phosphorylation Biocatalysts

The rational knowledge organization for a growing number of phosphorylation biocatalysts benefits from the unique classification of an enzymes according to its functions of catalyzing reactions from substrates to products, which was introduced more than 60 years ago and which is continuously developed further by the Enzyme Commission [18]. The description of enzymes by standardized nomenclature, classification and the four-digit Enzyme Commission number (EC number), which is well established for identifying an enzyme according to the reaction or transport catalyzed, is also very valuable for connecting information on phosphorylation biocatalysts, such as structures, functions and mechanisms.

Molecular information on phosphorylation biocatalysts can benefit greatly from resources accumulating structural data at a fast pace from sequencing work, from the first gene sequence to the present wealth of relevant sequences, from hypothetical or unreviewed annotations to well-characterized and experimentally validated genes. More than 2 billion nucleotide sequences [19] in the genetic sequence database GenBank in June 2022 and sequence data in the Universal Protein Resource Knowledgebase UniProtKB provide comprehensive sequence data of proteins [20], including phosphorylation biocatalysts. The UniProtKB Release 2022_02 contains in UniProtKB/TrEMBL more than 231 million entries, which are however largely predicted, automatically annotated and unreviewed, while 567,483 protein sequences have been manually annotated and reviewed in UniProtKB/SwissProt [20]. An impressive growth of nearly 10% per year can also be observed in the number of experimentally determined three-dimensional protein structures, from the first reported structure to the current number in the Protein Data Bank (PDB) [21]. The breakthrough of the artificial intelligence (AI) system Alphafold developed by Deepmind to predict protein structures with high accuracy [22,23] and the release of over 200 million protein structure predictions recently by the partnership of Deepmind with EMBL-EBI in

AlphaFold Protein Structure Database provide great opportunities for accelerating research on phosphorylation biocatalysts.

In addition to the information resources on the sequences and structures of phosphorylation biocatalysts, easy access to the current status of knowledge about their functional properties, such as activities and kinetic parameters, substrate scope or stabilities, as well as other information about their biological context and occurrence in nature is of major importance. By focusing for 35 years on the continuous and ongoing extraction of data on classified enzymes from the literature, the Braunschweig Enzyme Database (BRENDA) has become the most comprehensive and globally utilized information resource on enzymes, which also enables fast access to existing knowledge about classified phosphorylation biocatalysts [24], which has been extracted retrospectively from publications. With the growing number of publications on enzymes and as the functional descriptions of biocatalysts in publications may vary, guidelines on standardizing how biocatalytic reactions should be reported [25] facilitate the exchange of enzyme function data. The Standards for Reporting Enzymology Data (STRENDA) database, which has been launched as an enzyme function database [26] incorporating the STRENDA guidelines, enables authors to check the completeness and validity of enzymology datasets before submission to a journal and supports the quality of the whole workflow from discovery to publication and information retrieval. The elucidation of relationships between the structure and function of phosphorylation biocatalysts is not only of fundamental interest but also highly valuable for the discovery and development of novel phosphorylation biocatalysts with desired function and performance.

2.1. *Structures of Phosphorylation Biocatalysts*

Biocatalysts which catalyze phosphoryl group transfer reactions are mainly classified in the class of transferases, which have an Enzyme Commission (EC) four-digit number starting with two, but there are also various phosphorylation biocatalysts from the hydrolases of EC class 3 and the isomerases from EC class 5. Therefore, phosphotransferases and phosphorylases of EC class 1, phosphohydrolases of EC class 3 and phosphomutases of EC class 5 form the scope of this review. Beyond the scope of this review are biocatalysts which, in addition to phosphorylation, catalyze other reactions, such as D-glyceraldehyde 3-phosphate dehydrogenase from the oxidoreductases of EC class 1 catalyzing the oxidative phosphorylation of D-glyceraldehyde 3-phosphate to D-1,3-diphosphosphate glycerate, or carbamoyl phosphate synthases, which utilize ammonia or hydrolyze L-glutamine, from the ligases of EC class 6 catalyzing carbon–nitrogen bond formation in addition to phosphorylation.

2.1.1. Phosphotransferase Structures

The largest number of phosphorylation biocatalyst structures has been reported for phosphotransferases in EC class 2.7, with 36,705 reviewed protein sequences listed under EC class 2.7 in UniProtKB/SwissProt and 5,559,483 unreviewed protein sequences listed under EC class 2.7 in UniProtKB/TrEMBL [20]. The growing number of phosphotransferase/kinases has been classified according to sequences and different criteria such as structural folds, catalyzed reaction types, evolutionary relationships or biological organisms into protein families, which have also been assembled into fold groups according to similar structural folds [27–30]. A total of 213,201 experimentally determined three-dimensional structures of phosphotransferases, among which 1401 structures were obtained with a resolution of 1.5 A or better, have been deposited in PDB [21]. The first three-dimensional kinase structures, which were obtained more than 4 decades ago by x-ray diffraction, involved the metabolic enzymes hexokinase B from *Saccharomyces cerevisiae* [31], pyruvate kinase from cat muscle [32] and phosphoglycerate kinase from *Saccharomyces cerevisiae* [33]. In the meantime, the number of kinase structures with small molecules as substrates has been continuously growing, for example, for carbohydrate kinases, lipid kinases, nucleoside kinases, nucleoside monophosphate and nucleoside diphosphate kinases, hydroxyacid kinases and amino acid kinases [21]. More than 17,000 kinase sequences were classified based on the similarity of their sequences into 30 families, whereby 98% of all the sequences are represented by 19 families and fall into seven general structural folds [27]. These fold groups with known three-dimensional structure include the Rossmann fold, ferredoxin fold, ribonuclease H fold and TIM β/α-barrel, which are some of the most widespread folds [27]. Numerous structures of human and viral nucleoside kinases, nucleoside monophosphate kinases and nucleoside diphosphate kinases have been determined and are highly relevant for the activation cascade of nucleoside analogs to their active triphosphate forms at high concentrations and at the desired site [34]. Structures determined in different environments may differ, as described for example for the small integral membrane protein diacylglycerol kinase DgkA, which catalyzes the ATP-dependent phosphorylation of diacylglycerol to phosphatidic acid, where the domain swapping observed by NMR in the solution structure [35] was not observed by X-ray in the crystal structure [36] or by magic angle spinning solid-state NMR in phospholipid bilayers [37]. The analysis of protein kinase structures continues to attract major interest since the first report, describing the structure of a catalytic subunit of a protein kinase depending on cyclic adenosine monophosphate (cAMP), more than three decades ago [38], because of their fundamental importance in the post-translation modification of proteins, signaling and drug discovery [39–41]. The discovery of ribonucleic acid kinase ArkI and the determination of its structure opens a window into a whole new range of kinases and is of great interest in the post-transcriptional modification of tRNA [42].

2.1.2. Phosphohydrolase Structures

Phosphohydrolases or phosphatases in EC class 3.1, which in nature catalyze the hydrolytic reaction direction, can also be used for catalyzing the reverse reaction of phosphorylation. A large number of phosphatase sequences are known and have been deposited in UniProtKB/SwissProt and UniProt/TrEMBL: 4399 reviewed and 669,414 unreviewed protein sequences of the phosphoric-monoester hydrolases of enzyme class EC 3.1.3, 1043 reviewed and 247,424 unreviewed protein sequences of the phosphoric-diester hydrolases of enzyme class EC 3.1.4 and 34 reviewed and 7396 unreviewed protein sequences of the diphosphoric-monoester hydrolases of enzyme class EC 3.1.7 [20]. Numerous three-dimensional phosphatase structures have been determined experimentally and deposited in PDB: 2639 3D structures of the phosphoric-monoester hydrolases of enzyme class EC 3.1.3, 936 3D structures of the phosphoric-diester hydrolases of enzyme class EC 3.1.4 and 16 3D structures of the diphosphoric-monoester hydrolases of enzyme class EC 3.1.7 [21].

A wealth of structural information is now available for phosphatases, and the structural classification of proteins (SCOP) database [43] now lists 13 folds, 23 superfamilies and 48 families for phosphatases, enabling new insights about the role of the protein scaffold in the transfer of phosphoryl groups [44]. The example of the crystal structures determined for the enzyme encoded by the gene *Rv2131c* from *Mycobacterium tuberculosis*, which was originally assumed to be a bifunctional enzyme inositol monophosphatase/fructose-1,6-bisphosphatase and then discovered to be a 3′-phosphoadenosine-5′-phosphatase and showing 31% identity of its amino acid sequence with the regulator protein CysQ of sulfate assimilation in *E. coli* [45], illustrates the value of paying attention to molecular details and the experimental verification of annotations [46]. The CysQ structures have been determined in a ligand-free form; in the substrate-bound form, which contains phosphoadenosine-phosphate and is lithium-inhibited; and the product-bound form, in which AMP, phosphate and 3 Mg^{2+} ions are bound [46].

As the phosphotransferase versus phosphohydrolase balance of most phosphatases is unfavorable for phosphorylation reactions, the identification of structural determinants for shifting this balance towards high phosphotransferase activity and low hydrolases activity is desirable. The structures of nonspecific acid phosphatases have attracted much interest in order to understand the factors enabling their use in phosphorylation reactions with inexpensive phosphoryl donors for a broad range of substrates. The structure of the nonspecific acid phosphatase from *Escherichia blattae* [47] was important for the development of a phosphatase with enhanced phosphoryl group transfer activity over its inherent phosphatase activity. Thereby, the discovery that in the homologous nonspecific acid phosphatase from *Morganella morganii* the productivity of the 5′-phosphorylation of inosine was increased when the value for the Michaelis–Menten constant K_m for inosine was reduced, was important for guiding the structure-based engineering of the phosphatase for the inosine 5′-phosphorylation [48]. The replacement of Gly74 by Aspartic acid and Ile153 by Threonine in the wild-type enzyme of *Escherichia blattae* practically led to no tertiary structure change in the G74D/I153T mutant, but it reduced the K_m value for inosine [49]. Further mutations around the binding site of inosine have resulted in the S72F/G74D/I153T mutant enzyme, which showed V_{max} to be 2.7-fold higher and K_m to be 5.4-fold lower than the wild-type enzyme [49,50]. The same phosphotransferase level as in *Morganella morganii* was reached for the acid phosphatase by increasing the number of amino acid substitutions to 11 by site-directed mutagenesis, elucidating amino acid positions relevant for phosphotransferase activity [50].

The crystal structure of the acid phosphatase from *Pseudomonas aeruginosa*, containing in the asymmetric unit 3 identical units, each consisting of 10 α-helices, showed His132 acting as the key acid−base catalyst for phosphorylation reaction [51]. The introduction of charged residues near the active site has demonstrated an increase in the phosphorylation/hydrolysis ratio, for example with the optimal Asp135 → Arg135 mutation showing a 2.9-fold increase in the 2-phosphorylation of L-ascorbic acid [51].

2.1.3. Phosphorylase Structures

Phosphorylases in EC class 2.4, which catalyze the reversible phosphorolytic breakdown of an *O*- or an *N*-glycosidic bond using inorganic phosphate to generate a specific glycosyl phosphate, are listed with 1191 reviewed protein sequences in UniProtKB/SwissProt and with 282,371 unreviewed protein sequences in UniProtKB/TrEMBL [20]. Prominent phosphorylase biocatalysts are the glycoside phosphorylases, which have already been known for more than eight decades [52], and the nucleoside phosphorylases, which are involved in the reversible phosphorolysis of purine and pyrimidine nucleosides [53]. A total of 1319 experimentally determined three-dimensional structures of phosphorylases, among which 57 structures have been obtained with a resolution of 1.5 A or better, have been deposited in PDB [21].

The glycoside phosphorylases are mainly classified in the glycoside hydrolase family (GH family) of the carbohydrate-active enzymes database (CAZy database) [54] and more than one X-ray structure for each family of disaccharide phosphorylases, which are popular for phosphorolysis reactions, have been determined [55].

Recently, the crystal structures of a new class of nucleoside phosphorylases were determined, which showed a conserved dimeric Cupin fold with a high hydrophobic dimer interface [56], while the classical two families of nucleoside phosphorylases had a different structure. The nucleoside phosphorylase-I (NP-I) family enzymes are trimeric or hexameric and share a common α/β-subunit fold, while the nucleoside phosphorylase-II (NP-II) family enzymes have dimeric structures.

2.1.4. Phosphomutase Structures

Among the intramolecular transferases of EC class 5.4, the phosphotransferases or phosphomutases of EC class 5.4.2 are of much interest: 1933 reviewed and 156,197 unreviewed protein sequences of the phosphomutases can be found in UniProtKB/SwissProt and UniProtKB/TrEMBL, while 244 experimentally determined 3D structures of the phosphomutases of EC class 5.4.2, among which 107 structures have been obtained with a resolution of 1.5 A or better, have been deposited in PDB [21].

Phosphopentomutase from *Bacillus cereus*, the crystal structure of which was the first structure published of a procaryotic phosphopentomutase, has been shown to fold into a core domain organized around an alkaline phosphatase fold and a cap domain, with the active site housed by an electropositive cleft at the interface between the core and cap domains [57]. Phosphomannomutase HAD5 from *Plasmodium falciparum*, which is essential for this malaria-causing parasite and also has phosphoglucomutase activity, has recently been shown to have a similar structure to the human enzyme with two domains, although significant sequence variations have been found in an active site loop [58].

2.2. Functions of Phosphorylation Biocatalysts

The importance of powerful biocatalysts for the highly selective covalent introduction of polar and charged phosphoryl-groups to a large variety of small and large substrate molecules is reflected by the great functional diversity of phosphorylation biocatalysts. An overview of the major functions of the phosphoryl-group transferring biocatalysts of EC classes 1, 3 and 5 is summarized in Figure 2.

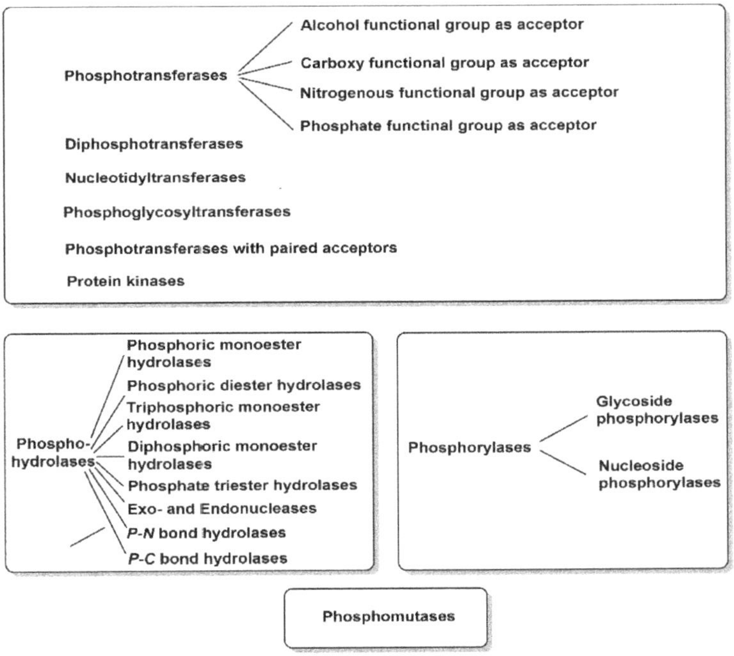

Figure 2. Overview of major functions of phosphoryl-group transferring biocatalysts.

2.2.1. Phosphotransferase Functions

The large functional diversity of phosphotransferases, illustrated also by the many different phosphorylation reactions and acceptor substrates ranging from metabolites and other small molecular weight compounds to large biomolecules such as proteins, enables the highly selective biocatalytic transfer of a defined number and the molecular nature of a phosphoryl-group-containing donor to one specific functional group of an acceptor molecule [59,60]. Thereby, the highly selective formation of new covalent bonds of phosphorus to oxygen, nitrogen, sulfur and carbon is very attractive, as a large number of phosphotransferases exist in nature and no protecting groups for donors and acceptors are needed. The exquisite selectivity of phosphotransferases in catalyzing the phosphorylation of a particular hydroxy group is illustrated in Figure 3 by the two ATP-dependent kinases AcbM and ValC. The kinase AcbM plays an important role in the biosynthetic pathway to the seven-carbon cyclitol unit of the antidiabetic drug acarbose in the *Actinoplanes* sp. and catalyzes the formation of 2-*epi*-5-*epi*-valiolone-7-phosphate [61]. For the phosphorylation of the closely related substrates valienone and validone to valienone-7-phosphate and validone-7-phosphate, respectively, the different kinase ValC is required, which has a critical function in the biosynthesis of the antifungal compound validamycin A [62]. Short chain fatty acid kinases are involved in central metabolic pathways catalyzing the reversible ATP-dependent phosphorylation of carboxylic acids leading to the formation of acyl-phosphates, such as acetate kinase (see Figure 3) catalyzing the phosphorylation of acetate to acetyl-phosphate [63,64]. Phosphagen kinases, such as creatine kinases (see Figure 3) catalyzing the ATP-dependent reversible *N*-phosphorylation of creatine, are key enzymes in catalyzing the formation of high energy phosphorus nitrogen bonds [65,66].

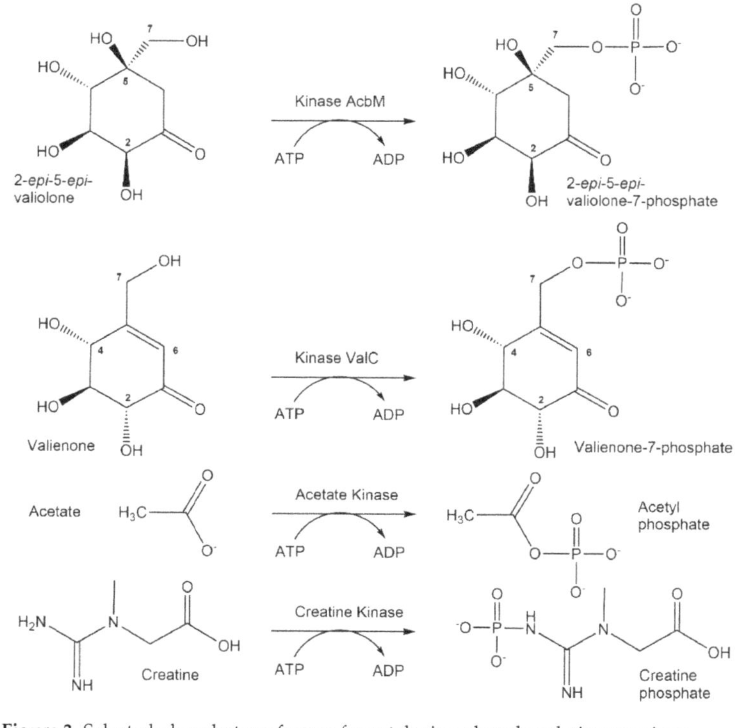

Figure 3. Selected phosphotransferases for catalyzing phosphorylation reactions.

The global analysis, identification and cataloguing of the functional diversity of all protein kinases in the genome of an organism, which was first introduced for the protein kinase complement of the human genome and was named kinome by Manning et al. [39], has been greatly expanded to animal, plant and microbial organisms, thus further extending the functional diversity. Although tremendous knowledge about protein kinase functions concerning the phosphorylations of serine, threonine and tyrosine residues in proteins exists, much less is known about the noncanonical phosphorylations of histidine, lysine, arginine, cysteine, aspartate and glutamate residues [67]. Evidence is emerging that protein kinase functions can also be exerted by metabolic enzymes, such as hexokinase, pyruvate kinase M2 and phosphoglycerate kinase 1 [68]. Despite the long history of metabolic kinases having small molecules as substrates, the functional diversity of kinases unrelated to protein kinases is understudied and is therefore of much interest [69,70].

2.2.2. Phosphohydrolase Functions

A broad functional diversity of EC class 3 enzymes is known to catalyze the hydrolysis of phosphorus–oxygen bonds in phosphoric-monoesters, phosphoric-diesters, diphosphoric-monoesters, triphosphoric acid monoesters, phosphoric acid triesters, RNA, DNA, phosphorus-containing anhydrides and the hydrolysis of phosphorus–nitrogen and phosphorus–carbon bonds.

The ability of a number of phosphatases to catalyze not only hydrolysis reactions but also phosphorylation reactions has been investigated for more than seven decades [71]. Although hydrolytic activities can counteract phosphorylation reactions by reducing the yield of phosphorylated products, the differential hydrolysis of enantiomers to chiral precursors of antiviral compounds by phosphotriesterase variants [72], increasing the transferase/hydrolase ratio of phosphatases by protein engineering and the combined use of suitable phosphatase, inexpensive phosphoryl donors and reaction engineering offer new opportunities for broadly applicable and scalable phosphorylation reactions. Acid phosphatase from *Morganella morganii* has been found to also catalyze the 5′-phosphorylation of nucleosides (see Figure 4) using not only the energy-rich carbamoyl phosphate and acetylphosphate, but also pyrophosphate as an inexpensive phosphoryl donor [73]. While the closely related acid phosphatase from *Providencia stuartii* exhibited phosphorylation activity similar to the *M. morganii* enzyme, the acid phosphatases from *Enterobacter aerogenes*, *Escherichia blattae* and Klebsiella *planticola* showed lower phosphorylation activities [74]. Nonspecific acid phosphatases from *Shigella flexneri* and *Salmonella enterica*, which have been shown to be selective for phosphorylate D-glucose to D-glucose-6-phosphate, without the formation of D-glucose-1-phosphate, using pyrophosphate [75], have also been demonstrated (see Figure 4), together with other acid phosphatases such as PiACP from *Prevotella intermedia*, to catalyze the selective phosphorylation of monoalcohols and diols [76].

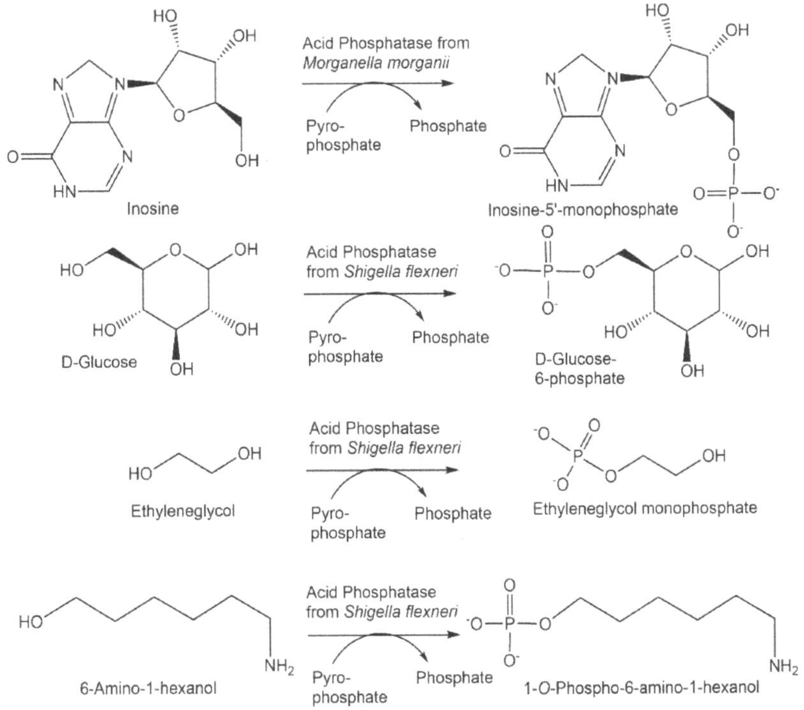

Figure 4. Selected phosphohydrolases for catalyzing phosphorylation reactions.

The transferase/hydrolase ratio is also of much fundamental interest in the regulation of the energy production and control of glycolysis by the bifunctional 6-phosphofructo-

2-kinase/fructose 2,6-bisphosphatase isoenzymes (PFKFBs), with the PFKFB3, which is overexpressed in many cancer types, having the highest kinase/phosphatase ratio [77].

2.2.3. Phosphorylase Functions

Retaining and inverting glycoside phosphorylases catalyze the reversible formation of glycosyl 1-phosphates with the anomeric configuration corresponding to both the enzyme type and the configuration of the substrate, for which a rather limited number of disaccharides, oligosaccharides and polysaccharides is known, although 544 glycoside phosphorylase entries from nine GH families and six GT families are listed in the CAZy database [54]. The reversible formation of the same α-D-ribose 1-phosphate and a purine base or a pyrimidine base from the corresponding purine or pyrimidine nucleoside is catalyzed by nucleoside phosphorylases of family NP-I or NP-II [53]. Improving the stability and broadening the substrate scope of nucleoside phosphorylases is of much interest for both the phosphorolysis as well as the synthesis direction [78–80].

Sucrose phosphorylase catalyzes the reversible phosphorolysis of sucrose using phosphate, whereby α-D-glucopyranose-1-phosphate is formed (see Figure 5) and D-fructose is obtained as a byproduct [81], while trehalose phosphorylase leads to the anomeric β-D-glucopyranose-1-phosphate using α,α-trehalose and phosphate as substrates [82]. A nearly quantitative phosphorolysis of 7-methyl- 2′-deoxyguanosine to α-D-2′-deoxyribofuranose-1-phosphate can be obtained when purine nucleoside phosphorylase from *E. coli* is used [83].

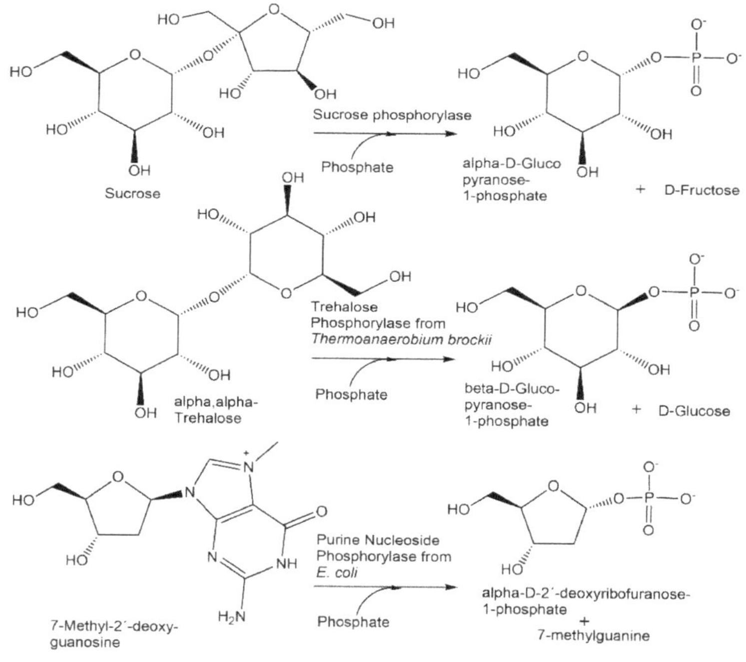

Figure 5. Selected phosphorylases for phosphorylation reactions.

2.2.4. Phosphomutase Functions

Phosphotransferases catalyzing the reversible intramolecular transfer of phosphoryl groups, also called phosphomutases, represent a metabolically important group of enzymes, not only for creating new phosphorus–oxygen bonds, but also for forming phosphorus–carbon bonds, while the phosphorus–oxygen bond of the starting substrate is cleaved. The reversible conversion of glyceric acids, pentoses and hexoses which are phosphorylated can thereby be achieved by the use of phosphoglycerate mutases, phosphopentomutases and phosphohexomutases with high selectivity. The reversible isomerization of 3-phospho-D-glycerate and 2-phospho-D-glycerate can be catalyzed (see Figure 6) by cofactor-independent as well as 2,3-bisphosphoglycerate-dependent phosphoglycerate mutases [84]. The phosphopentomutase function in catalyzing the reversible intramolecular phosphoryl group transfer is of much physiological and preparative interest, as shown for example in the isomerization of 2-deoxy-D-ribose-5-phosphate (see Figure 6) and 2-deoxy-alpha-D-ribose-1-phosphate [85]. Figure 6 also shows the enzymatic interconversion of D-mannose-6-phosphate and alpha-D-mannose-1-phosphate, which is catalyzed by phosphomannomutase and is essential in the activation of D-mannose and glycoconjugate biosynthesis in eukaryotes [86].

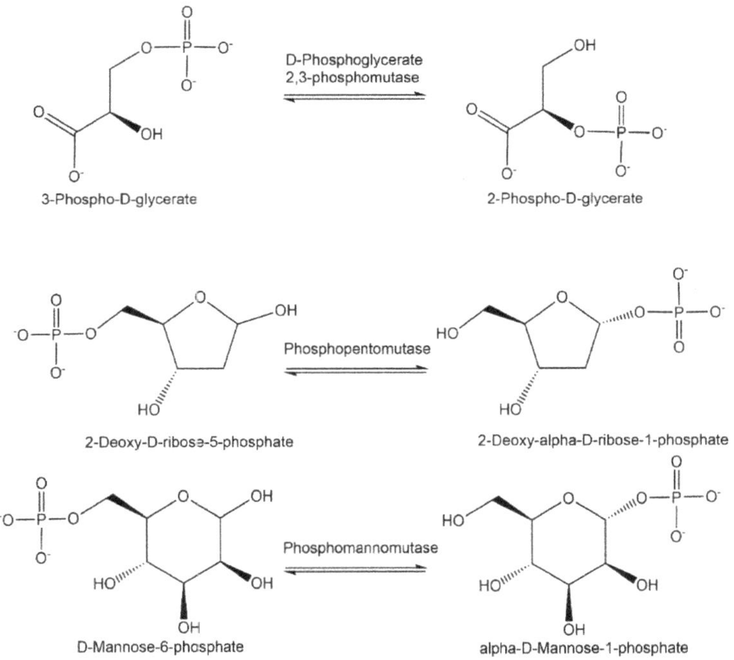

Figure 6. Selected phosphomutases for intramolecular phosphorylation reactions.

While D-glucose 6-phosphate interconversion catalyzed by α-D-phospho-glucomutases leads to α-D-glucose 1-phosphate [87], β-D-glucose 1-phosphate is obtained when its interconversion is catalyzed by β-D-phosphoglucomutases [88] in the thermodynamically unfavorable rearrangement of phosphoenolpyruvate to 3-phosphonopyruvate catalyzed by phosphoenolpyruvate mutase [89].

2.3. *Mechanisms of Phosphorylation Biocatalysts*

The mechanisms by which biocatalysts achieve the enormous rate accelerations by several orders of magnitude of the phosphoryl-group transfer reactions in comparison with the slow reaction rates without biocatalysts are of fundamental interest and continue to attract much attention [90–92]. Various distinct phosphoryl group transfer mechanisms have been elucidated in numerous enzyme families [93].

2.3.1. Phosphotransferases

The large 10^{12}–10^{14}-fold rate enhancements produced by hexokinase, homoserine kinase and N-acetylgalactosamine kinase has been demonstrated in a thermodynamic analysis to be due to two effects of these representative kinases compared with nonenzymatic phosphoryl group transfer: a more favorable entropy of activation and major reductions in the enthalpy of activation [94]. The catalytic mechanism by which adenylate kinase achieves this rate acceleration, which has been investigated in much detail by a combination of various techniques such as NMR and crystallography, involves the activation of phosphoryl transfer and lid opening, both by two orders of magnitude, by placing the charged cofactor Mg^{2+} in the active site organized before [95]. Based on structural and biochemical analyses of the ADP-phosphorylating class I polyphosphate kinase 2 from *Francisella tularensis* and the AMP- and ADP-phosphorylating class III polyphosphate kinase 2 from *Meiothermus ruber,* a mechanism of action has been proposed [96]. A key feature of the proposed mechanism is the in-line nucleophilic attack of the nucleotide on polyphosphate, which is activated by the active-site Lewis acidic Mg^{2+} upon the binding of polyphosphate [96].

2.3.2. Phosphohydrolases

The investigation of acid phosphatase α-D-glucose 1-phosphate phosphatase from *Escherichia coli* provided evidence for a 10^4-fold phosphatase efficiency advantage of a histidine compared to an aspartate nucleophile in position 18 and an additional 100-fold phosphatase efficiency advantage by the cooperative interaction of the catalytic nucleophile His18 with general acid catalytic groups, which is lost when His18 is replaced with Asp18 [97].

2.3.3. Phosphorylases

Structural and kinetic studies of hexameric purine nucleoside phosphorylase have provided detailed insights into the mechanism of purine nucleoside and phosphate binding, subunit conformations in open and closed forms, phosphate-induced conformational change, the sequence of nucleoside binding and subunit cooperation for effective catalysis [98].

2.3.4. Phosphomutases

Detailed NMR and X-ray investigations of the β-phosphoglucomutase-catalyzed isomerization of β-D-glucose 1-phosphate and D-glucose 6-phosphate via β-D-glucose 1,6-bisphosphate have revealed the importance of a conserved transition state organization by an invariant protein conformation, and priority for phosphate positioning over hexose in the substrate recognition [99]. The unusual mechanism of the phospho-enolpyruvate mutase-catalyzed intramolecular phosphoryl group transfer, which requires the cleavage of a low-energy oxygen–phosphorus bond and the formation of a high-energy carbon–phosphorus bond, is currently understood as a pericyclic rearrangement, which is retaining the configuration and placing the incoming nucleophile not strictly in line with the leaving group [100].

3. Discovery Methodologies and Tools for Phosphorylation Biocatalysts

The range of methodologies and tools for the discovery of novel biocatalysts and function–sequence relationships has been very much advanced from classical protein purification from a biological organism and its subsequent biochemical characterization of the purified protein, which also require the availability or synthesis of suitable enzyme substrates for both the activity-guided purification as well as for measuring enzyme kinetics. A number of methodologies, ranging from bioinformatics and genome mining to metagenomics, are available for finding phosphorylation biocatalysts with known functions that are based on sequence similarity. Important tools for the discovery of phosphorylation biocatalysts with novel functions are meaningful analytical assays of the enzyme-catalyzed substrate to product conversion. Robust and sensitive enzyme activity assays with a high information content, which enable the precise identification of the phosphorylation site in the substrate, continue to be a key prerequisite for the discovery of novel enzyme functions, or the experimental proof or disproof of automatically annotated protein sequences of phosphorylation biocatalysts. There are however many different approaches to utilizing enzyme activity assays for accelerating the assignment of the enzyme functions of the catalyzing substrate to product conversions to protein sequences and structures. Multiplexing selected enzyme activity assays in a single reaction vessel has been valuable for facilitating single protein kinase assays by using a kinase activity assay for kinome profiling in a single reaction, whereby 90 different peptide phosphorylation rates are obtained by mass spectrometry [101], requiring, however, stable isotope-labeled phosphorylated peptides as internal standards. The miniaturization and parallelization of enzyme activity assays have the advantages of saving costs, materials and time, but require higher sensitivity. This can be provided by highly informative and sensitive mass spectrometry methodologies, which have enabled the profiling of in vitro enzymatic activities of the 1,275 *Escherichia coli* proteins, which are functionally uncharacterized, to discover 241 enzymes which are potentially novel enzymes, of which 12 have been validated experimentally. This activity-based metabolite profiling (ABMP) of an overexpressed or purified protein is based on measuring by nontargeted mass spectrometry metabolites which are accumulated or depleted after the incubation of the protein of interest in a metabolome extract [102]. The identification in central enzymes of *Escherichia coli* of 34 new phosphorylation sites, which are functional in the regulation of enzyme activity, is of much interest for discovering novel protein kinases or for finding a relaxed specificity of the known serine/threonine or tyrosine kinases of *Escherichia coli* [103]. Chemical probes which are designed to specifically measure all proteins with the same characteristic activity in one class of phosphorylation biocatalysts can be used in activity-based protein profiling (ABPP), for example, for profiling a specific class of protein kinases, lipid kinases or other metabolic kinases [104].

A number of broadly applicable methodologies for the assignment of a novel or known enzyme function to a domain of unknown function (DUF) are provided by genomic enzymology [105], by the ligand-oriented screening for solute-binding transport proteins, by the analysis tools of sequence similarity networks (SSN) and genome neighborhood networks (GNN), enabling the identification of DUF1537 as a novel kinase family in acid sugar catabolic pathways and the discovery of four DUF1537 family members as novel C4-sugar kinases which depend on ATP [106].

The great methodological advances for the discovery and development of novel biocatalysts which have been provided by directed evolution, protein engineering and computational approaches to protein design have enabled the optimization of biocatalyst properties such as activities, selectivities or stabilities, and their adaptation to reaction conditions [107,108]. The power and potential of these tools and methodologies has attracted much interest as they provide the opportunity to develop numerous phosphorylation biocatalysts since the directed evolution of thymidine kinase for the phosphorylation of zidovudine [109], and their use in tackling challenging new-to-nature phosphorus chemistry is very interesting [110].

4. Novel Phosphorylation Biocatalysts

Based on the progress with biocatalysts for the formation of new *O-P* linkages, for example, in the synthesis of phosphorylated sugars or nucleotides, novel phosphorylation biocatalysts are highly desirable for efficient, highly selective and sustainable phosphorylation reactions and because biocatalytic *X-P* bond formation has been anticipated to lead to the faster development of nucleoside and nucleotide therapeutics in drug discovery and development [111].

4.1. Novel Phosphotransferases

Since the first demonstration 55 years ago that an enzyme preparation of an antibiotic-resistant *E. coli* strain could phosphorylate kanamycin and related antibiotics [112], the antibiotic resistome and assignments of enzyme functions to antibiotic resistance genes have attracted much interest [113,114]. Numerous novel phosphorylation biocatalysts have been discovered in microorganisms to inactivate antibiotics, such as aminoglycoside and macrolide antibiotics, for defense or self-protection by catalyzing their *O*-phosphorylation. A new antibiotic-resistance kinase AmiN from *Bacillus subtilis* has been discovered which inactivates amicoumacin by catalyzing the phosphorylation of the 2-hydroxy group of amicoumacin and was misannotated as a homoserine kinase [115]. From the recently discovered rifamycin inactivating phosphotransferase family, an unusual antibiotic resistance kinase, rifampicin phosphotransferase from *Listeria monocytogenes*, has been described to depend on ATP in phosphorylating the 21-hydroxy group of rifampicin, with the concomittant formation of AMP and inorganic phosphate [116]. The phosphotransferase Cpr17, which was discovered by cloning and characterizing the biosynthetic gene cluster of the capuramycin-type nucleoside antibiotic A-102395, was demonstrated to provide self-resistance and to prefer GTP as the phosphoryl donor [117].

Enzymes for the selective phosphorylation of hydroxycarboxylic acids are of much interest, and the recently discovered ATP-dependent pantoate kinase from *Thermococcus kodakaraensis*, which catalyzes the 4-phosphorylation of D-pantoate to D-4-phosphopantoate (see Figure 7), is not regulated by CoA and has been demonstrated to accept the nucleotides CTP, GTP and UTP [118].

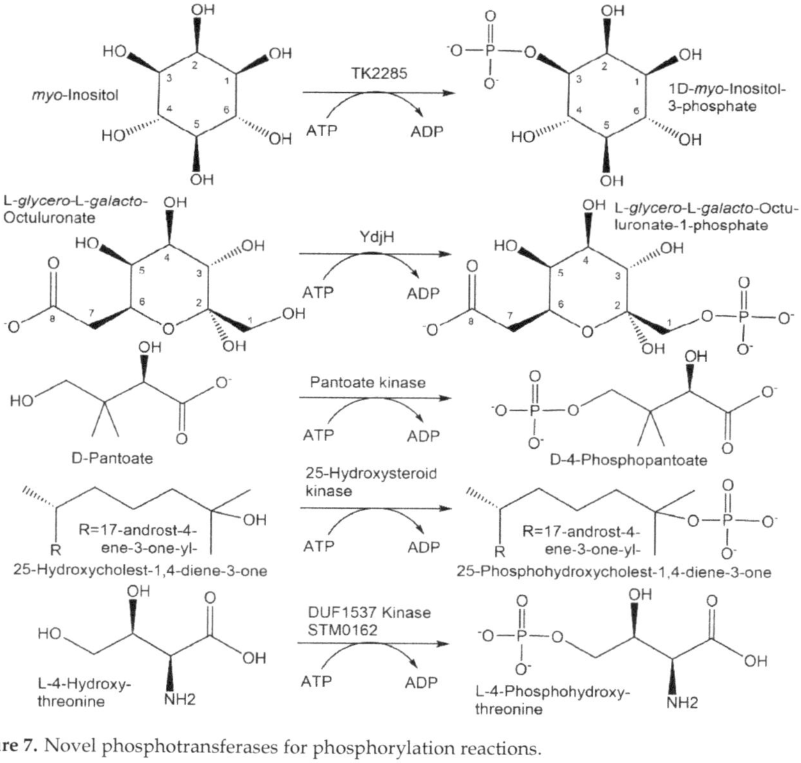

Figure 7. Novel phosphotransferases for phosphorylation reactions.

A novel flavonoid phosphotransferase from *Bacillus subtilis*, which has been discovered to catalyze the regioselective ATP-dependent phosphorylation of a broad range of flavonoids to the corresponding flavonoid monophosphates and the generation of AMP and phosphate, enables the efficient and sustainable conversion of poorly water-soluble natural flavones, isoflavones, flavonols, flavanones and flavonolignans into their monophosphosphorylated forms with improved water solubilities [119].

A highly active 25-hydroxysteroid kinase has been found to be involved in the microbial degradation of the side chain of cholesterol and sitosterol in *Sterolibacterium denitrificans* and catalyzed the selective and fully reversible phosphorylation of the tertiary hydroxy-group at the C25 of steroid alcohols such as the cholesterol metabolites 25-hydroxy-cholest-4-en-3-one and 25-hydroxy-cholest-1,4-diene-3-one (see Figure 7), the sitosterol metabolite 25-hydroxy-sitost-1,4-diene-3-one or 25-hydroxy vitamin D3 [120].

Numerous novel carbohydrate kinases have been discovered in microbial pathways of carbohydrate utilization, by expressing and characterizing DUFs, or by directed evolution. When directed evolution was used to broaden the substrate scope of galactokinase GalK from *Escherichia coli*, it was found that the Y371H variant, containing only a single amino acid exchange from tyrosine in the wild-type enzyme to histidine in the variant, already showed a wider substrate scope and accepted seven additional D-sugars, L-altrose and L-glucose as substrates [121]. The TK2285 protein from *Thermococcus kodakaraensis* has been discovered to function as an ATP-dependent *myo*-inositol kinase (see Figure 7) catalyzing the phosphorylation of the 3-hydroxy group of myo-inositol to 1D-*myo*-inositol 3-phosphate [122]. The same enzyme function could also be achieved by an engineered pyrophosphate-dependent myo-inositol 1-kinase by changing residues to recognize myo-inositol [123]. A highly active galactokinase from *Leminorella grimontii* has been found to catalyze the 1-phosphorylation of a number of galactose analogues [124]. From screening a series of wild-type enzymes, seven galacto- and six *N*-acetylhexosamine kinases, new enzyme phosphorylation activities towards fluorinated monosaccharides and four novel *N*-acetylhexosamine kinases have been discovered [125]. The novel sugar kinase YdhJ is the first kinase accepting C8-monosaccharides as substrates and has been discovered in the course of the functional characterization of the *ydj* gene cluster [126]. The 1-hydroxy group of a number of 2-keto-monosaccharides has been shown to be phosphorylated by YdhJ, and L-*glycero*-L-*galacto*-octuluronate was the best substrate (see Figure 7) being converted to L-*glycero*-L-*galacto*-octuluronate-1-phosphate [126]. The functional characterization of Cj1415, involved in the biosynthetic pathway to the *O*-methyl phosphoramidate modified capsular polysaccharide from *Campylobacter jejuni*, has led to its identification as cytidine diphosphoramidate kinase catalyzing 3′-phosphorylation [127].

The substrate scope of the ATP-dependent 5′-phosphorylation of all four 2-deoxynucleosides and analogues catalyzed by deoxynucleoside kinase from *Drosophila melanogaster* has been broadened to also include an unnatural nucleoside of an artificially expanded genetic information system by the Q81E variant [128].

The area of amino acid kinases has experienced a very interesting series of recent discoveries. Novel serine kinases catalyzing the *O*-phosphorylation of free serine have been discovered and were found to be dependent on ADP in *Thermococcus kodakaraensis* [129], while the serine kinase from *Staphylothermus marinus* was found to be ATP dependent [130]. An interesting novel amino acid kinase has been discovered as a DUF1537 family member STM0162, which catalyzes not only the phosphorylation of D-threonate, but also the phosphorylation of the toxic intermediate L-4-hydroxythreonine to the essential metabolite L-4-phosphohydroxythreonine (see Figure 7), providing thereby a recycling and detoxification path [131].

The protein Cj1418 has been discovered to be a unique ATP-dependent L-glutamine kinase catalyzing the *N*-phosphorylation of L-glutamine at its amide nitrogen, which is required in the already-mentioned biosynthetic pathway to the *O*-methyl phosphoramidate modification of the *Campylobacter jejuni* capsular polysaccharide [132,133].

4.2. Novel Phosphohydrolases

Novel phosphohydrolases which can be used in the reverse direction with inexpensive phosphoryl donors to catalyze phosphorylations have been found from microbial sources as well as by enzyme engineering. Nonspecific acid phosphatases from wild-type and genetically modified strains of *Enterobacter aerogenes* and *Raoultella planticola* have been found to also catalyze the phosphorylation of nucleosides and sugars (see Figure 8) using pyrophosphate as the phosphoryl donor [134].

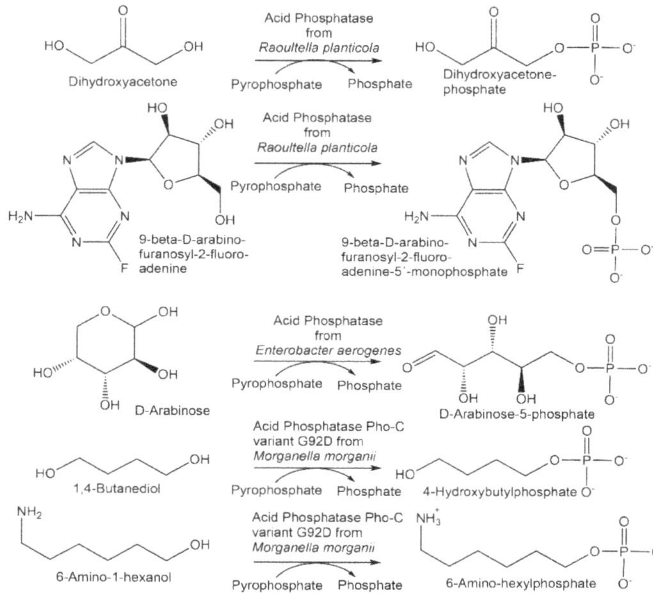

Figure 8. Novel phosphohydrolases for phosphorylation reactions.

The I171T-G92D mutant of the *M. morganii* acid phosphatase PhoC enabled the decrease in the K_m value for inosine in the phosphorylation reaction using pyrophosphate (see Figure 4) by two thirds of that of the wild-type enzyme, to a value well below the inosine solubility under the phosphorylation reaction conditions [135]. The double mutant I171T-G92D and the single mutants G92D, I171T, G92A and G92N of the *M. morganii* acid phosphatase PhoC and a nonspecific acid phosphatase variant from *E. blattae* have been investigated with respect to the phosphorylation of primary alcohols using pyrophosphate [135]. Beneficial mutations, for example G92D and I171T-G92D, have led to an increased affinity for alcohol substrates such as 1,4-butanediol, glycerol, ethylene glycol and 6-amino-1-hexanol (see Figure 8), as well as the decreased hydrolysis of phosphorylated products and the extension of the optimum pH towards a neutral pH [136]. The covalent immobilization of the nonspecific acid phosphatases PhoN from *Shigella flexneri* and *Salmonella enterica* ser. *typhimurium* LT2 enabled a better phosphorylation of alcohol substrates in an aqueous medium using pyrophosphate as the phosphoryl donor [137].

4.3. Novel Phosphorylases

The search for novel glycoside phosphorylases has attracted much interest for extending the rather limited functional diversity of known glycoside phosphorylases.

A high-throughput assay has been used to screen a metagenomic source library for glycoside phosphorylases, whereby seven new glycoside phosphorylases from the CAZy family GH94, among them two cellobiose phosphorylases, two cellodextrin phosphorylases and two laminaribiose phosphorylases, and a β-1,3-oligoglucan phosphorylase from the CAZy family GH149 [138]. Expressing and characterizing what was thought to be a sucrose phosphorylase from *Meiothermus silvanus*, residing in an unexplored branch of glycoside hydrolase family GH13, enabled the discovery of a glucosylglycerate phosphorylase (see Figure 9) [139]. By identifying and experimentally screening unknown clusters of the glycoside hydrolase family GH94 sequence space, using a combination of phylogenetic analysis and SSN, it was possible to discover a new 4-O-β-D-glucosyl-D-galactose phosphorylase (see Figure 9) from *Paenibacillus polymyxa* [140]. Efficient glucosaminide phosphorylases (see Figure 9), which have been engineered from GH84 *O*-GlcNacases by a single point mutation, have been shown to be 10-fold more active than their naturally occurring counterparts [141]. Widening the substrate spectrum is also of much interest for nucleoside phosphorylases, which has been achieved by engineering novel variants of purine nucleoside phosphorylases [142] and by the wider substrate spectrum of thermophilic nucleoside phosphorylases [143].

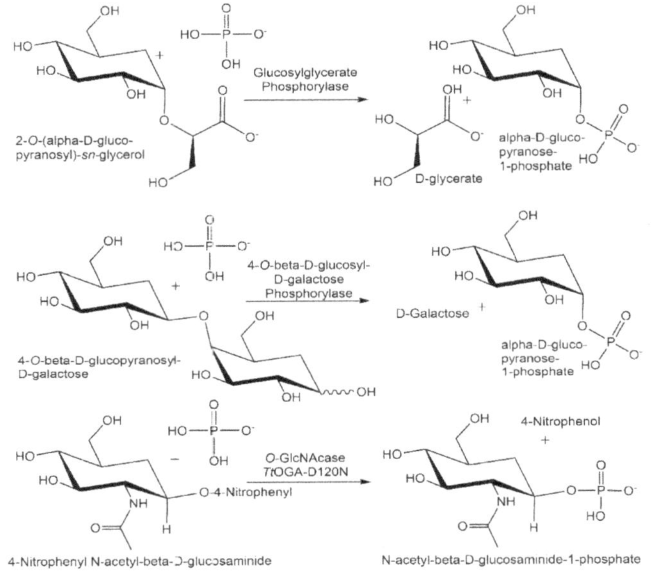

Figure 9. Novel phosphorylases for phosphorylation reactions.

4.4. Novel Phosphomutases

The discovery of novel enzymes catalyzing intramolecular phosphoryl-group transfer has been especially interesting in the area of the phosphopentomutases, for example, engineered phosphopentomutase variant enzymes (see Figure 10) catalyzing the isomerization of a deoxyribose-5-phosphate analog to the corresponding deoxyribose-1-phosphate analog [144]. The recombinant expression of the ST0242 protein from *Sulfolobus tokadaii* has led to the discovery of its phosphoglucosamine-mutase and phosphogalactosamine-mutase activity [145].

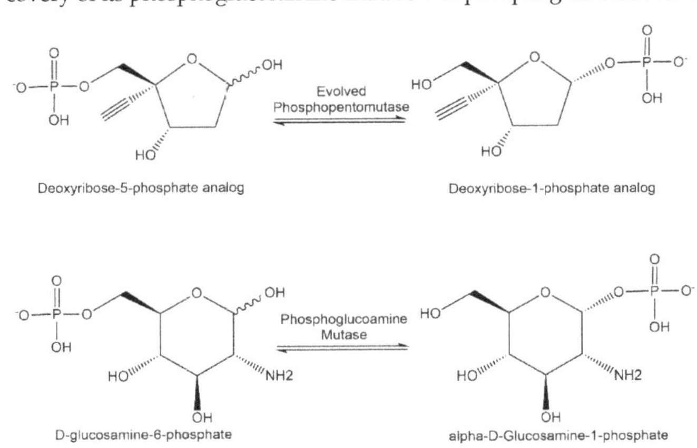

Figure 10. Novel phosphomutases for intramolecular phosphorylation reactions.

5. Analytical Applications of Phosphorylation Biocatalysts

The manufacturing of suitable and highly active phosphorylation biocatalysts which have been made available as fit-for-purpose enzymes and with negligible levels of undesired contaminating enzyme activities continue to be important prerequisites for analytical applications (see Figure 11), as already shown with the development of methods for enzymatic analysis several decades ago [146,147], for example, in the applications of glycerol kinase for the determination of glycerol or creatine kinase for the analysis of creatine. Significant improvements in the individual steps of the overall approach to enzymatic analysis, by efficient enzyme production using recombinant technologies, analytical method optimization, international standardization, miniaturization and automation have facilitated analytical applications and have led to its routine use in diagnostics. Building on the principles of enzymatic analysis established long ago and the power of phosphorylation biocatalysts, the highly selective determination of analytes in complex matrices can be achieved without time-consuming prior sample preparation.

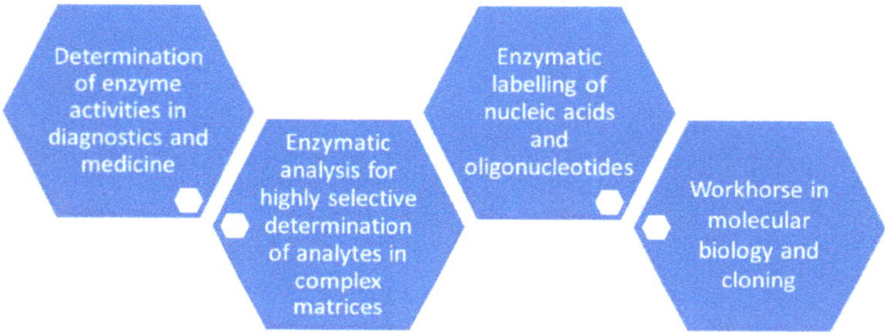

Figure 11. Major analytical application areas for phosphorylation biocatalysts.

Analytical applications connected with phosphorylation biocatalysts are also important with respect to the selective measurement of the corresponding enzyme activities, for example, in medicine as diagnostic tools for human health and disease and in the development of enzyme inhibitors by determining their characteristics for inhibiting a desired phosphorylation biocatalyst activity.

5.1. Analytical Applications of Phosphotransferases

The basic principle of the highly selective phosphorylation of glycerol to L-glycerol-3-phosphate catalyzed by ATP-dependent glycerol kinase continues to be a key reaction, which is then coupled to indicator reactions in modern quantitative assays of glycerol. The enzymatic determination of glycerol can be applied in the food, beverages and cosmetics process analysis and control in microbial fermentation and cell culture media, for example, in process analysis, or additionally following an enzymatic hydrolysis reaction for obtaining glycerol which is then also determined by applying glycerol kinase and an indicator reaction in triglycerides diagnostics in medicine. Phosphorylation biocatalysts are applied not only for the enzymatic analysis of substrates, but also for the determination of enzyme activities, and kinases are themselves target enzymes for biomedical analysis. An important application is the activity assay of creatine kinase and its isoenzymes for disease and emergency diagnosis, such as acute myocardial infarction and skeletal muscle diseases. The recommendation of the International Federation of Clinical Chemistry (IFCC) for the measurement of creatine kinase activity includes the utilization of hexokinase [148].

Phosphorylation biocatalysts have also found broad applications in the analysis of nucleic acids and oligonucleotides as well as for the analytical scale preparation of selectively phosphorylated nucleic acids. Recombinant T4 polynucleotide kinase has become a workhorse for molecular biology [149] and is used in molecular cloning for the 5′-phosphorylation of DNA or RNA. T4 polynucleotide kinase is also used for labeling nucleic acids and oligonucleotides at the 5′ end for generating analytical reagents for detection, size marker in electrophoresis or hybridization probes, which can be used in techniques for locating and binding the nucleic acids of a complementary sequence, such as Southern Blotting, Northern Blotting and in situ hybridization [150]. The acceptance of ATP-biotin as a substrate by T4 polynucleotide kinase is of much interest for the kinase-catalyzed biotinylation of single-stranded DNA [151].

5.2. Analytical Applications of Phosphohydrolases

Increased alkaline phosphatase activity in serum, if not caused by bone growth or pregnancy, occurs in bone and/or liver disease and therefore alkaline phosphatase is a target enzyme in clinical chemistry, with IFCC reference procedures available for measuring catalytic concentrations of alkaline phosphatase [152]. Alkaline phosphatase has also been widely used as a reporter enzyme in enzyme immunoassays due to its advantages, such as enzyme properties, easy preparation of enzyme–antibody conjugates and signal response for various detection types, making it a privileged label for enzyme immunoassays [153].

5.3. Analytical Applications of Phosphorylases

The substrate specificity of cellobiose phosphorylase has been attractive for the colorimetric quantification of cellobiose in the presence of other saccharides, whereby linear calibration curves have been obtained [154]. Maltose phosphorylase has been applied for the selective detection of phosphate without interference from other anions by using the immobilized enzyme in a biosensor with conductometric detection [155]. The activity of the glycogen phosphorylase isoenzyme BB is of much clinical interest in tissue hypoxia and ischemic myocardial damage [156].

5.4. Analytical Applications of Phosphomutases

As phosphoglycerate mutase 1 is the rate-limiting glycolytic enzyme in tumor cells, leukocytes and heart tissue, the analysis of its activity and its regulation is of much biomedical interest [157], for example, as an effector of the mammalian target of the rapamycin signaling pathway and as prognostic biomarker of non-small cell lung cancer [158].

6. Synthetic Applications of Phosphorylation Biocatalysts

Simple and broadly applicable process designs for synthetic phosphorylation reactions are attractive for reducing complexity when the required process targets with regard to selectivity and conversion can be achieved while meeting safety, health, environment and sustainability goals [159]. Synthetic applications of phosphorylation biocatalysts as well as the number and types of products have grown (see Figure 12), thus facilitating the further development of processes.

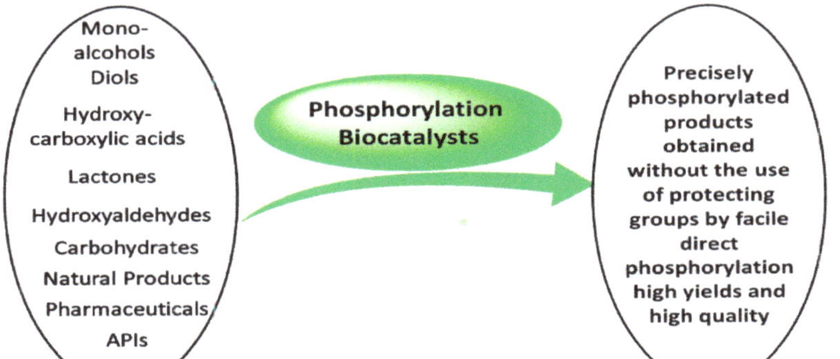

Figure 12. Compound classes of starting materials for preparative phosphorylations using phosphorylation biocatalysts.

As the applications, boundary conditions and approaches differ across the specific reactions catalyzed by the key classes of phosphorylation biocatalysts, the approaches towards the synthetic applications of phosphohydrolases, phosphotransferases, phosphorylases and phosphomutases are described individually within separate subsections. A separate subsection summarizes the most recent approaches in applying phosphorylation biocatalysts in cascade reactions.

6.1. Synthetic Applications of Phosphohydrolases

The application of simple and inexpensive phosphoryl donors such as pyrophosphate with phosphatases for catalyzing phosphorylation reactions for the synthesis of phosphorylated products (see Figure 13) requires, however, measures to counteract or prevent the favored phosphatase-catalyzed hydrolysis of the newly synthesized phosphorylated products.

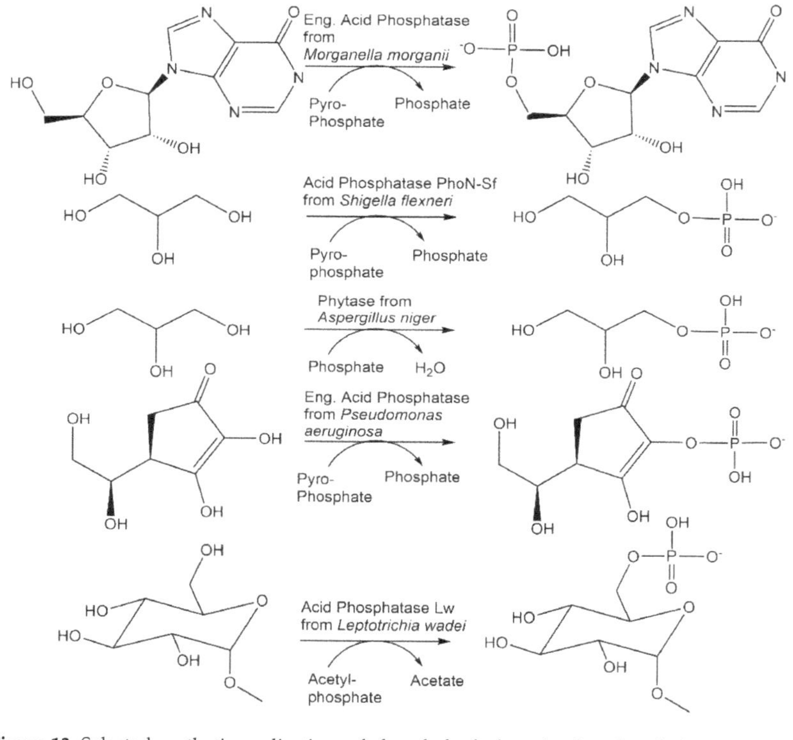

Figure 13. Selected synthetic applications of phosphohydrolases in phosphorylation reactions.

Reaction and enzyme engineering of the phosphatase-catalyzed 5′-phosphorylation of nucleosides using pyrophosphate as a phosphoryl donor has enabled the production of 5′-nucleotides at an industrial large scale, such as the production of inosine-5′-phosphate from inosine [160,161]. Bacterial nonspecific acid phosphatases have been exploited in the phosphorylation of a large range of monoalcohols and diols [162–167]. High product concentrations could be achieved in the phosphorylation of glycerol (see Figure 13), where 167 g per liter of racemic glycerol-1-phosphate was obtained when acid phosphatase PhoN-Sf from *Shigella flexneri* and pyrophosphate was used, and 104 g per liter of racemic glycerol-1-phosphate with the simple application of phytase from *Aspergillus niger* and monophosphate [167]. Even higher product concentrations of greater than 400 g per liter could be achieved in the 6-phosphorylation of maltotriose and methyl-α-D-glucopyranoside (see Figure 13) [166]. The phosphorylation efficiency of acid phosphatase from *Pseu-*

domonas aeruginosa towards 2-phosphorylation of L-ascorbic acid has been improved by protein engineering, and L-ascorbic acid-2-phosphate has been obtained with 48.6% conversion at a product concentration of 61.5 g per liter [168].

6.2. Synthetic Applications of Phosphotransferases

Another biocatalytic 5′-phosphorylation of the β-D-ribofuranoside substructure has been achieved in the synthesis of β-nicotinamide mononucleotide from the NAD^+ precursor nicotinamide riboside using nicotinamide riboside kinase from *Kluyveromyces marxianus* and the system acetyl phosphate/acetate kinase from *Bacillus stearothermophilus* for ATP regeneration (see Figure 14), whereby a product concentration of 93.5 g per liter and a productivity of 281 g per liter and per day could be achieved for NMN [169]. The synthesis of L-glyceraldehyde-3-phosphate has been performed by the enantioselective phosphorylation of glyceraldehyde catalyzed by glycerol kinase, whereby only the L-glyceraldehyde is taken as a substrate [170,171]. The dihydroxyacetone kinase-catalyzed phosphorylation has been shown to be enantioselective for the D-glyceraldehyde, which has enabled the synthesis of D-glyceraldehyde-3-phosphate [172].

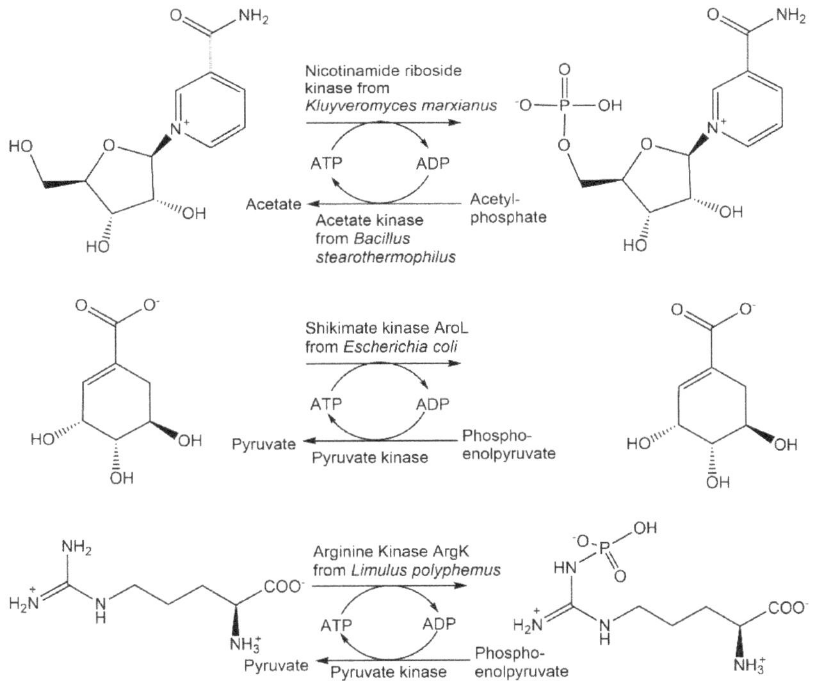

Figure 14. Synthetic applications of phosphotransferases in phosphorylation reactions.

Recombinant hydroxyacid kinases have been very valuable for selectively phosphorylating hydroxycarboxylic acids containing two or more hydroxy functional groups. Mevalonate-5-kinase from *Thermococcus kodakaraensis* has been applied in the synthesis of (*R*)-5-phosphomevalonate in >98% purity and a 65% isolated yield [173]. Glycerate-2-kinase from *Thermotoga maritima* was applied to manufacture D-glycerate-2-phosphate in excellent purity and with a 72% isolated yield [174]. The highly active shikimate kinase AroL from *Escherichia coli* illustrates the power of phosphorylation biocatalysts by its application in a highly efficient one-step synthesis of shikimate-3-phosphate (see Figure 14), which is superior to other approaches by avoiding lengthy synthetic routes, low yields, hazardous reagents or side product formation [175].

Carbohydrate kinases provide great advantages for the protecting group free and selective phosphorylation of a specific hydroxy functional group. Enantiocomplementary D- and L-xylulokinases have been applied for phosphorylating D- and L-xylulose with the complete conversion to the D- and L-enantiomers of xylulose-5-phosphate, respectively [176]. A facile one-step enzymatic synthesis of D-tagatose-1,6-phosphate using a D-tagatose-6-phosphate kinase-catalyzed 1-phosphorylation of D-tagatose-6-phosphate with ATP regeneration has been demonstrated to be highly efficient and scalable [177]. The selectivity of a series of carbohydrate kinases has been used for the synthesis of eight phosphorylated ketopentoses, where the high substrate selectivity towards the ketopentoses but not to the ketoses D-xylose and L-arabinose enabled the control of the coupling of the reversible isomerization/epimerization reactions with the subsequent phosphorylation i one pot [178,179].

The highly selective biocatalytic *O*-phosphorylation of psilocyn to psilocybin catalyzed by recombinant ATP-dependent *Psilocybe cubensis* 4-hydroxytryptamine kinase PsiK has been demonstrated at a gram scale without ATP regeneration, with an isolated yield of 88.5% [180]. Avoiding low-yield phosphorylation chemistry using protecting groups and their removal by hydrogenolysis depending on heavy metals, and optimization opportunities by bioprocess development, such as reducing ATP cofactor amounts by ATP regeneration, downstream processing and product recovery, illustrate the power of phosphorylation biocatalysts in natural product synthesis.

The application of the recombinant arginine kinase ArgK from *Limulus polyphemus* enabled a straightforward and highly efficient one-step synthesis of N_{ω}-phospho-L-arginine by the highly selective enzymatic *N*-phosphorylation of L-arginine at the ω-nitrogen (see Figure 14), thus avoiding lengthy routes with the introduction and removal of protecting groups [181].

6.3. Synthetic Applications of Phosphorylases

Sucrose phosphorylase from *Leuconostoc mesenteroides* has been covalently immobilized on Eupergit C for the continuous production in a packed bed reactor of α-D-glucose-1-phosphate from sucrose and phosphate (see Figure 5) at 0.6 M concentration each, whereby the reactor could be operated with a 91% conversion which remained stable up to a reaction time of 650 h [81]. When sucrose phosphorylase from *Bifidobacterium adolescentis* immobilized on Sepabeads for the continuous production at 60 °C of α-D-glucose-1-phosphate from sucrose and phosphate, a space–time yield of 179 g per liter and per hour was achieved [182]. A-D-glucose-1-phosphate has also been obtained in good yield from starch and phosphate at 0.7 M concentration each by using α-glucan phosphorylase from *Thermus caldophilus* [183]. Whole cells of *E. coli* expressing trehalose phosphorylase have been applied for the production of β-D-glucose-1-phosphate from trehalose and phosphate (see Figure 5) at 60 °C, with a 26% conversion [82]. The use of nucleoside phosphorylases for the synthesis of natural and modified α-D-pentofuranose-1-phosphates is of much interest for facilitated access to these key building blocks and metabolites. Highly efficient syntheses of α-D-ribose-1-phosphate and 2-deoxy-α-D-ribose-1-phosphate from 7-methylguanosine and 7-methly-2′-deoxyguanosine, respectively, and phosphate, have been achieved in a 74–96% yield after isolation and purification by enzymatic phosphorolysis (see Figure 5) using purine nucleoside phosphorylase [83].

Thermostable pyrimidine nucleoside phosphorylases have been applied in the facile and rapid enzymatic synthesis of a series of natural and modified α-D-pentofuranose-1-phosphates (see Figure 15) in gram quantities with a purity of greater than 95% from the corresponding nucleosides and phosphate [184]. After optimizing enzymatic phosphorolysis procedures and reaction conditions for the conversion of uridine and phosphate to α-D-ribofuranose-1-phosphate and uracil, and for the conversion of thymidine and phosphate to α-D-2′-deoxyribofuranose-1-phosphate and uracil, the protocols have been applied for the pyrimidine nucleoside phosphorylase-catalyzed synthesis of α-D-arabinofuranose-

1-phosphate, 2′-deoxy-2′-fluoro-α-D-ribofuranose-1-phosphate and 2′-deoxy-2′-fluoro-α-D-arabinofuranose-1-phosphate [184].

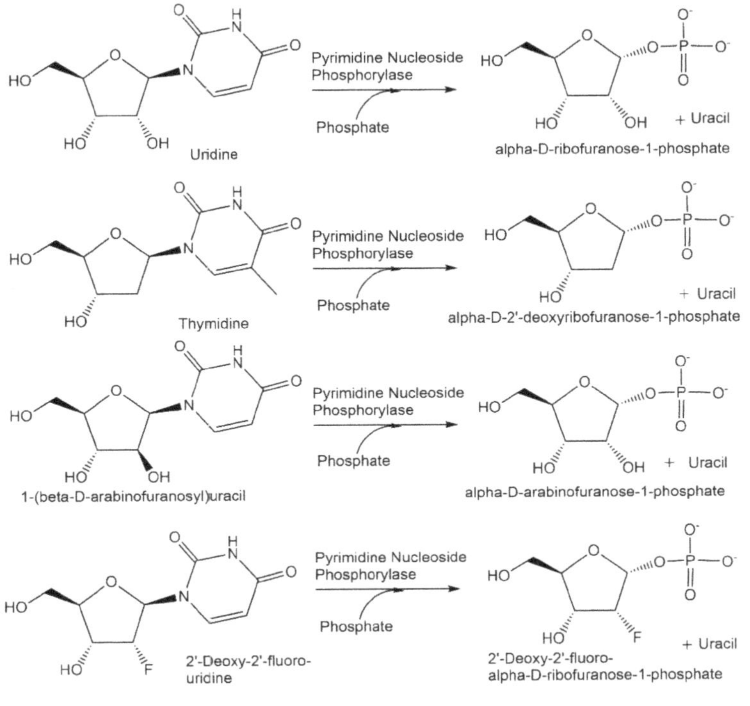

Figure 15. Synthetic applications of phosphorylases in phosphorylation reactions.

6.4. Synthetic Applications of Phosphomutases

The equilibrium thermodynamics of intramolecular isomerization reactions catalyzed by phosphomutases needs to be overcome in order to avoid challenging product recovery and purification. Coupling the phosphomutase-catalyzed reaction step with a subsequent reaction provides an elegant strategy, also used by nature, to shift the thermodynamic equilibrium towards the product side. An early application has been the use of crude preparations of phosphoacetylglucosamine mutase from *Neurospora crassa* for catalyzing the conversion of labeled *N*-acetylglucosamine-6-phosphate to labeled *N*-acetylglucosamine-1-phosphate coupled with the subsequent reaction catalyzed by uridinetriphosphate: *N*-acetylglucosamine 1-phosphate phosphotransferase for the synthesis of labeled uridinediphosphate-*N*-acetylglucosamine [185]. The recombinant expression of phosphoacetylglucosamine mutase Agm1 from *Saccharomyces cerevisiae* has been advantageous for the conversion of *N*-acetylglucosamine-6-phosphate to *N*-acetylglucosamine-1-phosphate (see Figure 16) and subsequently to uridinediphosphate-*N*-acetylglucosamine as part of a de novo pathway to legionaminic acid [186].

A phosphopentomutase from a *Bacillus sphaericus* strain, which was found by screening for an acetaldehyde- and phosphorylated-compound-tolerant enzyme, was applied in its recombinant version with the corresponding gene cloned and expressed in *E. coli*, for catalyzing the isomerization of 2-deoxyribose-5-phosphate to 2-deoxyribose-1-phosphate (see Figure 6), towards the enzymatic synthesis of 2′-deoxynucleoside [187]. Phosphomannomutase from *E. coli* has been applied to catalyze the conversion of 2-deoxy-D-glucose-6-phosphate to 2-deoxy-D-glucose-1-phosphate (see Figure 16) as part of the enzymatic synthesis of deoxythymidinediphosphate-2-deoxy-α-glucose [188].

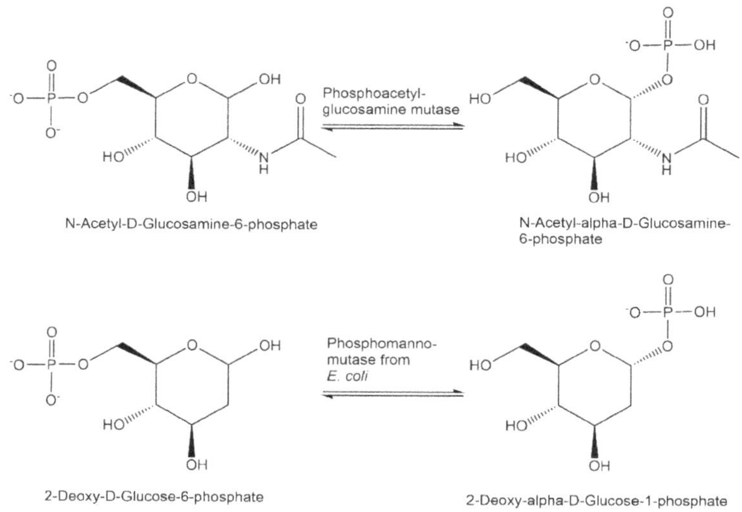

Figure 16. Synthetic applications of phosphomutases in phosphorylation reactions.

6.5. Phosphorylation Biocatalysts in Cascades

The five recombinant enzymes *N*–acetylhexosamine kinase NahK, glucose-1-phosphate uridyltransferase GalU, uridine monophosphate kinase URA6, polyphosphate kinase PPK3 and inorganic diphosphatase PmPpA have been applied in a cascade reaction in one pot for the production of uridine-5′-diphospho-*N*-acetylglucosamine (see Figure 17) from uridine-5′-monophosphate, N-acetylglucosamine and polyphosphate with a space–time yield of about 0.8 g per liter per hour [189]. The three enzymes uridine phosphorylase from *Clostridium perfringens*, purine nucleoside phosphorylase from *Aeromonas hydrophila* and deoxyadenosine kinase from *Dictyostelium discoideum* have been used for the synthesis of the antiviral arabinosyladenine-5′-monophosphate (see Figure 17) with 90% purity and 55% yield [190]. Kinases have also been successfully applied in synthetic reaction cascades for preparing key phosphorylated intermediates, for example dihydroxyacetone kinase in synthesizing phosphorylated D- and L-monosaccharides [191,192]. The most recent developments of various optimized phosphorylation biocatalysts in highly efficient and selective routes to antivirals are impressive examples of the fast move of biocatalytic phosphorylations into industrial processes of the pharmaceutical industry. Five phosphorylation biocatalysts, the evolved pantothenate kinase, phosphopentomutase, purine nucleoside phosphorylase and the two auxiliary enzymes acetate kinase from *Thermotoga maritima* and sucrose phosphorylase from *Alloscardovia omnicolens*, have been key for the success of the multistep enzymatic manufacturing process to the antiviral islatravir (see Figure 17) against HIV [193]. The advantages of biocatalytic cascade reactions for complexity reduction, faster delivery and higher yield are also important in the synthesis of stable isotope-labeled compounds, such as the enzymatic one-pot synthesis of ^{13}C- and ^{15}N-labeled ATP and GTP with up to 66% isolated yields using various kinases [194], due to the costs of the stable isotope-labeled starting materials. When radioactive isotopes are involved, the minimization of radioactive waste and the elimination of handling radioactive intermediates are further advantages, such as in the one-pot biocatalytic cascade reaction to the ^{14}C-labeled antiHIV nucleoside islatravir from 2-^{14}C-acetaldehyde, which gave a five-fold improvement in the overall radiochemical yield compared with the chemical route [195]. Evolving *S*-methyl-5-thioribose kinase activity for the 1-phosphorylation of the 5-isobutyryl-D-ribose with an α:β diastereomer ratio of >99:1, using propionyl-phosphate as phosphoryl donor, and engineering the uridine phosphorylase activity for the synthesis of 5′-isobutyryluridine from 5′-isobutyryl-D-ribose-1-phosphate and uracil have been key, together with other auxiliary enzymes, in a novel short reaction cascade to the antiviral molnupiravir, which decreased the 10 reaction steps of the initial chemical route from D-ribose and uracil to 3 steps, while increasing the overall yield from less than 10% to 69% [196].

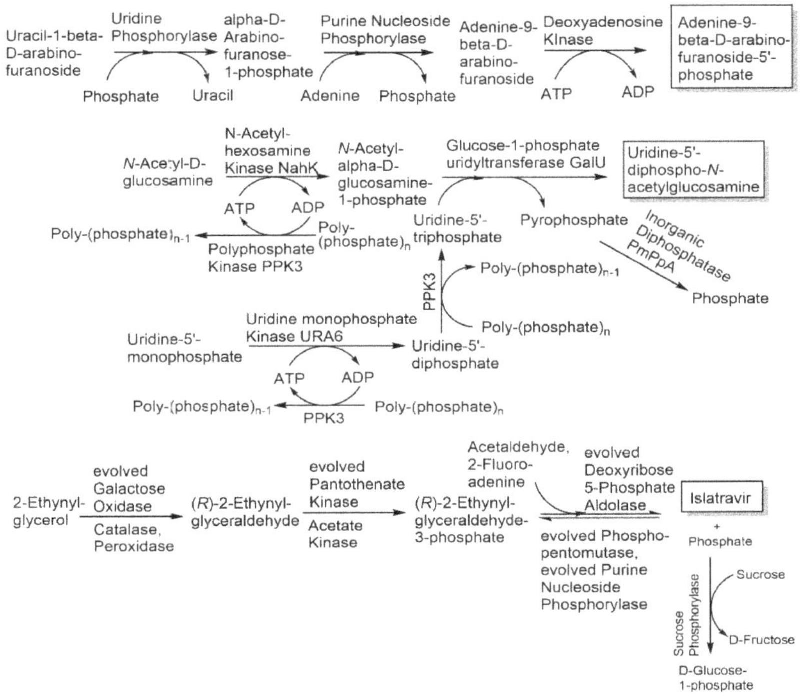

Figure 17. Synthetic reaction cascades involving phosphotransferases, phosphorylases and phosphomutases.

The chemically challenging diastereoselective synthesis of *P*-chiral nucleotide 1-thiotriphosphates has been achieved by a combination of the enzyme optimization of adenylate kinase, guanylate kinase and the phosphoryl donor regenerating acetate kinase by several rounds of directed evolution with reaction engineering [197].

6.6. Phosphoryl Donors and Systems for their Regeneration

Easily accessible, safe and inexpensive inorganic phosphates are attractive phosphoryl donors for a variety of biocatalytic phosphorylation reactions. Therefore, phosphorylation biocatalysts, which can accept inorganic phosphoryl donors, such as phosphorylases accepting phosphate, phosphatases accepting pyrophosphate or kinases accepting polyphosphates, are of much interest for phosphorylations at an industrial large scale. The full utilization of the phosphoryl donor is realized with simple phosphate as a donor in phosphorylase-catalyzed processes. The use of pyrophosphate for the phosphorylation of inosine using an engineered acid phosphatase (see Figure 4) has enabled an industrial large-scale production process for inosine-5-monophosphate with an 88% yield and a product concentration of greater than 100 g per liter [73,74]. As pyrophosphate can be easily prepared from phosphate, the phosphate byproduct formed in the phosphorylation reaction can also be recycled to pyrophosphate. Polyphosphate has not only received much attention as an easily accessible, inexpensive and stable phosphoryl donor for industrial phosphorylations but is also of fundamental interest [198,199], and early work on metaphosphate as a phosphoryl donor has led to the discovery of an NAD kinase [200] accepting metaphosphate, but not pyrophosphate or polyphosphate.

A great number of phosphorylation biocatalysts are however not dependent on inorganic phosphoryl donors but require high-energy organic compounds as phosphoryl donors [59], among which ATP is in wide use in natural and synthetic phosphorylation reactions. Therefore, systems for the synthesis and regeneration of ATP are of much interest, not only for phosphorylation biocatalysts with respect to possible enzyme inhibition, downstream processing aspects or costs of the phosphorylation process, but also in the wider context of ATP-dependent biocatalysts [201,202]. Regeneration systems utilizing sacrificial high energy phosphorylating agents together with the corresponding enzymes to catalyze the conversion of ADP to ATP, such as phosphoenolpyruvate and pyruvate kinase, or acetyl phosphate and acetate kinase, have been in wide use since the pioneering work of Whitesides and coworkers [203–205]. The further development of new ATP regeneration systems using easily accessible polyphosphates and polyphosphate kinases [206], high energy agents such as propionylphosphate [196] and phosphorus recycling enzymes in cascade reactions such as pyruvate oxidase [196,207], provides great opportunities for complexity reduction. A new electrochemical system using electricity instead of high energy phosphates or stoichiometric oxidants has been applied for ATP regeneration at the 20 g scale, whereby current and enzymatic processes need to be balanced [208].

6.7. Phosphorylation Reaction Engineering

The in-depth knowledge and characterization of the phosphorylation biocatalyst and the dependence of its properties, such as its activity, stability, activation and inhibition, on parameters such as temperature, pH, buffer or ionic strength for the reaction in focus is essential for achieving an optimized performance of a kinetically and thermodynamically feasible phosphorylation reaction. Kinetic as well as thermodynamic parameters and the determination of the Michaelis constant $K_{M,}$ k_{cat} or inhibitor constants, the equilibrium constant of the reaction and complexation constants, for example of ATP with a divalent cation such as Mg^{2+}, are of major importance not only for optimizing a single enzymatic reaction but also for a reaction cascade. The properties and stabilities of reaction components can be decisive for reaching optimal reaction engineering parameters, for example in the product half life of the glycerol kinase-catalyzed phosphorylation of L-glyceraldehyde to L-glyceraldehyde-3-phosphate [209] and the effect of the Mg/ATP-ratio on glycerol kinase [210].

6.8. Product Recovery and Purification

What is already being taken into account in the reaction design is how the phosphorylation can be driven to complete conversion and how a phosphorylated product is isolated from the reaction mixture is also essential for later operations. Facile product recovery and purification can be a significant factor for the economy and sustainability of an overall phosphorylation process. Therefore, making use of the most suitable and effective methodologies [211] and developing innovative new approaches for recovering and purifying the highly charged products from aqueous media is of major importance.

7. Opportunities and Outlook

The planetary boundary of the phosphorus biochemical flow at high risk and the essential and unique status of phosphorus bonds for many central features of life require resource-efficient use and the re-use of the nonrenewable phosphorus in order to close the phosphorus cycle. The characteristic and prevailing form of phosphorus occurs in nature in a stable oxidation state of five, which is what scholars have proposed chemists should use too [212], and in chemical bonds to oxygen in phosphates, with an impressive balance of stability and reactivity as pH-dependent ionized species.

The opportunities for discovering and utilizing the power of phosphorylation biocatalysts look exciting towards highly selective and efficient phosphorylation reactions, towards understanding fundamental and applied aspects and towards sustainable industrial chemistry [213]. Due to its benefits for selectivity, safety, health, the environment and sustainability, the use of phosphorylation biocatalysts enabling complete conversion to the phosphorylated product in one reaction step is a highly attractive and powerful synthetic strategy for replacing lengthy phosphorylation methodologies, which require protected phosphoryl group donors as well as the introduction of suitable protecting groups into the starting compound. The predominant use by nature of the phosphorus oxidation state +5 and of the phosphate group, which can be introduced from inorganic or organic phosphoryl donors and can occur in different ionization states carrying multiple negative charges, provides inspirations for making further valuable use of phosphorylation biocatalysts in various directions. Cost-efficient industrial large-scale processes can be envisaged by developing phosphorylation biocatalysts accepting inorganic phosphoryl donors. The preferential use of phosphate as a good leaving group in various biochemical reactions, such as carbon–carbon bond formation, decarboxylation, substitution and elimination reactions [7], is an opportunity to develop new reaction cascades involving phosphorylation biocatalysts and subsequent reactions, thereby replacing leaving groups traditionally used in synthetic chemistry, such as halides, tosylates or triflates. Finally, the biocatalytic introduction of the negatively charged phosphate group into poorly water-soluble compounds is of interest for improving the solubilities in aqueous media and for the retention of compounds within biological cells as negatively charged compounds after their poorly water-soluble precursors have passed the cell membranes and have undergone intracellular phosphorylation.

Therefore, the further exploration of the capabilities and power of natural and engineered phosphorylation biocatalysts from different angles and perspectives will be very beneficial, not only for highly selective and efficient phosphorylation reactions, but also for biocatalytic routes involving phosphate as a leaving group and for systems biocatalysis approaches.

Funding: This research received no external funding.

Conflicts of Interest: The author declares no conflict of interest.

References

1. Ruttenberg, K.C. The Global Phosphorus Cycle. *Treatise Geochem.* **2003**, *8*, 585–645. [CrossRef]
2. Steffen, W.; Richardson, K.; Rockström, J.; Cornell, S.E.; Fetzer, I.; Bennett, E.M.; Biggs, R.; Carpenter, S.R.; De Vries, W.; De Wit, C.A.; et al. Planetary boundaries: Guiding human development on a changing planet. *Science* **2015**, *347*, 1259855. [CrossRef] [PubMed]
3. Jupp, A.R.; Beijer, S.; Narain, G.C.; Schipper, W.; Slootweg, J.C. Phosphorus recovery and recycling–closing the loop. *Chem. Soc. Rev.* **2021**, *50*, 87–101. [CrossRef] [PubMed]
4. Geeson, M.B.; Cummins, C.C. Phosphoric acid as a precursor to chemicals traditionally synthesized from white phosphorus. *Science* **2018**, *359*, 1383–1385. [CrossRef]
5. Willey, N.; Timbs, P. Radioactivity in Future Phosphogypsum: New predictions based on estimates of 'Peak P' and rock phosphate resources. *J. Environ. Radioact.* **2022**, *244–245*, 106828. [CrossRef]
6. Pasek, M.A. Thermodynamics of Prebiotic Phosphorylation. *Chem. Rev.* **2020**, *120*, 4690–4706. [CrossRef]
7. Westheimer, F.H. Why Nature Chose Phosphates. *Science* **1987**, *235*, 1173–1178. [CrossRef]
8. Bowler, M.W.; Cliff, M.J.; Waltho, J.P.; Blackburn, G.M. Why did Nature select phosphate for its dominant roles in biology? *New J. Chem.* **2010**, *34*, 784–794. [CrossRef]
9. Kamerlin, S.C.; Sharma, P.K.; Prasad, R.B.; Warshel, A. Why Nature Really Chose Phosphate. *Q. Rev. Biophys.* **2013**, *246*, 1–132. [CrossRef]
10. Wohlgemuth, R. Key advances in biocatalytic phosphorylations in the last two decades: Biocatalytic syntheses in vitro and biotransformations in vivo (in humans). *Biotechnol. J.* **2021**, *16*, 2000090. [CrossRef]
11. Nam, I.; Lee, J.K.; Nam, H.G.; Zare, R.N. Abiotic production of sugar phosphates and uridine ribonucleoside in aqueous micro-droplets. *PNAS* **2017**, *114*, 12396–12400. [CrossRef] [PubMed]
12. Cramer, F. Neuere Methoden der präparativen organischen Chemie III 2. Darstellung von Estern, Amiden und Anhydriden der Phosphorsäure. *Angew. Chem.* **1960**, *72*, 236–249. [CrossRef]
13. Cramer, F.; Weimann, G. Imidoester, VII. Trichloracetonitril, ein Reagenz zur selektiven Veresterung von Phosphorsäuren. *Chem. Ber.* **1961**, *94*, 996–1007. [CrossRef]

14. Sakakura, A.; Katsukawa, M.; Ishihara, K. Selective synthesis of phosphate monoesters by dehydrative condensation of phosphoric acid and alcohols promoted by nucleophilic bases. *Org. Lett.* **2005**, *7*, 1999–2002. [CrossRef] [PubMed]
15. Dueymes, C.; Pirat, C.; Pascal, R. Facile synthesis of simple mono-alkyl phosphates from phosphoric acid and alcohols. *Tetrahedron Lett.* **2008**, *49*, 5300–5301. [CrossRef]
16. Lira, L.M.; Vasilev, D.; Pilli, R.A.; Wessjohann, L.A. One-pot synthesis of organophosphate monoesters from alcohols. *Tetrahedron Lett.* **2013**, *54*, 1690–1692. [CrossRef]
17. Domon, K.; Puripat, M.; Fujiyoshi, K.; Hatanaka, M.; Kawashima, S.A.; Yamatsugu, K.; Kanai, K. Catalytic Chemoselective *O*-Phosphorylation of Alcohols. *ACS Cent. Sci.* **2020**, *6*, 283–292. [CrossRef]
18. McDonald, A.G.; Tipton, K.F. Enzyme nomenclature and classification: The state of the art. *FEBS J.* **2022**. [CrossRef]
19. Sayers, E.W.; Cavanaugh, M.; Clark, K.; Pruitt, K.D.; Schoch, C.L.; Sherry, S.T.; Karsch-Mizrachi, I. GenBank. *Nucleic Acids Res.* **2021**, *49*, D92–D96. [CrossRef]
20. The UniProt Consortium. UniProt: The universal protein knowledgebase in 2021. *Nucleic Acids Res.* **2021**, *49*, D480–D489. [CrossRef]
21. Burley, S.K.; Bhikadiya, C.; Bi, C.; Bittrich, S.; Chen, L.; Crichlow, G.V.; Christie, C.H.; Dalenberg, K.; Di Costanzo, L.; Duarte, J.M.; et al. RCSB Protein Data Bank: Powerful new tools for exploring 3D structures of biological macromolecules for basic and applied research and education in fundamental biology, biomedicine, biotechnology, bioengineering and energy sciences. *Nucleic Acids Res.* **2021**, *49*, D437–D451. [CrossRef] [PubMed]
22. Jumper, J.; Evans, R.; Pritzel, A.; Green, T.; Figurnov, M.; Ronneberger, O.; Tunyasuvunakool, K.; Bates, R.; Žídek, A.; Potapenko, A.; et al. Highly accurate protein structure prediction with AlphaFold. *Nature* **2021**, *596*, 583–589. [CrossRef] [PubMed]
23. Varadi, M.; Anyango, S.; Deshpande, M.; Nair, S.; Natassia, C.; Yordanova, G.; Yuan, D.; Stroe, O.; Wood, G.; Laydon, A.; et al. AlphaFold Protein Structure Database: Massively expanding the structural coverage of protein-sequence space with high-accuracy models. *Nucleic Acids Res.* **2022**, *50*, D439–D444. [CrossRef]
24. Chang, A.; Jeske, L.; Ulbrich, S.; Hofmann, J.; Koblitz, J.; Schomburg, I.; Neumann-Schaal, M.; Jahn, D.; Schomburg, D. BRENDA, the ELIXIR core data resource in 2021: New developments and updates. *Nucleic Acids Res.* **2021**, *49*, D498–D508. [CrossRef] [PubMed]
25. Gardossi, L.; Poulsen, P.B.; Ballesteros, A.; Hult, K.; Švedas, V.K.; Vasić-Rački, Đ.; Carrea, G.; Magnusson, A.; Schmid, A.; Wohlgemuth, R.; et al. Guidelines for reporting of biocatalytic reactions. *Trends Biotechnol.* **2010**, *28*, 171–180. [CrossRef] [PubMed]
26. Swainston, N.; Baici, A.; Bakker, B.M.; Cornish-Bowden, A.; Fitzpatrick, P.F.; Halling, P.; Leyh, T.S.; O'Donovan, C.; Raushel, F.M.; Reschel, U.; et al. STRENDA DB: Enabling the validation and sharing of enzyme kinetics data. *FEBS J.* **2018**, *285*, 2193–2204. [CrossRef]
27. Cheek, S.; Zhang, H.; Grishin, N.V. Sequence and structure classification of kinases. *J. Mol. Biol.* **2002**, *320*, 855–881. [CrossRef]
28. Cheek, S.; Ginalski, K.; Zhang, H.; Grishin, N.V. A comprehensive update of the sequence and structure classification of kinases. *BMC Struct. Biol.* **2005**, *5*, 6. [CrossRef]
29. Blum, M.; Chang, H.; Chuguransky, S.; Grego, T.; Kandasaamy, S.; Mitchell, A.; Nuka, G.; Paysan-Lafosse, T.; Qureshi, M.; Raj, S.; et al. The InterPro protein families and domains database: 20 years on. *Nucleic Acids Res.* **2021**, *49*, D344–D354. [CrossRef]
30. Kannan, N.; Taylor, S.S.; Zhai, Y.; Venter, J.C.; Manning, G. Structural and Functional Diversity of the Microbial Kinome. *PloS Biology* **2007**, *5*, e17. [CrossRef]
31. Anderson, C.M.; Stenkamp, R.E.; Steitz, T.A. Sequencing a protein by X-ray crystallography: II. Refinement of yeast hexokinase B Co-ordinates and sequence at 2.1 Å resolution. *J. Mol. Biol.* **1978**, *123*, 15–33. [CrossRef]
32. Stuart, D.I.; Levine, M.; Muirhead, H.; Stammers, D.K. Crystal structure of cat muscle pyruvate kinase at a resolution of 2.6 Å. *J. Mol. Biol.* **1979**, *134*, 109–142. [CrossRef]
33. Watson, H.C.; Walker, N.P.; Shaw, P.J.; Bryant, T.N.; Wendell, P.L.; Fothergill, L.A.; Perkins, R.E.; Conroy, S.C.; Dobson, M.J.; Tuite, M.F. Sequence and structure of yeast phosphoglycerate kinase. *EMBO J.* **1982**, *1*, 1635–1640. [CrossRef] [PubMed]
34. Deville-Bonne, D.; El Amri, C.; Meyer, P.; Chen, Y.; Agrofoglio, L.A.; Janin, J. Human and viral nucleoside/nucleotide kinases involved in antiviral drug activation: Structural and catalytic properties. *Antivir. Res.* **2010**, *86*, 101–120. [CrossRef] [PubMed]
35. Van Horn, W.D.; Kim, H.-J.; Ellis, C.D.; Hadziselimovic, A.; Sulistijo, E.S.; Karra, M.D.; Tian, C.; Sönnichsen, F.D.; Sanders, C.R. Solution NMR Structure of Membrane-Integral Diacylglycerol Kinase. *Science* **2009**, *324*, 1726–1729. [CrossRef] [PubMed]
36. Li, D.; Lyons, J.A.; Pye, V.E.; Vogeley, L.; Aragao, D.; Kenyon, C.P.; Shah, S.T.A.; Doherty, C.; Aherne, M.; Caffrey, M. Crystal structure of the integral membrane diacylglycerol kinase. *Nature* **2013**, *497*, 521–524. [CrossRef]
37. Li, J.; Shen, Y.; Chen, Y.; Zhang, Z.; Ma, S.; Wan, Q.; Tong, Q.; Glaubitz, C.; Liu, M.; Yang, J. Structure of membrane diacylglycerol kinase in lipid bilayers. *Commun. Biol.* **2021**, *4*, 282. [CrossRef]
38. Knighton, D.R.; Zheng, J.; Ten Eyck, L.F.; Ashford, V.A.; Xuong, N.H.; Taylor, S.S. and Sowadski, J.M. Crystal structure of the catalytic subunit of cyclic adenosine monophosphate-dependent protein kinase. *Science* **1991**, *253*, 407–414. [CrossRef]
39. Manning, G.; Whyte, D.B.; Martinez, R.; Hunter, T.; Sudarsanam, S. The protein kinase complement of the human genome. *Science* **2002**, *298*, 1912–1934. [CrossRef]
40. Arter, C.; Trask, L.; Ward, S.; Yeoh, S. and Bayliss, R. Structural features of the protein kinase domain and targeted binding by small molecule inhibitors. *J. Biol. Chem.* **2022**, *298*, 102247. [CrossRef]
41. Modi, Y.; Dunbrack Jr, R.L. Kincore: A web resource for structural classification of protein kinases and their inhibitors. *Nucleic Acids Res.* **2022**, *50*, D654–D664. [CrossRef] [PubMed]

42. Ohira, T.; Minowa, K.; Sugiyama, K.; Yamashita, S.; Sakaguchi, Y.; Miyauchi, K.; Noguchi, R.; Kaneko, A.; Orita, I.; Fukui, T.; et al. Reversible RNA phosphorylation stabilizes tRNA for cellular thermotolerance. *Nature* **2022**, *605*, 372–379. [CrossRef] [PubMed]
43. SCOP 2 Database. Available online: http://scop.mrc-lmb.cam.ac.uk/ (accessed on 15 August 2022).
44. Allen, K.N.; Dunaway-Mariano, D. Catalytic scaffolds for phosphoryl group transfer. *Curr. Opin. Struct. Biol.* **2016**, *41*, 172–179. [CrossRef] [PubMed]
45. Hatzios, S.K.; Iavarone, A.T.; Bertozzi, C.R. Rv2131c from *Mycobacterium tuberculosis* Is a CysQ 3′-Phosphoadenosine-5′-phosphatase. *Biochemistry* **2008**, *47*, 5823–5831. [CrossRef]
46. Erickson, A.I.; Sarsam, R.D.; Fisher, A.J. Crystal Structures of *Mycobacterium tuberculosis* CysQ, with Substrate and Products Bound. *Biochemistry* **2015**, *54*, 6830–6841. [CrossRef]
47. Ishikawa, K.; Mihara, Y.; Gondoh, K.; Suzuki, E.; Asano, Y. X-ray structures of a novel acid phosphatase from *Escherichia blattae* and its complex with the transition-state analog molybdate. *EMBO J.* **2000**, *19*, 2412–2423. [CrossRef]
48. Asano, Y. Overview of screening for new microbial biocatalysts and their uses in organic synthesis – selection and optimization of biocatalysts. *J. Biotechnol.* **2002**, *94*, 65–72. [CrossRef]
49. Ishikawa, K.; Mihara, Y.; Shimba, N.; Ohtsu, N.; Kawasaki, H.; Suzuki, E.-i.; Asano, Y. Enhancement of nucleoside phosphorylation activity in an acid phosphatase. *Protein Eng.* **2002**, *15*, 539–543. [CrossRef]
50. Mihara, Y.; Ishikawa, K.; Suzuki, E.-i.; Asano, Y. Improving the Pyrophosphate-inosine Phosphotransferase Activity of *Escherichia blattae* Acid Phosphatase by Sequential Site-directed Mutagenesis. *Biosci. Biotechnol. Biochem.* **2004**, *68*, 1046–1050. [CrossRef]
51. Xu, X.; Yan, S.; Hou, X.; Song, W.; Wang, L.; Wu, T.; Qi, M.; Wu, J.; Rao, Y.; Wang, B.; et al. Local Electric Field Modulated Reactivity of *Pseudomonas aeruginosa* Acid Phosphatase for Enhancing Phosphorylation of L-Ascorbic Acid. *ACS Catal.* **2021**, *11*, 13397–13407. [CrossRef]
52. Puchart, V. Glycoside phosphorylases: Structure, catalytic properties and biotechnological potential. *Biotechnol. Adv.* **2015**, *33*, 261–276. [CrossRef] [PubMed]
53. Pugmire, M.J.; Ealick, S.E. Structural analyses reveal two distinct families of nucleoside phosphorylases. *Biochem. J.* **2002**, *361*, 1–25. [CrossRef] [PubMed]
54. Drula, E.; Garron, M.-L.; Dogan, S.; Lombard, V.; Henrissat, B.; Terrapon, N. The carbohydrate-active enzyme database: Func-tions and literature. *Nucleic Acids Res.* **2022**, *50*, D571–D577. [CrossRef]
55. Sun, S.; You, C. Disaccharide phosphorylases: Structure, catalytic mechanisms and directed evolution. *Synth. Syst. Biotechnol.* **2021**, *6*, 23–31. [CrossRef] [PubMed]
56. Wen, Y.; Li, X.; Guo, W.; Wu, B. Crystal structures of a new class of pyrimidine/purine nucleoside phosphorylase revealed a Cupin fold. *Proteins* **2022**, *90*, 1233–1241. [CrossRef]
57. Panosian, T.D.; Nannemann, D.P.; Watkins, G.R.; Phelan, V.V.; McDonald, W.H.; Wadzinski, B.E.; Bachmann, B.O.; Iverson, T.M. *Bacillus cereus* phosphopentomutase is an alkaline phosphatase family member that exhibits an altered entry point into the catalytic cycle. *J. Biol. Chem.* **2011**, *286*, 8043–8054. [CrossRef]
58. Frasse, P.M.; Miller, J.J.; Polino, A.J.; Soleimani, E.; Zhu, J.S.; Jakeman, D.L.; Jez, J.M.; Goldberg, D.E.; John, A.R.O. 2022. Enzymatic and structural characterization of HAD5, an essential phosphomannomutase of malaria-causing parasites. *J. Biol. Chem.* **2022**, *298*, 101550. [CrossRef]
59. Gauss, D.; Schönenberger, B.; Molla, G.S.; Kinfu, B.M.; Chow, J.; Liese, A.; Streit, W.R.; Wohlgemuth, R. Biocatalytic phosphorylation of metabolites. In *Applied Biocatalysis–From Fundamental Science to Industrial Applications*; Hilterhaus, L., Liese, A., Kettling, U., Antranikian, G., Eds.; Wiley-VCH: Weinheim, Germany, 2016; pp. 147–177.
60. Wohlgemuth, R.; Liese, A.; Streit, W. Biocatalytic phosphorylations of metabolites: Past, present, and future. *Trends Biotechnol.* **2017**, *35*, 452–465. [CrossRef]
61. Tsunoda, T.; Samadi, A.; Burade, S.; Mahmud, T. Complete biosynthetic pathway to the antidiabetic drug acarbose. *Nat. Commun.* **2022**, *13*, 3455. [CrossRef]
62. Minagawa, K.; Zhang, Y.; Ito, T.; Bai, L.; Deng, Z.; Mahmud, T. ValC, a New Type of C7-Cyclitol Kinase Involved in the Biosynthesis of the Antifungal Agent Validamycin, A. *ChemBioChem* **2007**, *8*, 632–641. [CrossRef]
63. Smith, C.-I.; Martin, S.R.; and Smith, K.S. Acetate kinase: Not just a bacterial enzyme. *Trends Microbiol.* **2006**, *14*, 249–253. [CrossRef] [PubMed]
64. Bachochin, M.J.; Van Allen, M.; Barber, R.D. Characterization of a *Rhodobacter sphaeroides* primary fatty acid kinase. *Arch. Microbiol.* **2021**, *203*, 861–864. [CrossRef] [PubMed]
65. Wallimann, T.; Tokarska-Schlattner, M.; Schlattner, U. The creatine kinase system and pleiotropic effects of creatine. *Amino Acids* **2011**, *40*, 1271–1296. [CrossRef] [PubMed]
66. Gao, Y.-S.; Wang, Y.; Li, C.; Chen, Z.; Yan, Y.-B.; Zhou, H.-M. Dissecting the key residues crucial for the species-specific thermostability of muscle-type creatine kinase. *Int. J. Biol. Macromol.* **2010**, *47*, 366–370. [CrossRef]
67. Hunter, T. A journey from phosphotyrosine to phosphohistidine and beyond. *Mol. Cell* **2022**, *82*, 2190–2220. [CrossRef]
68. Lu, Z.; Hunter, T. Metabolic kinases moonlighting as protein kinases. *Trends Biochem. Sci.* **2018**, *43*, 301–310. [CrossRef]
69. Berginski, M.E.; Moret, N.; Liu, C.; Goldfarb, D.; Sorger, P.K.; Gomez, S.M. The Dark Kinase Knowledgebase: An online compendium of knowledge and experimental results of understudied kinases. *Nucleic Acids Res.* **2021**, *49*, D529–D535. [CrossRef]

70. Moret, N.; Liu, C.; Gyori, B.M.; Bachman, J.A.; Steppi, A.; Hug, C.; Taujale, R.; Huang, L.C.; Berginski, M.E.; Gomez, S.M.; et al. A resource for exploring the understudied human kinome for research and therapeutic opportunities. *BioRxiv* **2021**. Available online: https://www.biorxiv.org/content/10.1101/2020.04.02.022277v3 (accessed on 15 August 2022).
71. Morton, R.K. The Phosphotransferase Activity of Phosphatases. *Biochem. J.* **1958**, *70*, 139–155. [CrossRef]
72. Holden, H.M.; Raushel, F.M. From the Three-Dimensional Structure of Phosphotriesterase. *Biochemistry* **2021**, *60*, 3413–3415. [CrossRef]
73. Asano, Y.; Mihara, Y.; Yamada, H. A novel selective nucleoside phosphorylation enzyme from *Morganella morganii*. *J. Biosci Bioeng.* **1999**, *87*, 732–738. [CrossRef]
74. Mihara, Y.; Utagawa, T.; Yamada, H.; Asano, Y. Acid phosphatase/phosphotransferases from enteric bacteria. *J. Biosci. Bioeng* **2001**, *92*, 50–54. [CrossRef]
75. Tanaka, N.; Hasan, Z.; Hartog, A.F.; van Herk, T.; Wever, R. Phosphorylation and dephosphorylation of polyhydroxy compounds by class A bacterial acid phosphatases. *Org. Biomol. Chem.* **2003**, *1*, 2833–2839. [CrossRef] [PubMed]
76. Tasnádi, G.; Lukesch, M.; Zechner, M.; Jud, W.; Hall, M.; Ditrich, K.; Baldenius, K.; Hartog, A.F.; Wever, R.; Faber, K. Exploiting acid phosphatases in the synthesis of phosphorylated monoalcohols and diols. *Eur. J. Org. Chem.* **2016**, *1*, 45–50. [CrossRef]
77. Ros, S.; Schulze, A. Balancing glycolytic flux: The role of 6-phosphofructo-2-kinase/fructose 2,6-bisphosphatases in cancer metabolism. *Cancer Metab.* **2013**, *1*, 8. [CrossRef] [PubMed]
78. Nannemann, D.P.; Kaufmann, K.W.; Meiler, J.; Bachmann, B.O. Design and directed evolution of a dideoxy purine nucleoside phosphorylase. *Protein Eng. Des. Sel.* **2010**, *23*, 607–616. [CrossRef]
79. Birmingham, W.R.; Starbird, C.A.; Panosian, T.D.; Nannemann, D.P.; Iverson, T.M.; Bachmann, B.O. Bioretrosynthetic construction of a didanosine biosynthetic pathway. *Nat. Chem. Biol.* **2014**, *10*, 392–399. [CrossRef]
80. Zhou, X.; Szeker, K.; Janocha, B.; Böhme, T.; Albrecht, D.; Mikhailopulo, I.A.; Neubauer, P. Recombinant purine nucleoside phosphorylases from thermophiles: Preparation, properties and activity towards purine and pyrimidine nucleosides. *FEBS J* **2013**, *280*, 1475–1490. [CrossRef]
81. Goedl, C.; Schwarz, A.; Minani, A.; Nidetzky, B. Recombinant sucrose phosphorylase: Characterization, kinetic studies of transglucosylation, and application of immobilized enzyme for production of alpha-D-glucose 1-phosphate. *J. Biotechnol.* **2007**, *129*, 77–86. [CrossRef]
82. Van der Borght, J.; Desmet, T.; Soetaert, W. Enzymatic production of β-d-glucose-1-phosphate from trehalose. *Biotechnol. J.* **2010**, *5*, 986–993. [CrossRef]
83. Kulikova, I.V.; Drenichev, M.S.; Solyev, P.N.; Alexeev, C.S.; Mikhailov, C.N. Enzymatic Synthesis of 2-Deoxyribose 1-phosphate and Ribose 1-phosphate and Subsequent Preparation of Nucleosides. *Eur. J. Org. Chem.* **2019**, 6999–7004. [CrossRef]
84. Fothergill-Gilmore, L.A.; Watson, H.C. The phosphoglycerate mutases. *Adv. Enzymol. Relat. Areas Mol. Biol* **1989**, *62*, 227–313. [CrossRef]
85. Tozzi, M.G.; Camici, M.; Mascia, L.; Sgarrella, F.; Ipata, P.L. Pentose phosphates in nucleoside interconversion and catabolism. *FEBS.J.* **2006**, *273*, 1089–1101. [CrossRef] [PubMed]
86. Kedzierski, L.; Malby, R.L.; Smith, B.J.; Perugini, M.A.; Hodder, A.N.; Ilg, T.; Colman, P.M.; Handman, E. Structure of *Leishmania mexicana* phosphomannomutase highlights similarities with human isoforms. *J. Mol. Biol.* **2006**, *363*, 215–227. [CrossRef] [PubMed]
87. Stiers, K.M.; Muenks, A.G.; Beamer, L.J. Biology, mechanism, and structure of enzymes in the α-D-phosphohexomutase superfamily. *Adv. Protein Chem. Struct. Biol.* **2017**, *109*, 265–304. [CrossRef] [PubMed]
88. Dai, J.; Finci, L.; Zhang, C.; Lahiri, S.; Zhang, G.; Peisach, E.; Allen, K.N.; Dunaway-Mariano, D. Analysis of the structural determinants underlying discrimination between substrate and solvent in β-phosphoglucomutase catalysis. *Biochemistry* **2009**, *48*, 1984–1995. [CrossRef] [PubMed]
89. Liu, S.; Lu, Z.; Han, Y.; Jia, Y.; Howard, A.; Dunaway-Mariano, D.; Herzberg, O. Conformational flexibility of PEP mutase. *Biochemistry* **2004**, *43*, 4447–4453. [CrossRef] [PubMed]
90. Knowles, J.R. Enzyme-Catalyzed Phosphoryl Transfer Reactions. *Annu. Rev. Biochem.* **1980**, *49*, 877–919. [CrossRef]
91. Cleland, W.W.; Hengge, A.C. Enzymatic Mechanisms of Phosphate and Sulfate Transfer. *Chem. Rev.* **2006**, *106*, 3252–3278. [CrossRef]
92. Lassila, J.K.; Zalatan, J.G.; Herschlag, D. Biological Phosphoryl-Transfer Reactions: Understanding Mechanism and Catalysis. *Ann. Rev. Biochem.* **2011**, *80*, 669–702. [CrossRef]
93. Allen, K.N.; Dunaway-Mariano, D. Phosphoryl group transfer: Evolution of a catalytic scaffold. *Trends Biochem. Sci.* **2004**, *29*, 495–503. [CrossRef] [PubMed]
94. Stockbridge, R.B.; Wolfenden, R. The intrinsic reactivity of ATP and the catalytic proficiencies of kinases acting on glucose, N-acetylgalactosamine, and homoserine: A thermodynamic analysis. *J. Biol. Chem.* **2009**, *284*, 22747–22757. [CrossRef] [PubMed]
95. Kerns, S.J.; Agafonov, R.V.; Cho, Y.-J.; Pontiggia, F.; Otten, R.; Pachov, D.V.; Kutter, S.; Phung, L.A.; Murphy, P.N.; Thai, V.; et al. The energy landscape of adenylate kinase during catalysis. *Nat. Struct. Mol. Biol.* **2015**, *22*, 124–131. [CrossRef] [PubMed]
96. Parnell, A.E.; Mordhorst, S.; Kemper, F.; Giurrandino, M.; Prince, J.P.; Schwarzer, N.J.; Hofer, A.; Wohlwend, D.; Jessen, H.J.; Gerhardt, S.; et al. Substrate recognition and mechanism revealed by ligand-bound polyphosphate kinase 2 structures. *Proc. Natl. Acad. Sci. USA* **2018**, *115*, 3350–3355. [CrossRef] [PubMed]

97. Pfeiffer, M.; Crean, R.M.; Moreira, C.; Parracino, A.; Oberdorfer, G.; Brecker, L.; Hammerschmidt, F.; Lynn Kamerlin, S.C.; Nidetzky, B. Essential Functional Interplay of the Catalytic Groups in Acid Phosphatase. *ACS Catal.* **2022**, *12*, 3357–3370. [CrossRef]
98. Štefanić, Z.; Narczyk, M.; Mikleušević, G.; Kazazić, S.; Bzowska, A.; Luić, M. Crystallographic snapshots of ligand binding to hexameric purine nucleoside phosphorylase and kinetic studies give insight into the mechanism of catalysis. *Sci. Rep.* **2018**, *8*, 15427. [CrossRef]
99. Jin, Y.; Bhattasali, D.; Pellegrini, E.; Forget, S.M.; Baxter, N.J.; Cliff, M.J.; Bowler, M.W.; Jakeman, D.L.; Blackburn, G.M.; Waltho, J.P. α-Fluorophosphonates reveal how a phosphomutase conserves transition state conformation over hexose recognition in its two-step reaction. *Proc. Natl. Acad. Sci. USA* **2014**, *111*, 12384–12389. [CrossRef]
100. Ramos-Figueroa, J.S.; Palmer, D.R.; Horsman, G.P. Phosphoenolpyruvate mutase-catalyzed C-P bond formation: Mechanistic ambiguities and opportunities. *ChemBioChem* **2022**, *23*, e202200285. [CrossRef]
101. Kubota, K.; Anjum, R.; Yu, Y.; Kunz, R.C.; Andersen, J.N.; Kraus, M.; Keilhack, H.; Nagashima, K.; Krauss, S.; Paweletz, C.; et al. Sensitive multiplexed analysis of kinase activities and activity-based kinase identification. *Nat. Biotechnol.* **2009**, *27*, 933–940. [CrossRef]
102. Sévin, D.C.; Fuhrer, T.; Zamboni, N.; Sauer, U. Nontargeted in vitro metabolomics for high-throughput identification of novel enzymes in *Escherichia coli*. *Nat. Methods* **2017**, *14*, 187–194. [CrossRef]
103. Schastnaya, E.; Raguz Nakic, Z.; Gruber, C.H.; Doubleday, P.F.; Krishnan, A.; Johns, N.I.; Park, J.; Wang, H.H.; Sauer, U. Extensive regulation of enzyme activity by phosphorylation in *Escherichia coli*. *Nat. Commun.* **2021**, *12*, 5650. [CrossRef] [PubMed]
104. Cravatt, B.F.; Wright, A.T.; Kozarich, J.W. Activity-based protein profiling: From enzyme chemistry to proteomic chemistry. *Annu. Rev. Biochem.* **2008**, *77*, 383–414. [CrossRef] [PubMed]
105. Zallot, R.; Oberg, N.; Gerlt, J.A. Discovery of new enzymatic functions and metabolic pathways using genomic enzymology web tools. *Curr. Opin. Biotechnol.* **2021**, *69*, 77–90. [CrossRef]
106. Zhang, X.; Carter, M.S.; Vetting, M.W.; San Francisco, B.; Zhao, S.; Al-Obaidi, N.F.; Solbiati, J.O.; Thiaville, J.J.; de Crécy-Lagard, V.; Jacobson, M.P.; et al. Assignment of function to a domain of unknown function: DUF1537 is a new kinase family in catabolic pathways for acid sugars. *Proc. Natl. Acad. Sci. USA* **2016**, *113*, E4161–E4169. [CrossRef] [PubMed]
107. Arnold, F.H. Directed evolution: Bringing new chemistry to life. *Angew. Chem. Int. Ed.* **2018**, *57*, 4143–4148. [CrossRef] [PubMed]
108. Molina, R.S.; Rix, G.; Mengiste, A.A.; Álvarez, B.; Seo, D.; Chen, H.; Hurtado, J.E.; Zhang, Q.; García-García, J.D.; Heins, Z.J.; et al. In vivo hypermutation and continuous evolution. *Nat. Rev. Methods Prim.* **2022**, *2*, 1–22. [CrossRef]
109. Christians, F.C.; Scapozza, L.; Crameri, A.; Folkers, G.; Stemmer, W.P. Directed evolution of thymidine kinase for AZT phosphorylation using DNA family shuffling. *Nat. Biotechnol.* **1999**, *17*, 259–264. [CrossRef]
110. Miller, D.C.; Athavale, S.V.; Arnold, F.H. Combining chemistry and protein engineering for new-to-nature biocatalysis. *Nat. Synth.* **2022**, *1*, 18–23. [CrossRef]
111. Fryszkowska, A.; Devine, P.N. Biocatalysis in drug discovery and development. *Curr. Opin. Chem. Biol.* **2020**, *55*, 151–160. [CrossRef]
112. Umezawa, H.; Okanishi, M.; Kondo, S.; Hamana, K.; Utahara, R.; Maeda, K.; Mitsuhashi, S. Phosphorylative inactivation of aminoglycosidic antibiotics by *Escherichia coli* carrying R factor. *Science* **1967**, *157*, 1559–1561. [CrossRef]
113. Crofts, T.S.; Gasparrini, A.J.; Dantas, G. Next-generation approaches to understand and combat the antibiotic resistome. *Nat. Rev. Microbiol.* **2017**, *15*, 422–434. [CrossRef] [PubMed]
114. Surette, M.D.; Spanogiannopoulos, P.; Wright, G.D. The enzymes of the rifamycin antibiotic resistome. *Acc. Chem. Res.* **2021**, *54*, 2065–2075. [CrossRef] [PubMed]
115. Terekhov, S.S.; Mokrushina, Y.A.; Nazarov, A.S.; Zlobin, A.; Zalevsky, A.; Bourenkov, G.; Golovin, A.; Belogurov Jr, A.; Osterman, I.A.; Kulikova, A.A.; et al. A kinase bioscavenger provides antibiotic resistance by extremely tight substrate binding. *Sci. Adv.* **2020**, *6*, eaaz9861. [CrossRef] [PubMed]
116. Stogios, P.J.; Cox, G.; Spanogiannopoulos, P.; Pillon, M.C.; Waglechner, N.; Skarina, T.; Koteva, K.; Guarné, A.; Savchenko, A.; Wright, G.D. Rifampin phosphotransferase is an unusual antibiotic resistance kinase. *Nat. Commun.* **2016**, *7*, 11343. [CrossRef]
117. Cai, W.; Goswami, A.; Yang, Z.; Liu, X.; Green, K.D.; Barnard-Britson, S.; Baba, S.; Funabashi, M.; Nonaka, K.; Sunkara, M.; et al. The Biosynthesis of Capuramycin-type Antibiotics. *J. Biol. Chem.* **2015**, *290*, 13710–13724. [CrossRef]
118. Kita, A.; Kishimoto, A.; Shimosaka, T.; Tomita, H.; Yokooji, Y.; Imanaka, T.; Atomi, H.; Miki, K. Crystal structure of pantoate kinase from *Thermococcus kodakarensis*. *Proteins* **2020**, *88*, 718–724. [CrossRef]
119. Hsu, C.; Tsai, H.Y.; Chang, C.F.; Yang, C.C.; Su, N.W. Discovery of a novel phosphotransferase from *Bacillus subtilis* that phos-phorylates a broad spectrum of flavonoids. *Food Chem.* **2022**, *400*, 134001. [CrossRef]
120. Jacoby, C.; Goerke, M.; Bezold, D.; Jessen, H.; Boll, M. A fully reversible 25-hydroxy steroid kinase involved in oxygen-independent cholesterol side-chain oxidation. *J. Biol. Chem.* **2021**, *297*, 101105. [CrossRef]
121. Hoffmeister, D.; Yang, J.; Liu, L.; Thorson, J.S. Creation of the first anomeric D/L-sugar kinase by means of directed evolution. *Proc. Natl. Acad. Sci. USA* **2003**, *100*, 13184–13189. [CrossRef]
122. Nagata, R.; Fujihashi, M.; Sato, T.; Atomi, H.; Miki, K. Crystal Structure and Product Analysis of an Archaeal myo-Inositol Kinase Reveal Substrate Recognition Mode and 3-OH Phosphorylation. *Biochemistry* **2015**, *54*, 3494–3503. [CrossRef]
123. Tashiro, R.; Sato, T.; Atomi, H.; Miki, K.; Fujihashi, M. Altering the phosphorylation position of pyrophosphate-dependent *myo*-inositol-1-kinase based on its crystal structure. *ACS Chem. Biol.* **2021**, *16*, 794–799. [CrossRef]

124. Huang, H.; Parmeggiani, F.; Pallister, E.; Huang, C.-J.; Liu, F.-F.; Li, Q.; Birmingham, W.R.; Both, P.; Thomas, B.; Liu, L.; et al. Characterisation of a Bacterial Galactokinase with High Activity and Broad Substrate Tolerance for Chemoenzymatic Synthesis of 6-Aminogalactose-1-Phosphate and Analogues. *ChemBioChem* **2018**, *19*, 388–394. [CrossRef] [PubMed]
125. Keenan, T.; Parmeggiani, F.; Malassis, J.; Fontenelle, C.Q.; Vendeville, J.B.; Offen, W.; Both, P.; Huang, K.; Marchesi, A.; Heyam, A.; et al. Profiling substrate promiscuity of wild-type sugar kinases for multi-fluorinated monosaccharides. *Cell Chem. Biol.* **2020**, *27*, 1199–1206. [CrossRef] [PubMed]
126. Huddleston, J.P.; Raushel, F.M. Functional Characterization of YdjH, a Sugar Kinase of Unknown Specificity in *Escherichia coli* K12. *Biochemistry* **2019**, *58*, 3354–3364. [CrossRef] [PubMed]
127. Taylor, Z.W.; Raushel, F.M. Cytidine Diphosphoramidate Kinase: An Enzyme Required for the Biosynthesis of the *O*-Methyl Phosphoramidate Modification in the Capsular Polysaccharides of *Campylobacter jejuni*. *Biochemistry* **2018**, *57*, 2238–2244. [CrossRef] [PubMed]
128. Matsuura, M.F.; Winiger, C.B.; Shaw, R.W.; Kim, M.-J.; Kim, M.-S.; Daugherty, A.B.; Chen, F.P.; Moses, J.D.; Lutz, S.; Benner, S.A. A Single Deoxynucleoside Kinase Variant from *Drosophila melanogaster* Synthesizes Monophosphates of Nucleosides That Are Components of an Expanded Genetic System. *ACS Synth. Biol.* **2017**, *6*, 388–394. [CrossRef]
129. Makino, Y.; Sato, T.; Kawamura, H.; Hachisuka, S.I.; Takeno, R.; Imanaka, T.; Atomi, H. An archaeal ADP-dependent serine kinase involved in cysteine biosynthesis and serine metabolism. *Nat. Commun.* **2016**, *7*, 13446. [CrossRef]
130. Mori, Y.; Kawamura, H.; Sato, T.; Fujita, T.; Nagata, R.; Fujihashi, M.; Miki, K.; Atomi, H. Identification and Enzymatic Analysis of an Archaeal ATP-Dependent Serine Kinase from the Hyperthermophilic Archaeon *Staphylothermus marinus*. *J. Bact.* **2021**, *203*, e00025-21. [CrossRef]
131. Thiaville, J.J.; Flood, J.; Yurgel, S.; Prunetti, L.; Elbadawi-Sidhu, M.; Hutinet, G.; Forouhar, F.; Zhang, X.; Ganesan, V.; Reddy, P.; et al. Members of a Novel Kinase Family (DUF1537) Can Recycle Toxic Intermediates into an Essential Metabolite. *ACS Chem. Biol.* **2016**, *11*, 2304–2311. [CrossRef]
132. Taylor, Z.W.; Brown, H.A.; Narindoshvili, T.; Wenzel, C.Q.; Szymanski, C.M.; Holden, H.M.; Raushel, F.M. Discovery of a glutamine kinase required for the biosynthesis of the O-methyl phosphoramidate modifications found in the capsular polysaccharides of *Campylobacter jejuni*. *J. Am. Chem. Soc.* **2017**, *139*, 9463–9466. [CrossRef]
133. Taylor, Z.W.; Chamberlain, A.R.; Raushel, F.M. Substrate specificity and chemical mechanism for the reaction catalyzed by glutamine kinase. *Biochemistry* **2018**, *57*, 5447–5455. [CrossRef] [PubMed]
134. Médici, R.; Garaycoechea, J.I.; Valino, A.I.; Pereira, C.A.; Lewkowicz, E.S.; Iribarren, A.M. A comparative study on phosphotransferase activity of acid phosphatases from *Raoultella planticola* and *Enterobacter aerogenes* on nucleosides, sugars and related compounds. *Appl. Microbiol. Biotechnol.* **2014**, *98*, 3013–3022. [CrossRef] [PubMed]
135. Mihara, Y.; Utagawa, T.; Yamada, H.; Asano, Y. Phosphorylation of Nucleosides by the Mutated Acid Phosphatase from *Morganella morganii*. *Appl. Environ. Microbiol.* **2000**, *66*, 2811–2816. [CrossRef] [PubMed]
136. Tasnádi, G.; Zechner, M.; Hall, M.; Baldenius, K.; Ditrich, K.; Faber, K. Investigation of acid phosphatase variants for the synthesis of phosphate monoesters. *Biotechnol. Bioeng.* **2017**, *114*, 2187–2195. [CrossRef]
137. Nagy, F.; Tasnádi, G.; Balogh-Weiser, D.; Bell, E.; Hall, M.; Faber, K.; Poppe, L. Smart nanoparticles for selective immobilization of acid phosphatases. *ChemCatChem* **2018**, *10*, 3490–3499. [CrossRef]
138. Macdonald, S.S.; Armstrong, Z.; Morgan-Lang, C.; Osowiecka, M.; Robinson, K.; Hallam, S.J.; Withers, S.G. Development and Application of a High-Throughput Functional Metagenomic Screen for Glycoside Phosphorylases. *Cell Chem. Biol.* **2019**, *26*, 1001–1012. [CrossRef]
139. Franceus, J.; Pinel, D.; Desmet, T. Glucosylglycerate phosphorylase, an enzyme with novel specificity involved in compatible solute metabolism. *Appl. Env. Microbiol.* **2017**, *83*, e01434-17. [CrossRef]
140. De Doncker, M.; De Graeve, C.; Franceus, J.; Beerens, K.; Křen, V.; Pelantová, H.; Vercauteren, R.; Desmet, T. Exploration of GH94 sequence space for enzyme discovery reveals a novel Glucosylgalactose phosphorylase specificity. *ChemBioChem* **2021**, *22*, 3319–3325. [CrossRef]
141. Teze, D.; Coines, J.; Raich, L.; Kalichuk, V.; Solleux, C.; Tellier, C.; Andre-Miral, C.; Svensson, B.; Rovira, C. A single point mutation converts GH84 O-GlcNac hydrolases into phosphorylases: Experimental and theoretic evidence. *J. Am. Chem. Soc.* **2020**, *142*, 2120–2124. [CrossRef]
142. Novick, S.J.; Dellas, N.; Mitchell, V.; Duan, D.; Nazor, J.; Alvizo, O.; Sowell-Kantz, A.A.; Moore, J.C.; Huffman, M.; Rodriguez-Granillo, A.; et al. Engineered Purine Nucleoside Phosphorylase Variant Enzymes. US 2022/0010316 A1, 2 November 2021.
143. Kamel, S.; Thiele, I.; Neubauer, P.; Wagner, A. Thermophilic nucleoside phosphorylases: Their properties, characteristics and applications. *BBA Proteins Proteom.* **2020**, *1868*, 140304. [CrossRef]
144. Vroom, J.; Sivaramakrishnan, S.; Hurtak, J.A. Engineered Phosphopentomutase Variant Enzymes. WO 2022/076454 A1, 15 June 2021.
145. Dadashipour, M.; Iwamoto, M.; Hossain, M.M.; Akutsu, J.-I.; Zhang, Z.; Kawarabayasi, Y. Identification of a Direct Biosyn-thetic Pathway for UDP-*N*-Acetylgalactosamine from Glucosamine-6-phosphate in Thermophilic Crenarchaeon *Sulfolobus tokodaii*. *J. Bacteriol.* **2018**, *200*, e00048-18. [CrossRef] [PubMed]
146. Bergmeyer, H.-U. *Methods of Enzymatic Analysis*; Academic Press: New York, NY, USA, 1965; pp. 407–410.
147. Wieland, O. An enzymic method for estimating glycerol. *Biochem. Ztschr.* **1957**, *329*, 313–319.

148. Hørder, M.; Elser, R.C.; Gerhardt, W.; Mathieu, M.; Sampson, E.J. International Federation of Clinical Chemistry (IFCC): Scientific Division, Committee on Enzymes. IFCC methods for the measurement of catalytic concentration enzymes. Part 7. IFCC method for creatine kinase (creatine *N*-phosphotransferase, EC 2.7.3.2). IFCC Recommendation. *J. Autom. Chem.* **1990**, *12*, 22–40. [CrossRef]
149. Galburt, E.A.; Pelletier, J.; Wilson, G.; Stoddard, B.L. Structure of a tRNA repair enzyme and molecular biology workhorse: T4 polynucleotide kinase. *Structure* **2002**, *10*, 1249–1260. [CrossRef]
150. Green, M.R.; Sambrook, J. Preparation of Labeled DNA, RNA, and Oligonucleotide Probes. *Cold Spring Harb. Protoc.* **2022**, *1*, 1–26. [CrossRef] [PubMed]
151. Anthony, T.M.; Pflum, M.K.H. Kinase-catalyzed biotinylation of DNA. *Bioorg. Med. Chem.* **2018**, *26*, 2331–2336. [CrossRef]
152. Schumann, G.; Klauke, R.; Canalias, F.; Bossert-Reuther, S.; Franck, P.F.H.; Gella, F.-J.; Jørgensen, P.J.; Kang, D.; Lessinger, J.-M.; Panteghini, M.; et al. IFCC primary reference procedures for the measurement of catalytic activity concentrations of enzymes at 37 °C. Part 9: Reference procedure for the measurement of catalytic concentration of alkaline phosphatase. *Clin. Chem. Lab. Med.* **2011**, *49*, 1439–1446. [CrossRef]
153. Shaban, S.M.; Jo, S.B.; Hafez, E.; Cho, J.H.; Kim, D.-H. A comprehensive overview on alkaline phosphatase targeting and reporting assays. *Coord. Chem. Rev.* **2022**, *465*, 214567. [CrossRef]
154. Kitaoka, M.; Aoyagi, C.; Hayashi, K. Colorimetric Quantification of Cellobiose Employing Cellobiose Phosphorylase. *Anal. Biochem.* **2001**, *292*, 163–166. [CrossRef]
155. Zhang, Z.; Jaffrezic-Renault, N.; Bessueille, F.; Leonard, D.; Xia, S.; Wang, X.; Chen, L.; Zhao, J. Development of a conductometric phosphate biosensor based on tri-layer maltose phosphorylase composite films. *Ana. Chim. Acta* **2008**, *615*, 73–79. [CrossRef]
156. Apple, F.S.; Wu, A.H.B.; Mair, J.; Ravkilde, J.; Panteghini, M.; Tate, J.; Pagani, F.; Christenson, R.H.; Mockel, M.; Danne, O.; et al. Future Biomarker for Detection of Ischemia and Risk Stratification in Acute Coronary Syndrome. *Clin. Chem* **2005**, *51*, 810–824. [CrossRef] [PubMed]
157. Hallows, W.C.; Yu, W.; Denu, J.M. Regulation of Glycolytic Enzyme Phosphoglycerate Mutase-1 by Sirt1 Protein-mediated Deacetylation. *J. Biol. Chem.* **2012**, *287*, 3850–3856. [CrossRef] [PubMed]
158. Sun, Q.; Li, S.; Wang, Y.; Peng, H.; Zhang, X.; Zheng, Y.; Li, C.; Li, L.; Chen, R.; Chen, X.; et al. Phosphoglyceric acid mutase-1 contributes to oncogenic mTOR-mediated tumor growth and confers non-small cell lung cancer patients with poor prognosis. *Cell Death Diff.* **2018**, *25*, 1160–1173. [CrossRef]
159. Wohlgemuth, R.; Littlechild, J. Complexity Reduction and Opportunities in the Design, Integration and Intensification of Biocatalytic Processes for Metabolite Synthesis. *Front. Bioeng. Biotechnol. Bioprocess. Eng.* **2022**, *10*, 958606. [CrossRef] [PubMed]
160. Asano, Y.; Mihara, Y.; Yamada, H. A new enzymatic method of selective phosphorylation of nucleosides. *J. Mol. Catal. B Enzym.* **1999**, *6*, 271–277. [CrossRef]
161. Suzuki, E.; Ishikawa, K.; Mihara, Y.; Shimba, N.; Asano, Y. Structural-based engineering for transferases to improve the industrial production of 5′-nucleotides. *Bull. Chem. Soc. Jpn.* **2007**, *80*, 276–286. [CrossRef]
162. Van Herk, T.; Hartog, A.F.; van der Burg, A.M.; Wever, R. Regioselective phosphorylation of carbohydrates and various alcohols by bacterial acid phosphatases; Probing the substrate specificity of the enzyme from *Shigella flexneri*. *Adv. Synth. Catal.* **2005**, *347*, 1155–1162. [CrossRef]
163. Babich, L.; Hartog, A.F.; van der Horst, M.A.; Wever, R. Continuous-flow reactor-based enzymatic synthesis of phosphorylated compounds on a large scale. *Chem. Eur. J.* **2012**, *18*, 6604–6609. [CrossRef]
164. Van Herk, T.; Hartog, A.F.; Schoemaker, H.E.; Wever, R.; Faber, K. Simple Enzymatic in situ Generation of Dihydroxyacetone phosphate and its Use in a Cascade Reaction for the Production of Carbohydrates: Increased Efficiency by Phosphate Cycling. *J. Org. Chem.* **2006**, *71*, 6244–6247. [CrossRef]
165. Tasnádi, G.; Hall, M.; Baldenius, K.; Ditrich, K.; Faber, K. Biocatalytic functionalization of hydroxyalkyl acrylates and phenoxyethanol via phosphorylation. *J. Biotechnol.* **2016**, *233*, 219–227. [CrossRef]
166. Tasnádi, G.; Jud, W.; Hall, M.; Baldenius, K.; Ditrich, K.; Faber, K. Evaluation of Natural and Synthetic Phosphate Donors for the Improved Enzymatic Synthesis of Phosphate Monoesters. *Adv. Synth. Catal.* **2018**, *360*, 2394–2401. [CrossRef] [PubMed]
167. Tasnádi, G.; Staśko, M.; Ditrich, K.; Hall, M.; Faber, K. Preparative-Scale Enzymatic Synthesis of *rac*-Glycerol-1-phosphate from Crude Glycerol Using Acid Phosphatases and Phosphate. *ChemSusChem* **2020**, *13*, 1759–1763. [CrossRef] [PubMed]
168. Song, S.; Zheng, K.; Xu, X.; Gao, C.; Guo, L.; Liu, J.; Chen, X.; Liu, L.; Hu, G.; Wu, J. Enzymatic Production of Ascorbic Acid-2-Phosphate by Engineered *Pseudomonas aeruginosa* Acid Phosphatase. *J. Agric. Food Chem.* **2021**, *69*, 14215–14221. [CrossRef] [PubMed]
169. Qian, X.-L.; Dai, Y.-S.; Li, C.X.; Pan, J.; Xu, J.-H.; Mu, B. Enzymatic synthesis of high-titer nicotinamide mononucleotide with a new nicotinamide riboside kinase and an efficient ATP regeneration system. *Bioresour. Bioprocess.* **2022**, *9*, 26. [CrossRef]
170. Gauss, D.; Schoenenberger, B.; Wohlgemuth, R. Chemical and enzymatic methodologies for the synthesis of enantiomerically pure glyceraldehyde 3-phosphates. *Carbohydr. Res.* **2014**, *389*, 18–24. [CrossRef]
171. Wong, C.H.; Whitesides, G.M. Synthesis of Sugars by Aldolase-Catalyzed Condensation Reactions. *J. Org. Chem.* **1983**, *48*, 3199–3205. [CrossRef]
172. Gauss, D.; Sánchez-Moreno, I.; Oroz-Guinea, I.; García-Junceda, E.; Wohlgemuth, R. Phosphorylation catalyzed by dihydroxyacetone kinase. *Eur. J. Org. Chem.* **2018**, *2018*, 2892–2895. [CrossRef]

173. Matsumi, R.; Hellriegel, C.; Schoenenberger, B.; Milesi, T.; Van Der Oost, J.; Wohlgemuth, R. Biocatalytic asymmetric phosphorylation of mevalonate. *RSC Adv.* **2014**, *4*, 12989–12994. [CrossRef]
174. Hardt, N.; Kinfu, B.M.; Chow, J.; Schoenenberger, B.; Streit, W.R.; Obkircher, M.; Wohlgemuth, R. Biocatalytic Asymmetric Phosphorylation Catalyzed by Recombinant Glycerate-2-Kinase. *ChemBioChem* **2017**, *18*, 1518–1522. [CrossRef]
175. Schoenenberger, B.; Wszolek, A.; Meier, R.; Brundiek, H.; Obkircher, M.; Wohlgemuth, R. Recombinant AroL-Catalyzed Phosphorylation for the Efficient Synthesis of Shikimic Acid 3-Phosphate. *Biotechnol. J.* **2018**, *13*, 1700529. [CrossRef]
176. Hardt, N.; Kind, S.; Schoenenberger, B.; Eggert, T.; Obkircher, M.; Wohlgemuth, R. Facile synthesis of D-xylulose-5-phosphate and L-xylulose-5-phosphate by xylulokinase-catalyzed phosphorylation. *Biocat. Biotrans.* **2020**, *38*, 35–45. [CrossRef]
177. Schoenenberger, B.; Kind, S.; Meier, R.; Eggert, T.; Obkircher, M.; Wohlgemuth, R. Efficient biocatalytic synthesis of D-tagatose 1,6-diphosphate by LacC-catalysed phosphorylation of D-tagatose 6-phosphate. *Biocat. Biotrans.* **2020**, *38*, 53–63. [CrossRef]
178. Wen, L.; Huang, K.; Liu, Y.; Wang, P.G. Facile enzymatic synthesis of phosphorylated ketopentoses. *ACS Catal.* **2016**, *6*, 1649–1654. [CrossRef]
179. Wen, L.; Huang, K.; Wei, M.; Meisner, J.; Liu, Y.; Garner, K.; Zang, L.; Wang, X.; Li, X.; Fang, J.; et al. Facile Enzymatic Synthesis of Ketoses. *Angew. Chem., Int. Ed.* **2015**, *54*, 12654–12658. [CrossRef] [PubMed]
180. Fricke, J.; Kargbo, R.; Regestein, L.; Lenz, C.; Peschel, G.; Rosenbaum, M.A.; Sherwood, A.; Hoffmeister, D. Scalable Hybrid Synthetic/Biocatalytic Route to Psilocybin. *Chem. Eur. J.* **2020**, *26*, 8281–8285. [CrossRef] [PubMed]
181. Schoenenberger, B.; Wszolek, A.; Milesi, T.; Brundiek, H.; Obkircher, M.; Wohlgemuth, R. Synthesis of Nω-Phospho-L-arginine by Biocatalytic Phosphorylation of L-Arginine. *ChemCatChem* **2017**, *9*, 121–126. [CrossRef]
182. De Winter, K.; Cerdobbel, A.; Soetaert, W.; Desmet, T. Operational stability of immobilized sucrose phosphorylase: Continuous production of α-glucose-1-phosphate at elevated temperatures. *Proc. Biochem.* **2011**, *46*, 2074–2078. [CrossRef]
183. Bae, J.; Lee, D.; Kim, D.; Cho, S.-J.; Park, J.E.; Koh, S.; Kim, J.; Park, B.-H.; Choi, Y.; Shin, H.-J.; et al. Facile synthesis of glucose-1-phosphate from starch by *Thermus thermophilus* GK24 α-glucan phosphorylase. *Proc. Biochem.* **2005**, *40*, 3707–3713. [CrossRef]
184. Kamel, S.; Weiss, M.; Klare, H.F.T.; Mikhailopoulo, I.A.; Neubauer, P.; Wagner, A. Chemo-enzymatic synthesis of α-D-pentofuranose-1-phosphates using thermostable pyrimidine nucleoside phosphorylases. *Mol. Catal.* **2018**, *458*, 52–59. [CrossRef]
185. Heptinstall, J.; Ward, P.J.; Hancock, I.C. The enzymic synthesis of [^{32}P]-N-acetylglucosamine. *Anal. Biochem.* **1978**, *91*, 158–165. [CrossRef]
186. Hassan, M.I.; Lundgren, B.R.; Chaumun, M.; Whitfield, D.M.; Clark, B.; Schoenhofen, I.C.; Boddy, C.N. Total Biosynthesis of Legionaminic Acid, a Bacterial Sialic Acid Analogue. *Angew. Chem. Intl. Ed.* **2016**, *55*, 12018–12021. [CrossRef] [PubMed]
187. Horinouchi, N.; Kawano, T.; Sakai, T.; Matsumoto, S.; Sasaki, M.; Mikami, Y.; Ogawa, J.; Shimizu, S. Screening and characterization of a phosphopentomutase useful for enzymatic production of 2′-deoxynucleoside. *New Biotechnol.* **2009**, *26*, 75–82. [CrossRef] [PubMed]
188. Yang, Y.-H.; Kang, Y.-B.; Kim, D.-H.; Lee, T.-H.; Park, S.-H.; Lee, K.; Yoo, D.; Liou, K.-K.; Lee, H.-C.; Sohng, J.-K.; et al. One-pot enzymatic synthesis of deoxy-thymidine-diphosphate (TDP)-2-deoxy-α-D-glucose using phosphomannomutase. *J. Mol. Catal. B Enzym.* **2010**, *62*, 282–287. [CrossRef]
189. Mahour, R.; Klapproth, J.; Rexer, T.F.T.; Schildbach, A.; Klamt, S.; Pietzsch, M.; Rapp, E.; Reichl, U. Establishment of a five-enzyme cell-free cascade for the synthesis of uridine diphosphate *N*-acetylglucosamine. *J. Biotechnol.* **2018**, *283*, 120–129. [CrossRef]
190. Robescu, M.S.; Serra, I.; Terreni, M.; Ubiali, D.; Bavaro, T. A Multi-Enzymatic Cascade Reaction for the Synthesis of Vidarabine-5′-Monophosphate. *Catalysts* **2020**, *10*, 60. [CrossRef]
191. Sánchez-Moreno, I.; Hélaine, V.; Poupard, N.; Charmantray, F.; Légeret, B.; Hecquet, L.; García-Junceda, E.; Wohlgemuth, R.; Guérard-Hélaine, C.; Lemaire, M. One-pot cascade reactions using fructose-6-phosphate aldolase: Efficient synthesis of D-arabinose 5-phosphate, D-fructose 6-phosphate and analogues. *Adv. Synth. Catal.* **2012**, *354*, 1725–1730. [CrossRef]
192. Hélaine, V.; Mahdi, R.; Sudhir Babu, G.V.; de Berardinis, V.; Wohlgemuth, R.; Lemaire, M.; Guérard-Hélaine, C. Straightforward synthesis of terminally phosphorylated L-sugars via multienzymatic cascade reactions. *Adv. Synth. Catal.* **2015**, *357*, 1703–1708. [CrossRef]
193. Huffman, M.A.; Fryszkowska, A.; Alvizo, O.; Borra-Garske, M.; Campos, K.R.; Canada, K.A.; Devine, P.N.; Duan, D.; Forstater, J.H.; Grosser, S.T.; et al. Design of an in vitro biocatalytic cascade for the manufacture of islatravir. *Science* **2019**, *366*, 1255–1259. [CrossRef]
194. Schultheisz, H.L.; Szymczyna, B.R.; Scott, L.G.; Williamson, J.R. Pathway Engineered Enzymatic de Novo Purine Nucleotide Synthesis. *ACS Chem. Biol.* **2008**, *3*, 499–511. [CrossRef]
195. Ren, S.; Huffman, M.A.; Whittaker, A.M.; Yang, H.; Nawrat, C.C.; Waterhouse, D.J.; Maloney, K.M.; Strotman, N.A. Synthesis of Isotopically Labeled Anti-HIV Nucleoside Islatravir though a One-Pot Biocatalytic Cascade Reaction. *Org. Process Res. Dev.* **2021**, *25*, 516–521. [CrossRef]
196. McIntosh, J.A.; Benkovics, T.; Silverman, S.M.; Huffman, M.A.; Kong, J.; Maligres, P.E.; Itoh, T.; Yang, H.; Verma, D.; Pan, W.; et al. Engineered ribosyl-1-kinase enables concise synthesis of molnupiravir, an antiviral for COVID-19. *ACS Centr. Sci.* **2021**, *7*, 1980–1985. [CrossRef] [PubMed]
197. McIntosh, J.A.; Liu, Z.; Andresen, B.M.; Marzijarani, N.S.; Moore, J.C.; Marshall, N.M.; Borra-Garske, M.; Obligacion, J.V.; Fier, P.S.; Peng, F.; et al. A kinase-cGAS cascade to synthesize a therapeutic STING activator. *Nature* **2022**, *603*, 439–444. [CrossRef] [PubMed]

198. Murata, K.; Uchida, T.; Kato, J.; Chibata, I. Polyphosphate kinase: Distribution, Some Properties and Its Application as an ATP Regeneration System. *Agric. Biol. Chem.* **1988**, *52*, 1471–1477.
199. Murata, K. Polyphosphate-dependent nicotinamide adenine dinucleotide (NAD) kinase: A novel missing link in human mitochondria. *Proc. Jpn. Acad., Ser. B* **2021**, *97*, 479–498. [CrossRef] [PubMed]
200. Murata, K.; Uchida, T.; Kato, J.; Chibata, I. A Metaphosphate-dependent Nicotinamide Adenine Dinucleotide Kinase from *Brevibacterium ammoniagenes*. *Agric. Biol. Chem.* **1980**, *44*, 1165–1172. [CrossRef]
201. Andexer, J.N.; Richter, M. Emerging Enzymes for ATP Regeneration in Biocatalytic Processes. *ChemBioChem* **2015**, *16*, 380–386. [CrossRef]
202. Mordhorst, S.; Andexer, J.N. Round, round we go – strategies for enzymatic cofactor regeneration. *Nat. Prod. Rep.* **2020**, *37*, 1316–1333. [CrossRef]
203. Hirschbein, B.L.; Mazenod, F.P.; Whitesides, G.M. Synthesis of Phosphoenolpyruvate and Its Use in Adenosine Triphosphate Cofactor Regeneration. *J. Org. Chem.* **1982**, *47*, 3765–3766. [CrossRef]
204. Crans, D.C.; Whitesides, G.M. A Convenient Synthesis of Disodium Acetyl Phosphate for Use in in Situ ATP Cofactor Regeneration. *J. Org. Chem.* **1983**, *48*, 3130–3132. [CrossRef]
205. Kazlauskas, R.J.; Whitesides, G.M. Synthesis of Methoxycarbonyl Phosphate, a New Reagent Having High Phosphoryl Donor Potential for Use in ATP Cofactor Regeneration. *J. Org. Chem.* **1985**, *50*, 1069–1076. [CrossRef]
206. Tavanti, M.; Hosford, J.; Lloyd, R.C.; Brown, M.J.B. Recent Developments and Challenges for the Industrial Implementation of Polyphosphate Kinases. *ChemCatChem* **2021**, *13*, 3565–3580. [CrossRef]
207. Kim, D.-M.; Swartz, J.R. Prolonging cell-free protein synthesis with a novel ATP regeneration system. *Biotech. Bioeng.* **1999**, *66*, 180–188. [CrossRef]
208. Ruccolo, S.; Brito, G.; Christensen, M.; Itoh, T.; Mattern, K.; Stone, K.; Strotman, N.A.; Sun, A.C. Electrochemical Recycling of Adenosine Triphosphate in Biocatalytic Reaction Cascades. *Chemrxiv* **2022**. [CrossRef]
209. Molla, G.S.; Kinfu, B.M.; Chow, J.; Streit, W.; Wohlgemuth, R.; Liese, A. Bioreaction Engineering Leading to Efficient Synthesis of L-Glyceraldehyd-3-phosphate. *Biotechnol. J.* **2017**, *12*, 1600625. [CrossRef] [PubMed]
210. Molla, G.S.; Himmelspach, A.; Wohlgemuth, R.; Haupt, E.T.K.; Liese, A. Mechanistic and kinetic elucidation of Mg^{2+}/ATP molar ratio effect on glycerol kinase. *Mol. Catal.* **2018**, *445*, 36–42. [CrossRef]
211. Wohlgemuth, R. Product Recovery. *Compr. Biotechnol.* **2011**, *2*, 591–601. [CrossRef]
212. Knouse, K.W.; Flood, D.T.; Vantourout, J.C.; Schmidt, M.A.; Mcdonald, I.M.; Eastgate, M.D.; Baran, P.S. Nature chose phosphates and chemists should too: How emerging P (V) methods can augment existing strategies. *ACS Cent. Sci.* **2021**, *7*, 1473–1485. [CrossRef]
213. Alcántara, A.R.; Domínguez de María, P.; Littlechild, J.A.; Schürmann, M.; Sheldon, R.A.; Wohlgemuth, R. Biocatalysis as Key to Sustainable Industrial Chemistry. *ChemSusChem* **2022**, *15*, e202102709. [CrossRef]

catalysts

MDPI

Article

Synthesis and Laccase-Mediated Oxidation of New Condensed 1,4-Dihydropyridine Derivatives

Jelena Milovanovic [1,†], Miyase Gözde Gündüz [2,†], Anastasia Zerva [3], Milos Petkovic [4], Vladimir Beskoski [5], Nikolaos S. Thomaidis [6], Evangelos Topakas [3] and Jasmina Nikodinovic-Runic [1,*]

[1] Institute of Molecular Genetics and Genetic Engineering, University of Belgrade, Vojvode Stepe 444a, 11221 Belgrade, Serbia; jelenaradivojevic@imgge.bg.ac.rs
[2] Department of Pharmaceutical Chemistry, Faculty of Pharmacy, Hacettepe University, Sıhhiye, 06100 Ankara, Turkey; miyasegunduz@yahoo.com
[3] Industrial Biotechnology and Biocatalysis Group, Biotechnology Laboratory, School of Chemical Engineering, National Technical University of Athens, 5 Iroon Polytechniou Str., Zografou Campus, 15780 Athens, Greece; anazer@chemeng.ntua.gr (A.Z.); vtopakas@chemeng.ntua.gr (E.T.)
[4] Faculty of Pharmacy, University of Belgrade, Vojvode Stepe 450, 11221 Belgrade, Serbia; milosp@pharmacy.bg.ac.rs
[5] Faculty of Chemistry, University of Belgrade, Studentski trg 16, 11158 Belgrade, Serbia; vbeskoski@chem.bg.ac.rs
[6] Laboratory of Analytical Chemistry, Department of Chemistry, National and Kapodistrian University of Athens, Panepistimioupolis Zografou, 15771 Athens, Greece; ntho@chem.uoa.gr
* Correspondence: jasmina.nikodinovic@gmail.com or jasmina.nikodinovic@imgge.bg.ac.rs; Tel.: +381-11-397-60-34
† Equal contribution.

Citation: Milovanovic, J.; Gündüz, M.G.; Zerva, A.; Petkovic, M.; Beskoski, V.; Thomaidis, N.S.; Topakas, E.; Nikodinovic-Runic, J. Synthesis and Laccase-Mediated Oxidation of New Condensed 1,4-Dihydropyridine Derivatives. *Catalysts* **2021**, *11*, 727. https://doi.org/10.3390/catal11060727

Academic Editor: Antonio Zuorro

Received: 25 May 2021
Accepted: 10 June 2021
Published: 12 June 2021

Abstract: We describe herein the synthesis and laccase mediated oxidation of six novel 1,4-dihydropyridine (DHP)-based hexahydroquinolines (DHP1-DHP3) and decahydroacridines (DHP4-DHP6). We employed different laccase enzymes with varying redox potential to convert DHP1-DHP3 and DHP4-DHP6 to the corresponding pyridine-containing tetrahydroquinoline and octahydroacridine derivatives, respectively. Intensively coloured products were detected in all biocatalytic reactions using laccase from *Trametes versicolor* (*Tv*Lacc), possibly due to the presence of conjugated chromophores formed in products after oxidation. The NMR assessment confirmed that the oxidation product of DHP1 was its corresponding pyridine-bearing tetrahydroquinoline derivative. Laccase from *Bacillus subtillis* (*Bacillus*Lacc) was the most efficient enzyme for this group of substrates using HPLC assessment. Overall, it could be concluded that DHP2 and DHP5, bearing catecholic structures, were easily oxidized by all tested laccases, while DHP3 and DHP6 containing electron-withdrawing nitro-groups are not readily oxidized by laccases. DHP4 with decahydroacridine moiety consisting of three condensed six-membered rings that contribute not only to the volume but also to the higher redox potential of the substrate rendered this compound not to be biotransformed with any of the mentioned enzymes. Overall, we showed that multiple analytical approaches are needed in order to assess biocatalytical reactions.

Keywords: dihydropyridine; hexahydroquinoline; acridinedione; laccase; biotransformation

1. Introduction

1,4-Dihydropyridine (DHP) scaffold occupies a significant position amongst all heterocycles, especially due to its therapeutic value [1]. Commercial DHPs, represented by nifedipine, target the L-type calcium channel and are widely prescribed for the treatment of cardiovascular conditions [2]. This prominent ring is also suitable for various chemical modifications. Introducing DHP into fused ring systems such as hexahydroquinoline and decahydroacridine has led to the discovery of novel molecules with a vast variety of biological activities [3,4]. DHPs are under investigation not only for medical purposes but also for their physicochemical properties. The oxidation of DHP ring to pyridine is one of the most examined characteristics of DHPs also providing an approach for the preparation of pyridines starting from DHPs [5]. The oxidation of DHPs is of great interest since DHP-based calcium channel blockers are biologically converted to the corresponding pyridine derivatives by the action of cytochrome P-450 in the liver [6,7].

Although a number of catalytic oxidative aromatization of DHPs using molecular oxygen as the oxidant have been reported, many of these methods rely on transition metals, employ organic solvents and require high reaction temperatures [8–10]. On the other hand, laccases (EC 1.10.3.2) are multi-copper oxidoreductases (MCOs), and as such, use the distinctive redox ability of copper ions to catalyze the oxidation of a wide range of aromatic substrates, in parallel with the reduction of molecular oxygen to two water molecules [11].

The rate of a laccase-oxidized reaction has been shown to depend on the redox potential difference (ΔE^0) between the enzyme and the substrate. However, other factors such as the size, shape and hydrophobicity of the substrate-binding site near the T1 copper apparently also affect the reaction rate [12]. All laccases can be divided into three groups according to their redox potential (E°) at the T1 Cu site: (a) low E° enzymes with E° ≈ 430 mV referred to a standard normal hydrogen electrode (NHE); (b) a middle E° group including enzymes with potentials 470-710 mV; (c) high E° enzymes with redox potentials of 730–800 mV [13].

Only a few reports on the enzyme-mediated oxidation of DHPs are available, using purified *Trametes versicolor* laccase (*Tv*Lacc) and aerial oxygen, under mild conditions [6,14]. Abdel-Mohsen and coworkers showed that ABTS (2,2′-Azino-bis(3-ethylbenzothiazoline-6-sulfonic acid) diammonium salt) is the best mediator for the oxidation of DHPs [6]. A cooperative catalytic system of *Tv*Lacc and 4-phenyl urazole has been applied for the aerobic oxidative aromatization of DHPs [15]. In our previous work, Simic et al. showed that recombinant bacterial laccase, BliLacc from *Bacillus licheniformis* was successfully applied in the preparative oxidation of some DHPs, but the oxidation of DHPs with bulkier side groups resulted in poor yields of the corresponding products [16].

As laccases oxidize phenolic substrates [17], we aimed to synthesize six novels DHPs carrying hydroxyl group(s) on the phenyl ring (Figure 1). Unlike commercial DHPs, we introduced this scaffold into two condensed ring systems; hexahydroquinoline (DHP1-DHP3) and decahydroacridine (DHP4-DHP6). We investigated their laccase-mediated oxidation to the corresponding pyridine-containing derivatives using laccase enzymes with different redox potential. To examine the influence of redox potential on the efficiency of biocatalytic reaction, we have used four laccases: *Trametes versicolor* laccase (*Tv*Lacc) (E^0 = 800 mV) [18], *Myceliophthora thermophila* (Novozym®51003) laccase (E^0 = 460 mV) [19], *Bacillus subtilis* laccase (*Bacillus*Lacc) (E^0 = 525 mV) [20] and *Thermothelomyces thermophila* laccase-like multicopper oxidase (*Tt*LMCO1) (E^0 = 317 mV) [21]. Molecular docking analysis was employed for a better understanding of substrate-enzyme interactions.

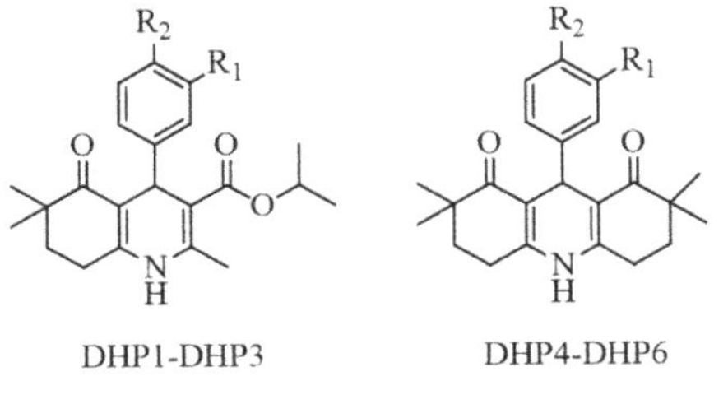

DHP	R_1	R_2
DHP1	OH	H
DHP2	OH	OH
DHP3	NO_2	OH
DHP4	OH	H
DHP5	OH	OH
DHP6	NO_2	OH

Figure 1. Chemical structures of the synthesized compounds.

2. Results and Discussion

We have successfully synthesized and structurally characterized two small subseries of new condensed DHPs carrying hexahydroquinoline (DHP1-DHP3) and decahydroacridine (DHP4-DHP6) scaffolds. The compounds were synthesized from moderate to good yields via modified Hantzsch reaction using 1,3-dicarbonyl compounds, appropriate benzaldehyde and ammonium acetate. The synthetic route to obtain the target compounds is depicted in Figure 2.

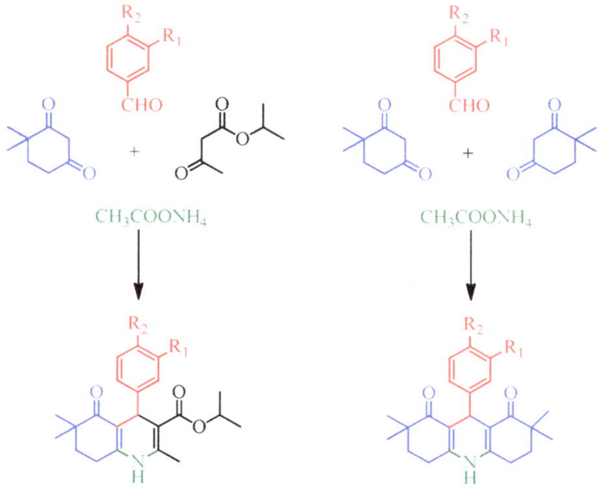

Figure 2. Synthetic scheme for the preparation of DHP-based compounds.

With this approach, we aimed to introduce DHP scaffold into condensed ring systems unlike the structures of the commercial ones such as nifedipine, isradipine and amlodipine. Additionally, we introduced at least one hydroxyl group on the phenyl ring of the compounds since laccases oxidize phenolic substrates. To see the effect of the phenyl substitution pattern on the laccase-mediated oxidation of the compounds, we synthesized compounds carrying one (DHP1, DHP4) or two phenolic hydroxyl groups (DHP2, DHP5). Moreover, we introduced an electron-withdrawing substituent, nitro group, to determine its impact on the oxidation of DHP3 and DHP6.

2.1. Oxidation of DHPs by Laccases from *Trametes versicolor* (*Tv*Lacc)

We have recently described the oxidation of a series of DHP derivatives by recombinant bacterial laccase, *Bli*Lacc from *B. licheniformis* [16]. Encouraged with previous results this study was initiated to explore if the commercially available laccase *Tv*Lacc could oxidize DHP1-DHP6. It was noticed that all the compounds were insoluble in most of the tested solvents and poorly soluble in warm ethanol and methanol; however, all reaction mixtures, after 18 h incubation, resulted in the colour change of all reactions: DHP1 and DHP4 turned to purple, DHP2 and DHP5 to orange and DHP3 and DHP6 to green (Figure 3). Notably, despite the colour change, the additional products were detected by TLC only in reaction mixtures containing DHP1 and DHP5 as substrates (Figure 3). The observed colour change indicated that there was a change in the chromophore structure of the starting compounds. Indeed, UV–VIS spectra of both substrate and reaction product confirmed this (Figures S1–S6). We expected that the oxidation of the dihydropyridine ring led to aromatization and formation of a pyridine ring which together with the phenyl group forms a conjugated chromophore. In the control reactions without enzyme and in the presence of ABTS, colour changed to green, due to oxidation of ABTS to the green cation radical (ABTS•+) [22]. In the control reactions without the ABTS as mediator and *Tv*Lacc, a slight change of colour in reactions with substrate DHP2 and DHP5 was notable (Figure S7), while TLC and HPLC analysis showed very low conversion rate but the same pattern to reactions in the presence of ABTS (Figures S8–S13, panel f). The addition of

$CuSO_4$ did not have any influence on resulting products compared to the same reactions without Cu ions (data not shown).

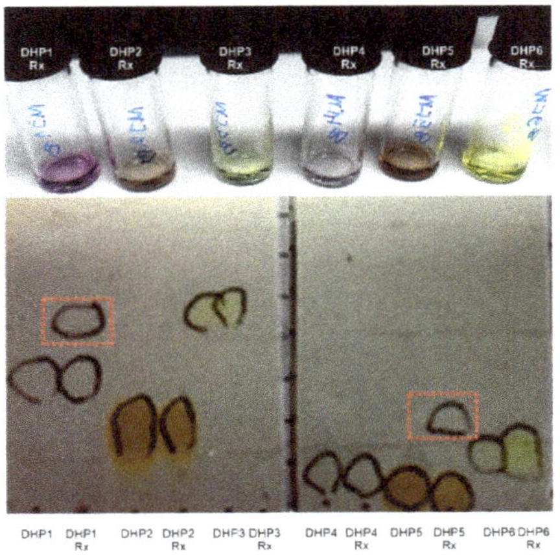

Figure 3. Biocatalytic reactions of DHP1-DHP6 using *Tv*Lacc as biocatalyst. Reactions after 18 h incubation and extraction with ethyl acetate are depicted in the upper panel and thin-layer chromatography analysis of DHP substrates and corresponding biotransformation products (eluent hexane/ethyl-acetate= 1/1, UV visualization). Red rectangles indicate DHP1 and DH5 reaction products.

Reaction mixtures were analysed by HPLC, as well (Figures S8–S13). We found that the results obtained on HPLC (with a UV detector) are consisted with TLC analysis results. The reactions using only DHP1 and DHP5 gave an additional significant signal, belonging to the product of the biocatalytic reaction, in the corresponding chromatograms.

Reaction product from DHP1 oxidation was successfully isolated and characterized by NMR spectroscopy. The proton NMR spectrum of the starting compound DHP1 contains signals derived from nitrogen-bound hydrogen and C4-hydrogen in the pyridine nucleus at about $\delta \sim 9$ ppm and $\delta = 4.74$, respectively. These signals do not exist in the ^{1}H NMR spectrum of the product (Figure 4, Figure S14). On the other hand, in the ^{13}C NMR spectrum of the product, an additional signal originating from the carbon C4 in the aromatic pyridine ring appears in the range from 140 ppm to 170 ppm (Figure S15). Taking into account these differences in the spectra the structure of the obtained reaction product was determined (Figure 4). According to the appearance of the HPLC chromatogram and the retention time (Figure S12, data b), the same transformation was also expected during the reaction with the substrate DHP5. Noteworthy, when DHP5-R was purified on a silica column, hydrolysis of the product occurred, preventing the representative NMR spectrum; however, signature peaks of the expected oxidation product were present (data not shown). When liquid chromatography coupled with a mass detector was used to analyze products of the biocatalytic transformation, it was quite difficult to determine the identity and the amount of the products, as the expected mass difference was only a single H atom (Figure 4).

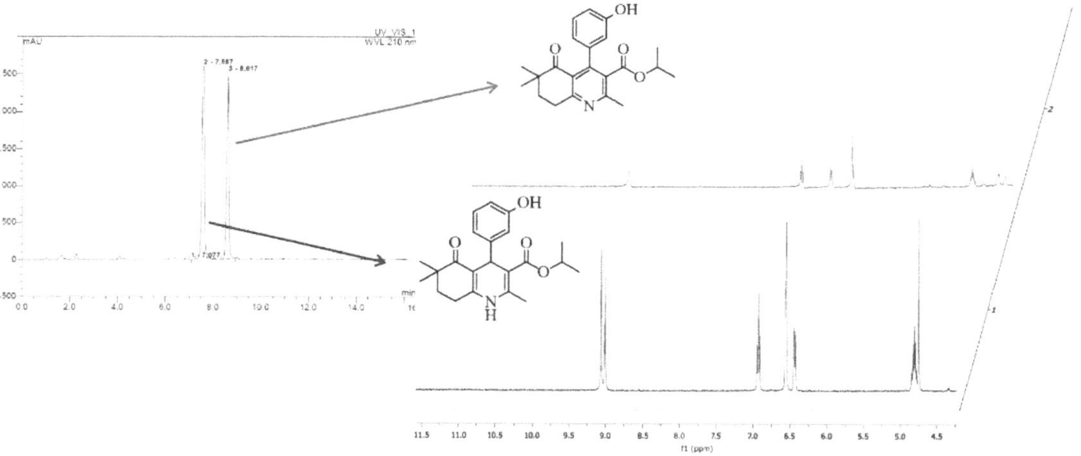

Figure 4. HPLC chromatogram of the laccase mediated transformation of DHP1 showing both DHP1 and DHP1-R as reaction product and ^{1}H NMR spectra of the DHP1 (bottom) and its tetrahydroquinoline derivative as the isolated oxidation product (top).

It has been well established that in the mediator-redox-system, oxidation occurs through a mediator, and it is highly dependent on the type of mediator [23]. Recently, it was shown that, in the presence of ABTS as a mediator, the substrate oxidations by the laccase and by the ABTS cation radical take place simultaneously, at least for phenolic substrates [24]. In order to better understand why oxidation products did not occur in reactions with other substrates, we calculated the relative energies of all substrates and expected pyridines (Table 1). All substrates have higher stability (lower absolute energy) in comparison with the corresponding oxidation products. Substrate-product pair DHP1-DHP1R has the lowest energy barrier in comparison with others; this could be the reason for the successful oxidation of DHP1. On the other hand, the reaction product occurs in the oxidation of DHP5 although the relative energy difference is similar to the other substrate-product pairs (Table 1).

Table 1. Relative energies of substrates DHP1-DHP6 and expected reaction products (DHP1-R–6-R).

Compound	Relative Energy (kcal/mol)
DHP1	0.0 [a]
DHP1-R	750.8
DHP2	0.0 [b]
DHP2-R	763.2
DHP3	0.0 [c]
DHP3-R	784.4
DHP4	0.0 [d]
DHP4-R	763.6
DHP5	0.0 [e]
DHP5-R	765.8
DHP6	0.0 [f]
DHP6-R	763.3

Absolute energy baselines (in Hartree). *a*: −1210.607757755; *b*: −1285.85580863; *c*: −1415.17774797; *d*: −1212.73819921; e: −1287.97793577; *f*: −1417.28900326.

2.2. Oxidation of DHPs by Laccases with Different Redox Potential

After the first set of biocatalytic oxidation of a series of DHP compounds with *Tv*Lacc, we found that there was a difference in the success of the reaction with various substrates (Figure 5A). As the progression of the reaction is influenced by many factors, among them enzyme redox potential, we decided to examine biocatalytic oxidation with three additional laccases having lower redox potential compared to *Tv*Lacc (E^0 = 800 mV), Novozym 51003 laccase (E^0 = 460 mV) [19], *Bacillus*Lacc (E^0 = 525 mV) [20], *Tt*LMCO1 (E^0 = 317 mV). *Tt*LMCO1 is a novel thermostable laccase-like multicopper oxidase with low redox potential (E^0 = 317 mV vs. NHE at 30 °C) [21]. *Tt*LMCO1 has been used previously for the cyclization of chalcones to the respective aurones with satisfactory yields [25]. The low redox potential of this particular LMCO leads to higher selectivity and therefore could be an advantage in synthetic applications, compared to high- E^0 laccases, which can quickly polymerize their substrate to insoluble products.

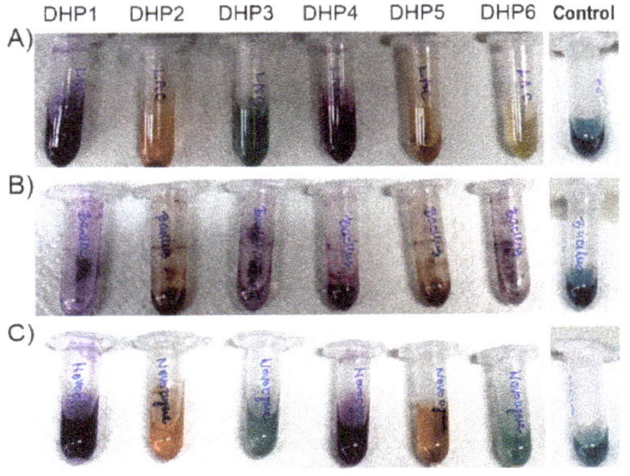

Figure 5. Colours of reactions after 18 h incubation with different types of laccases (**A**) *Tv*Lacc; (**B**) *Bacillus*Lac; (**C**) Novozym 51003; Control reactions with enzymes and ABTS mediator without DHP substrate.

Reactions were set by the same general procedure on small scale (using the same ratio of enzyme U/mg of a substrate; Table S1) and analyzed by HPLC after 18 h incubation (Table 2)). Using DHP4 and DHP6 as substrates, no detectable products were observed with any of the enzymes, although DHP4 changed colour to purple. Similarly, DHP3 was also very difficult to oxidize by laccases and only about 5–6% of the substrate was converted using *Bacillus*Lacc and Novozym 51003 laccases (Table 2). Among all tested enzymes, *Bacillus*Lacc was the most successful in the oxidation of DHP1 with more than 45% of this substrate converted to products, followed by *Tv*Lacc>Novozym 51003>*Tt*LMCO1. Judging by the HPLC assessment, the easiest to transform was DHP5 with *Tt*LMCO1 utilising about 80% of the substrate in the tested time-frame. The biocatalytic reactions without the presence of the ABTS mediator were also assessed (Table S2). Overall, lower conversion has been detected, but substrates DHP1, DHP2 and DHP5 were still the most readily oxidized. On the other hand, it was quite difficult to assess the biocatalytic reaction products using mass spectral analysis as the major expected product differed only in single hydrogen from the substrate, but MS analysis detected dimers in reaction with DHP5 as substrate (Figure S17). Since the detected dimer co-eluted with the respective monomer, it is possible that it could be formed during the ionization process itself (i.e., the dimerization took place after elution from the column).

Table 2. DHP1-DHP6 biotransformation with four laccases assessed by HPLC with the amount of substrate remaining after 18 h reaction expressed as %. Chromatograms are given in supporting material.

Substrate * Enzyme	DHP1	DHP2	DHP3	DHP5
*Tv*Lacc	61.8	45.9	100	52.8
*Bacillus*Lacc	54.7	63.8	94.1	43.6
Novozym 51003	67.2	79.4	94.1	37.7
*Tt*LMCO1	79.4	76.6	100	21.2

* For reactions DHP4 and DHP6 the results are not shown, since there was no reaction.

DHP3 and DHP6 containing electron-withdrawing nitro-groups are not readily oxidized by laccases [26]; therefore, the absence of activity against $-NO_2$–bearing DHPs could be expected, despite the use of ABTS as mediator. NO_2-bearing DHPs were oxidized at very low yield only by *Bacillus*Lacc and Novozym 51003, but not *Tv*Lacc. This is in accordance with the findings of Tadesse and coworkers, where NO_2-bearing phenols were readily oxidized by Novozym laccase, but not *Tv*Lacc, despite the difference of the enzymes in E^0 and the high E^0 of the substrates [19]. The authors attributed this apparent discrepancy to steric issues related to the active site architecture of *Tv*Lacc: the existence of two phenylalanines (Phe332 and Phe337), forming a 'gate' for the entrance of the substrate, could prohibit access to bulky substrates. Phe162 also plays an important role in the shaping of the hydrophobic cavity: replacement by Ala resulted in more efficient oxidation of bulky substrates [27]. Small size residues in these positions result in a more relaxed entrance to the T1 copper centre (Figure 6 and Figure S18), which is reflected in more efficient oxidation of bulky substrates by *Bacillus*Lacc, compared to the other laccases. On the other hand, Novozym 51003 contains a Trp residue in the position of Phe337 of *Tv*Lacc laccase, which is a rather bulky amino acid (Table S2, Figures S12 and S19). This may contribute to the lower oxidation efficiency towards DHP2, compared to *Bacillus*Lacc.

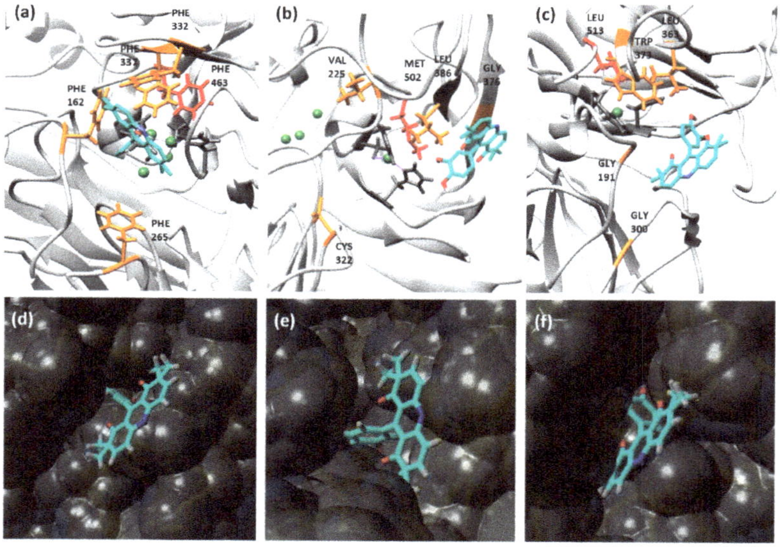

Figure 6. Models of DHP5 interacting with laccases *Tv*Lacc (**a**,**d**), *Bacillus*Lacc (**b**,**e**) and Novozym 51003 (**c**,**f**). (**a**–**c**) The interaction model of DHP5 with the tested laccases. Ligand is coloured in cyan, copper atoms are coloured in green, equatorial T1 copper ligands are coloured in dark grey, an axial copper ligand is coloured in red, binding pocket residues are coloured in orange. (**d**–**f**) The surface binding model of DHP5 with the tested laccases.

To better understand the obtained results, we explored the interaction of compounds DHP5 and DHP2 and laccases using standard docking protocols. *Tt*LMCO1 was not included in docking analyses due to a lack of available crystal structure. The architecture of the substrate-binding pocket, together with the respective models are shown in Figure 6 for DHP5 substrate and in Figure S12 for DHP2 substrate. The results of the molecular docking analysis for DHP5 and DHP2 are shown in Tables S3 and S4, respectively.

Some conclusions can be drawn from the correlation of the size of the substrates with the architecture of the substrate pocket for each laccase: for example, according to structure-based sequence alignments, *Bacillus*Lacc contains Val, Leu and Gly in the positions Phe 162, Phe337 and Phe332 of *Tv*Lacc laccase, respectively, as shown by structure-based sequence alignments (Figure S18).

The redox potential of laccases in some cases can be correlated with one residue, acting as a weakly coordinating axial ligand to the T1 copper centre. This amino acid is usually a Phe in high-redox, white-rot fungal laccases, while in *Bacillus*Lacc and Novozym 51003 laccase this position is occupied by a Met and a Leu, respectively (Figure 6) [28]. Even though DHP1 and DHP4 have the same phenol ring, their reactivity with all tested enzymes is completely different. DHP1 is oxidized by all tested laccases in good yield while DHP4 did not react at all. This is expected because of decahydroacridine moiety in DHP4, which consists of three condensed six-membered rings, which contribute to the volume but more importantly to the higher redox potential of the substrate resulting in no detectable oxidation using laccases.

The addition of –OH groups in the phenol ring lowers the redox potential (E^0) of the compound (for example the E^0 of catechol is 0.53 V vs. NHE, compared to phenol with E^0 >0.8 V vs. NHE), and therefore, the oxidation of catecholic compounds is easier by laccases [29]. This is reflected by our results, since DHP2 and DHP5, bearing catecholic structures, were easily oxidized by almost all tested laccases. Among the catecholic substrates tested, significant differences can be observed in the oxidation efficiency by different laccases: For example, the oxidation of DHP2 follows the E^0 of the tested enzymes (less substrate remaining with increasing E^0), but the exact opposite happens in the case of DHP5 substrate. This is surprising: in the case of *Tt*LMCO1, which is the enzyme with the lowest E^0 tested, the maximum activity was observed. Nonetheless, it is in line with previous studies, reporting that substrate specificity does not always correlate well with the redox potential of the enzyme, and kinetic factors also play a significant role. For example, Xu and co-workers tested the kinetic properties of several multicopper oxidases and concluded that enzyme affinity is independent of the redox potential, although the turnover number is increasing with increased E^0 [30]. More recently, Mehra and co-workers concluded that the apparent K_m of a given laccase is the result of both affinity and electronic properties of the substrate [31]. Glazunova and co. also reported that high E^0 does not necessarily lead to more efficient oxidation, and they attributed the observed discrepancies to the architecture of the substrate-binding pocket [32]. According to this study, both the size and flexibility of the substrate pocket, together with the presence of specific amino acids, could determine the affinity for a specific substrate, and thus the architecture of the substrate pocket can be engineered properly to yield tailored biocatalysts for specific applications.

3. Materials and Methods

3.1. Reagents and Chemicals

All chemicals and solvents were purchased from commercial sources and were used without further purification.

*Tv*Lacc was purchased from Sigma (Munich, Germany), provided in the form of powder. Recombinant *Bacillus subtilis* CotA laccase preparation was obtained as described [33]. *Tt*LMCO1 was heterologously produced in *Pichia pastoris* and purified with metal affinity chromatography according to our previous study [25]. Laccase from *Myceliophthora thermophila* (Novozym 51003) was a kind gift from Novozymes A/S® (Bagsværd, Denmark) and was used as supplied.

3.2. Analytical Methods

The reactions were monitored and the purity of the obtained compounds was initially checked by thin-layer chromatography (TLC) performed on silica gel 60 F254 precoated aluminium sheets (Merck, (Darmstadt, Germany) using mobile phase ethyl acetate-hexane (1:1) and UV light for visualization. Melting points were determined using Thomas Hoover Capillary Melting Point Apparatus (Philadelphia, PA, USA) and were uncorrected. ^{1}H and ^{13}C NMR spectra were recorded at 400 and 100 MHz, respectively, using High-Performance Digital FT-NMR Spectrometer, Varian Mercury 400, (Palo Alto, California, USA) in dimethylsulfoxide (DMSO-d_6) solutions. Chemical shifts are expressed in ppm (δ) using tetramethylsilane as an internal standard, and coupling constants (J) are measured in hertz (Hz). The ESI-MS spectra were obtained on a micro mass ZQ-4000 single quadrupole mass spectrometer (Waters, Eschborn, Germany). Elemental analyses were carried out on a Leco CHNS 932 Elemental Analyzer (St. Joesph, MI, USA). UV–VIS spectra were recorded at spectrophotometer Ultrospec 3300pro, Amersham Biosciences (Amersham, Buckinghamshire, UK). Chromatographic analyzes were performed on HPLC (Thermo Scientific Ultimate 3000, Thermo Scientific, Waltham, Massachusetts, USA) with a UV detector (UltiMate 3000 DAD) .

3.3. General Procedure for the Synthesis and Structure Elucidation of DHPs

DHP-based compounds (DHP1-DHP6) were obtained according to modified Hantzsch synthesis. DHP-based hexahydroquinolines (DHP1-DHP3) and decahydroacridines (DHP4-DHP6) were synthesized as follows:

DHP1-DHP3: Equimolar amount (1 mmol) of 4,4-dimethyl-1,3-cyclohexanedione, substituted benzaldehyde and isopropyl acetoacetate were dissolved in 15 mL absolute ethanol and heated for 8 h under reflux conditions. The reaction mixture was cooled and poured into ice water. The obtained precipitate was filtered, and this crude solid was purified by recrystallization from ethanol-water.

DHP4-DHP6: 2 mmol 4,4-dimethyl-1,3-cyclohexanedione, 1 mmol substituted benzaldehyde and an excess amount of ammonium acetate (5 mmol) were heated in 15 mL absolute ethanol under reflux conditions for 8 h. After the reaction mixture was cooled down, the forming crystals were filtered and washed with cold ethanol, or the mixture was poured into ice-water, the obtained precipitate was filtered and recrystallized from the ethanol-water mixture.

Isopropyl 4-(3-hydroxyphenyl)-2,6,6-trimethyl-5-oxo-1,4,5,6,7,8-hexahydroquinoline-3- carboxylate (DHP1): Yield: 34%. M.p. 229–231 °C. ^{1}H-NMR (δ, DMSO-d_6): 0.88 (3H; s), 0.95 (3H; s), 1.05 (3H; d; J = 6.4 Hz), 1.15 (3H; d; J = 6.4 Hz), 1.67–1.71 (2H; m), 2.22 (3H; s), 2.44–2.48 (2H; m), 4.74 (1H; s), 4.77–4.83 (1H; m), 6.41–6.93 (4H; m), 8.97 (1H; s), 9.02 (1H; s). ESI-MS (m/z): 392 $[M+Na]^+$. Anal. Calcd. for $C_{22}H_{27}NO_4$; C, 71.52; H, 7.37; N, 3.79. Found: C, 71.25; H, 7.44; N, 3.82.

Isopropyl 4-(3,4-dihydroxyphenyl)-2,6,6-trimethyl-5-oxo-1,4,5,6,7,8-hexahydro quinoline-3-carboxylate (DHP2): Yield: 32%. M.p. 242–244 °C. ^{1}H-NMR (δ, DMSO-d_6): 0.90 (3H; s), 0.97 (3H; s), 1.08 (3H; d; J = 6.4 Hz), 1.17 (3H; d; J = 6.4 Hz), 1.69–1.73 (2H; m), 2.22 (3H; s), 2.45–2.49 (2H; m), 4.66 (1H; s), 4.79–4.85 (1H; m), 6.38 (1H; dd; J = 2.0/8.0 Hz), 6.49 (1H; d; J = 8.0 Hz), 6.54 (1H; d; J = 2.0 Hz), 8.45 (1H; s), 8.53 (1H; s), 8.92 (1H; s). ESI-MS (m/z): 408 $[M+Na]^+$. Anal. Calcd. for $C_{22}H_{27}NO_5$; C, 68.55; H, 7.06; N, 3.63. Found: C, 68.38; H, 7.11; N, 3.62.

Isopropyl 4-(4-hydroxy-3-nitrophenyl)-2,6,6-trimethyl-5-oxo-1,4,5,6,7,8-hexahydro quinoline-3-carboxylate (DHP3): Yield: 65%. M.p. 256–258 °C. ^{1}H-NMR (δ, DMSO-d_6): 0.86 (3H; s), 0.95 (3H; s), 1.02 (3H; d; J = 6.4 Hz), 1.16 (3H; d; J= 6.4 Hz), 1.66–1.72 (2H; m), 2.24 (3H; s), 2.45–2.49 (2H; m), 4.75 (1H; s), 4.77–4.83 (1H; m), 6.95 (1H; d; J = 8.8 Hz), 7.28 (1H; dd; J = 2.0/8.8 Hz), 7.57 (1H; d; J= 2.0 Hz), 9.09 (1H; s). ESI-MS (m/z): 437 $[M+Na]^+$. Anal. Calcd. for $C_{22}H_{26}N_2O_6$; C, 63.76; H, 6.32; N, 6.76. Found: C, 63.31; H, 6.29; N, 6.76.

9-(3-hydroxyphenyl)-2,2,7,7-tetramethyl-3,4,6,7,9,10-hexahydroacridine-1,8(2H,5H) -dione (DHP4): Yield: 25%. M.p. 294–295 °C. ^{1}H-NMR (δ, DMSO-d_6): 0.91 (6H; s), 0.97 (6H; s),

1.71–1.75 (4H; m), 2.49–2.53 (4H; m), 4.79 (1H; s), 6.37–6.90 (4H; m), 9.01 (1H; s), 9.29 (1H; s). ESI-MS (*m/z*): 388 $[M+Na]^+$. Anal. Calcd. for $C_{23}H_{27}NO_3$; C, 75.59; H, 7.45; N, 3.83. Found: C, 75.43; H, 7.39; N, 3.80.

9-(3,4-dihydroxyphenyl)-2,2,7,7-tetramethyl-3,4,6,7,9,10-hexahydroacridine-1,8 (2H,5H)-dione (DHP5): Yield: 39%. M.p. 260–261 °C. ^{1}H-NMR (δ, DMSO-d_6): 0.89 (6H; s), 0.94 (6H; s), 1.70 (4H; t), 2.47 (4H; t), 4.68 (1H; s), 6.32 (1H; dd; *J* = 2.0/8.0 Hz), 6.44 (1H; d; *J* = 8.0 Hz), 6.52 (1H; d; *J* = 2.0 Hz), 8.34 (1H; s), 8.50 (1H; s), 9.19 (1H; s). ESI-MS (*m/z*): 404 $[M+Na]^+$. Anal. Calcd. for $C_{23}H_{27}NO_4$; C, 72.42; H, 7.13; N, 3.67. Found: C, 72.20; H, 7.07; N, 3.69.

9-(4-hydroxy-3-nitrophenyl)-2,2,7,7-tetramethyl-3,4,6,7,9,10-hexahydroacridine-1,8 (2H,5H)-dione (DHP6): Yield: 58%. M.p. 267–268 °C. ^{1}H-NMR (δ, DMSO-d_6): 0.87 (6H; s), 0.95 (6H; s), 1.69-1.73 (4H; m), 2.47–2.52 (4H; m), 4.77 (1H; s), 6.92 (1H; d; *J* = 8.4 Hz), 7.30 (1H; dd; *J* = 2.4/8.4 Hz), 7.52 (1H; d; *J* = 2.4 Hz), 9.38 (1H; s). ESI-MS (*m/z*): 433 $[M+Na]^+$. Anal. Calcd. for $C_{23}H_{26}N_2O_5$; C, 67.30; H, 6.38; N, 6.82. Found: C, 67.43; H, 6.43; N, 6.89.

3.4. *Biocatalytic Reaction Setup and Analytical Procedures*

3.4.1. General Procedure for the Laccase-Catalyzed Oxidation of DHPs -Analytical Scale

In 2 mL Eppendorf tubes, 1 mg of DHP substrate was suspended in 50 µL methanol and 850 µL sodium acetate buffer (0.1 M, pH = 4.5). Subsequently, 5 µL of ABTS diammonium salt (final concentration 0.2 mM) dissolved in acetate buffer, and 100 µL *Tv*Lacc laccase (4–5 U) dissolved in acetate buffer, was added to the suspension. In some cases, a water solution of $CuSO_4$ to a final concentration of 0.09 mM was also added to the reaction mixture. Reaction mixtures were set up in Thermomixer (Eppendorf, Hamburg, Germany) (rpm 500), 50 °C overnight. After extraction with ethyl acetate (3 × 300 µL), the solvent was removed under reduced pressure. The crude products were dissolved in methanol, filtered and analyzed with HPLC.

3.4.2. General Procedure for the Laccase-Catalyzed Oxidation of DHPs -Medium Scale

In 100 mL flasks, 5 mg of DHP substrate was suspended in 300 µL methanol and 4.5 mL sodium acetate buffer (0.1 M, pH = 4.5). Subsequently, 87.5 µL ABTS diammonium salt (final concentration 0.2 mM) dissolved in acetate buffer and 200 µL *Tv*Lacc (5 U per 1 mg of the substrate) dissolved in acetate buffer was added to the suspension. When all different laccases were used, the amount of enzyme per mg of the substrate was kept constant (the activity units of laccases were assessed against catechol in the sodium acetate buffer (0.1 M, pH = 4.5)). Reaction mixtures were set up in a water bath at 150 rpm, 50 °C overnight. (New Brunswick Scientific, Edison, New JerseyUSA). Reaction products were extracted with ethyl acetate (3 × 10 mL); the combined organic extracts were dried over anhydrous $MgSO_4$; the solvent was removed under reduced pressure.

3.4.3. HPLC Analysis of Reaction Mixtures

Prepared crude products of all reaction mixtures were dissolved in methanol, filtered and analyzed by HPLC (Thermo Scientific Ultimate 3000, Vertex plus C18A 150 × 4.6 mm column, Thermo Scientific, Waltham, Massachusetts, USA) using eluent acetonitrile/water in gradient (acetonitrile 30% to 90% for 13 min, 90% constant for 0.5 min, return 3 min to initial conditions). The eluted products were detected at 210 nm, and quantitative analysis was performed by calculating the peak areas into a percentage using Chromeleon software. Reaction products were isolated on a semi-preparative column (Vertex plus C18, 250 × 8 mm) with the same solvent system.

3.4.4. NMR Analysis of Isolated Product from Biocatalytic Oxidation of DHP1

The reaction product was isolated using HPLC, as we described in the previous section. NMR spectra were recorded in DMSO-d_6 solution.

Isopropyl 4-(3-hydroxyphenyl)-2,6,6-trimethyl-5-oxo-5,6,7,8-tetrahydroquinoline-3- carboxylate: ^{1}H-NMR (δ, DMSO-d_6): 9.45 (s, 1H), 7.10 (t, *J* = 7.7 Hz, 1H), 6.71 (d, *J* = 7.3 Hz, 1H), 6.40 (d, *J* = 7.9 Hz, 1H), 4.79 (dd, *J* = 12.3, 6.1 Hz, 1H), 3.10 (t, *J* = 6.2 Hz, 2H), 2.45 (s,

3H), 1.94 (t, J = 6.4 Hz, 2H), 1.06 (s, 3H), 0.94 (d, J = 6.2 Hz, 3H). ^{13}C NMR (δ, DMSO-d_6): 201.74, 166.81, 163.39, 157.16, 156.86, 148.45, 138.91, 129.94, 129.17, 123.73, 118.69, 115.02, 114.81, 69.19, 42.58, 34.59, 29.51, 24.09, 22.96, 21.38.

3.5. Molecular Docking Simulations

The protein database (PDB) structures selected for docking analyses were 1KYA for *Tv*Lacc, 4Q8B for *Bacillus*Lacc and 6F5K for Novozym 51003 laccase. *Tt*LMCO1 was not included in docking analyses due to a lack of available crystal structure. The structures were optimized, and co-crystallized ligands were removed. Molecular docking analyses were performed with YASARA (Yet Another Scientific Artificial Reality Application), with a simulation cell of 13.0 Å filled with water around T1 copper [34]. The receptor was considered rigid, and the ligand was considered flexible. All other parameters were set to defaults. After 25 AutoDock VINA runs, the resulting clusters with RMSD > 5.0 Å were visually inspected. The complex conformations with the highest binding energies were selected, and visualization of the resulting models was performed with UCSF Chimera 1.14, which was also used for structure-based sequence alignments. Structure-based sequence alignments were visualized by ESPRIPT 3.0.

Molecular Energies Calculations

Optimization of geometry and calculation of relative energies (kcal/mol) of compounds DHP1-DHP6 and expected reaction products were obtained by Gaussian with B3LYP method using a 6-311++G(d,p) basis set.

4. Conclusions

Laccase oxidation usually leads to the formation of multiple products, and thus, laccase-mediated biocatalytic reactions require the use of multiple approaches for their characterization. In the present work, the laccase-mediated biocatalytic transformation of new DHP molecules was described. DHP2 and DHP5, which contain catechol structures, were easily oxidized by all tested laccases, while DHP3 and DHP6, which contain electron-withdrawing nitro-groups, are not readily oxidized, even with the addition of ABTS as mediator. The presence of a voluminous acridine ring, as well as the reportedly higher redox potential of the phenolic hydroxyl group in DHP4 contributed to the small reactivity of this compound. From the results of the present study, it can be concluded that molecular descriptors, such as size and electronic properties, have the greatest influence on the susceptibility of DHP-based substrates to laccase oxidation. Molecular docking analysis suggests that the architecture of the substrate-binding pocket is important for efficient oxidation. *Bacillus*Lacc has small gate-forming residues compared to the other tested laccases, and this enzyme was shown to be the most versatile regarding DHP oxidation, since it showed satisfactory activity against all tested DHP substrates. Overall, the present work reports the synthesis and laccase-mediated oxidation of new DHP compounds. The results of the present study highlight the complexity of laccase-mediated oxidation, and complement previous studies, supporting that the redox and kinetic properties of a given laccase may not always be adequate predictors of the activity on a given substrate, since other factors also may play a defining role, such as the size and side groups.

Supplementary Materials: The following are available online at https://www.mdpi.com/article/10.3390/catal11060727/s1. Figure S1: DHP1 (a) wave scan 220–800 nm, starting compound from 290–410 nm (concentrated sample dissolved in MeOH, recorded at spectrophotometer); (b) wave scan 200–400 nm, starting compound from 313–400 nm (sample conc. 1 mg/mL, dissolved in MeOH, recorded at HPLC); (c) wave scan 200–400 nm, reaction product -peak of product (blue line) (sample conc. 1 mg/mL, dissolved in MeOH). Figure S2: DHP2 (a) wave scan 220–800 nm, starting compound from 290–410 nm (concentrated sample dissolved in MeOH, recorded at spectrophotometer); (b) wave scan 200–400 nm, starting compound from 313–400 nm (sample conc. 1 mg/mL, dissolved in MeOH, recorded at HPLC); (c) wave scan 200–400 nm, reaction product (sample conc. 1 mg/mL, dissolved in MeOH, recorded at HPLC). Figure S3: DHP3 (a) wave scan 220–800 nm, starting compound from 290–410 nm (concentrated sample dissolved in MeOH, recorded at spectrophotometer); (b) wave scan 200–400 nm, starting compound from 313–400 nm (sample conc. 1 mg/mL, dissolved in MeOH, recorded at HPLC); (c) wave scan 200–400 nm, reaction product (sample conc. 1 mg/mL, dissolved in MeOH). Figure S4: DHP4 (a) wave scan 220–800 nm, starting compound from 290–410 nm (concentrated sample dissolved in MeOH, recorded at spectrophotometer); (b) wave scan 200–400 nm, starting compound from 313–400 nm (sample conc. 1 mg/mL, dissolved in MeOH, recorded at HPLC); (c) wave scan 200–400 nm, reaction product (sample conc. 1 mg/mL, dissolved in MeOH, recorded at HPLC). Figure S5: DHP5 (a) wave scan 220–800 nm, starting compound from 290–410 nm (concentrated sample dissolved in MeOH, recorded at spectrophotometer); (b) wave scan 200–400 nm, starting compound from 313–400 nm (sample conc. 1 mg/mL, dissolved in MeOH, recorded at HPLC); (c) wave scan 200–400 nm, reaction product (sample conc. 1 mg/mL, dissolved in MeOH, recorded at HPLC). Figure S6: DHP6 a) wave scan 220–800 nm, starting compound from 290–410 nm (concentrated sample dissolved in MeOH, recorded at spectrophotometer); (b) wave scan 200–400 nm, starting compound from 313–400 nm (sample conc. 1 mg/mL, dissolved in MeOH, recorded at HPLC); (c) wave scan 200–400 nm, reaction product (sample conc. 1 mg/mL, dissolved in MeOH, recorded at HPLC). Figure S7: Colors of reactions after 18 h incubation with *Tv*Lacc without ABTS mediator using six different DHP substrates. Figure S8: HPLC chromatograms of DHP1 before (a) and after oxidation with different enzymes (b) *Tv*Lacc, (c) *Bacillus*Lacc, (d) Novozym51003, (e) *Tt*LMCO1 and (f) *Tv*Lacc without ABTS. Figure S9: HPLC chromatograms of DHP2 before (a) and after oxidation with different enzymes (b) *Tv*Lacc, (c) *Bacillus*Lacc, (d) Novozym51003 (e) *Tt*LMCO1 and (f) *Tv*Lacc without ABTS. Figure S10: HPLC chromatograms of DHP3 before (a) and after oxidation with different enzymes (b) *Tv*Lacc, (c) *Bacillus*Lacc, (d) Novozym51003, (e) *Tt*LMCO1 and (f) *Tv*Lacc without ABTS. Figure S11: HPLC chromatograms of DHP4 before (a) and after oxidation with different enzymes (b) *Tv*Lacc, (c) *Bacillus*Lacc, (d) Novozym51003, (e) *Tt*LMCO1 and f) *Tv*Lacc without ABTS. Figure S12: HPLC chromatograms of DHP5 before (a) and after oxidation with different enzymes b) *Tv*Lacc, (c) *Bacillus*Lacc, (d) Novozym51003, (e) *Tt*LMCO1 and (f) *Tv*Lacc without ABTS. Figure S13: HPLC chromatograms of DHP6 before (a) and after oxidation with different enzymes (b) *Tv*Lacc, (c) *Bacillus*Lacc, (d) Novozym51003, (e) *Tt*LMCO1 and (f) *Tv*Lacc without ABTS. Figure S14: ^{1}H NMR spectra of isolated product from biocatalytic oxidation of DHP1. Figure S15: ^{13}C NMR spectra of isolated product from biocatalytic oxidation of DHP1. Figure S16: 2D COSY NMR spectra of isolated product from biocatalytic oxidation of DHP1. Figure S17: (a) Overlaid chromatograms of DHP5 (in blue, for the m/z value 382.2013 ($\pm$5 mDa) and its corresponding dimer (in orange, for the m/z value 763.3952 ($\pm$5 mDa) eluted at 6.80 min in positive ionization mode. (b) MS/MS spectra of the dimer of DHP5 in autoMS mode. Figure S18: Structure-based sequence alignment of laccases from *T. versicolor* (1kya), *B. subtilis* (4q8b) and Novozym 51003 (6f5k). Green arrows indicate the gate-forming phenylalanines in laccase from *T. versicolor*. Conserved residues are shown in red. α- helices, β-sheets and turns are indicated in black. Figure S19: Models of DHP2 interacting with laccases *Tv*Lacc (a, d), *Bacillus*Lacc (b, e) and Novozym 51003 (c, f). a–c: The interaction model of DHP2 with the tested laccases. Ligand is coloured in cyan, copper atoms are coloured in green, equatorial T1 copper ligands are coloured in dark grey, an axial copper ligand is coloured in red, binding pocket residues are coloured in orange. d-f: The surface binding model of DHP2 with the tested laccases. Table S1: Specific activities of the different laccases used in this study. Table S2: DHP1-DHP6 biotransformation with four laccases, without mediator ABTS, assessed by HPLC with the amount of substrate remaining after 18 h reaction expressed as %. (Chromatograms for *Tv*Lacc are provided in Figures S2–S7 as panel (f)). Table S3: The results of molecular docking for DHP5 in the active site of the tested laccases. Table S4: The results of molecular docking for DHP2 in the active site of the tested laccases.

Author Contributions: Conceptualization, J.M., E.T. and J.N.-R. methodology, J.M, M.G.G., A.Z. and N.S.T.; investigation, J.M., M.G.G., A.Z., M.P. and V.B.; resources, V.B., N.S.T., E.T. and J.N.-R.; data curation, J.M., M.G.G., A.Z.; writing—original draft preparation, J.M., A.Z. and M.P.; writing—review and editing, M.G.G., V.B., N.S.T., E.T. and J.N.-R.; funding acquisition, J.N.-R. All authors have read and agreed to the published version of the manuscript.

Funding: This research was funded by Ministry of Education, Science and Technological Development of the Republic of Serbia, 451-03-68/2020-14/200042, 2020.

Acknowledgments: The authors are grateful to Novozymes A/S for the generous gift of Novozym® 51003. M.G.G. gratefully acknowledges the financial support provided by the BAGEP Award of the Science Academy.

Conflicts of Interest: The authors declare no conflict of interest.

References

1. Mishra, A.P.; Bajpai, A.; Rai, A.K. 1,4-Dihydropyridine: A Dependable Heterocyclic Ring with the Promising and the Most Anticipable Therapeutic Effects. *Mini Rev. Med. Chem.* **2019**, *19*, 1219–1254. [CrossRef] [PubMed]
2. Aygün Cevher, H.; Schaller, D.; Gandini, M.A.; Kaplan, O.; Gambeta, E.; Zhang, F.X.; Çelebier, M.; Tahir, M.N.; Zamponi, G.W.; Wolber, G.; et al. Discovery of Michael acceptor containing 1,4-dihydropyridines as first covalent inhibitors of L-/T-type calcium channels. *Bioorganic Chem.* **2019**, *91*, 103187. [CrossRef] [PubMed]
3. Ranjbar, S.; Edraki, N.; Firuzi, O.; Khoshneviszadeh, M.; Miri, R. 5-Oxo-hexahydroquinoline: An attractive scaffold with diverse biological activities. *Mol. Divers.* **2019**, *23*, 471–508. [CrossRef]
4. Sahiba, N.; Sethiya, A.; Soni, J.; Agarwal, S. Acridine-1,8-diones: Synthesis and Biological Applications. *ChemistrySelect* **2021**, *6*, 2210–2251. [CrossRef]
5. Bai, C.-B.; Wang, N.-X.; Wang, Y.-J.; Lan, X.-W.; Xing, Y.; Wen, J.-L. A new oxidation system for the oxidation of Hantzsch-1,4-dihydropyridines and polyhydroquinoline derivatives under mild conditions. *RSC Adv.* **2015**, *5*, 100531–100534. [CrossRef]
6. Abdel-Mohsen, H.T.; Conrad, J.; Beifuss, U. Laccase-catalyzed oxidation of Hantzsch 1,4-dihydropyridines to pyridines and a new one pot synthesis of pyridines. *Green Chem.* **2012**, *14*, 2686–2690. [CrossRef]
7. Triggle, D.J. Calcium channel antagonists: Clinical uses—Past, present and future. *Biochem. Pharmacol.* **2007**, *74*, 1–9. [CrossRef]
8. Wang, B.; Hu, Y.; Hu, H. The Aromatization of Hantzsch 1,4-Dihydropyridines by Tetrakis-Pyridine Cobalt (II) Dichromate (TPCD). *Synth. Commun.* **1999**, *29*, 4193–4199. [CrossRef]
9. Kumar, P.; Kumar, A. An Expeditious Oxidative Aromatization of Hantzsch 1,4-Dihydropyridines to Pyridines Using Cetyltrimethylammonium Peroxodisulfate: A Phase Transferring Oxidant. *Bull. Korean Chem. Soc.* **2010**, *31*, 2299–2303. [CrossRef]
10. Mirza-Aghayan, M.; Boukherroub, R.; Nemati, M.; Rahimifard, M. Graphite oxide mediated oxidative aromatization of 1,4-dihydropyridines into pyridine derivatives. *Tetrahedron Lett.* **2012**, *53*, 2473–2475. [CrossRef]
11. Giardina, P.; Faraco, V.; Pezzella, C.; Piscitelli, A.; Vanhulle, S.; Sannia, G. Laccases: A never-ending story. *Cell Mol. Life Sci.* **2010**, *67*, 369–385. [CrossRef]
12. Frasconi, M.; Favero, G.; Boer, H.; Koivula, A.; Mazzei, F. Kinetic and biochemical properties of high and low redox potential laccases from fungal and plant origin. *Biochim. Biophys. Acta Proteins Proteom.* **2010**, *1804*, 899–908. [CrossRef] [PubMed]
13. Cambria, M.T.; Minniti, Z.; Librando, V.; Cambria, A. Degradation of Polycyclic Aromatic Hydrocarbons by Rigidoporus lignosus and its Laccase in the Presence of Redox Mediators. *Appl. Biochem. Biotechnol.* **2008**, *149*, 1–8. [CrossRef] [PubMed]
14. Shariati, M.; Imanzadeh, G.; Rostami, A.; Ghoreishy, N.; Kheirjou, S. Application of laccase/DDQ as a new bioinspired catalyst system for the aerobic oxidation of tetrahydroquinazolines and Hantzsch 1,4-dihydropyridines. *Comptes Rendus Chim.* **2019**, *22*, 337–346. [CrossRef]
15. Khaledian, D.; Rostami, A.; Zarei, S.A.; Mohammadi, B. Aerobic oxidative aromatization of Hantzsch 1,4-dihydropyridines via an anomeric-based oxidation in the presence of Laccase enzyme/4-Phenyl urazole as a cooperative catalytic oxidation system. *J. Iran. Chem. Soc.* **2019**, *16*, 1871–1878. [CrossRef]
16. Simić, S.; Jeremic, S.; Djokic, L.; Božić, N.; Vujčić, Z.; Lončar, N.; Senthamaraikannan, R.; Babu, R.; Opsenica, I.M.; Nikodinovic-Runic, J. Development of an efficient biocatalytic system based on bacterial laccase for the oxidation of selected 1,4-dihydropyridines. *Enzym. Microb. Technol.* **2020**, *132*, 109411. [CrossRef]
17. Kurniawati, S.; Nicell, J.A. Efficacy of mediators for enhancing the laccase-catalyzed oxidation of aqueous phenol. *Enzym. Microb. Technol.* **2007**, *41*, 353–361. [CrossRef]
18. Piontek, K.; Antorini, M.; Choinowski, T. Crystal structure of a laccase from the fungus Trametes versicolor at 1.90-A resolution containing a full complement of coppers. *J. Biol. Chem.* **2002**, *277*, 37663–37669. [CrossRef]
19. Tadesse, M.A.; D'Annibale, A.; Galli, C.; Gentili, P.; Sergi, F. An assessment of the relative contributions of redox and steric issues to laccase specificity towards putative substrates. *Org. Biomol. Chem.* **2008**, *6*, 868–878. [CrossRef]
20. Durão, P.; Chen, Z.; Fernandes, A.T.; Hildebrandt, P.; Murgida, D.H.; Todorovic, S.; Pereira, M.M.; Melo, E.P.; Martins, L.O. Copper incorporation into recombinant CotA laccase from Bacillus subtilis: Characterization of fully copper loaded enzymes. *J. Biol. Inorg. Chem. JBIC Publ. Soc. Biol. Inorg. Chem.* **2008**, *13*, 183–193. [CrossRef]
21. Zouraris, D.; Kiafi, S.; Zerva, A.; Topakas, E.; Karantonis, A. FTacV study of electroactive immobilized enzyme/free substrate reactions: Enzymatic catalysis of epinephrine by a multicopper oxidase from Thermothelomyces thermophila. *Bioelectrochemistry* **2020**, *134*, 107538. [CrossRef] [PubMed]
22. Gramss, G. Reappraising a Controversy: Formation and Role of the Azodication (ABTS2+) in the Laccase-ABTS Catalyzed Breakdown of Lignin. *Fermentation* **2017**, *3*, 27. [CrossRef]
23. Morozova, O.V.; Shumakovich, G.P.; Shleev, S.V.; Yaropolov, Y.I. Laccase-mediator systems and their applications: A review. *Appl. Biochem. Microbiol.* **2007**, *43*, 523–535. [CrossRef]
24. Hilgers, R.; Vincken, J.-P.; Gruppen, H.; Kabel, M.A. Laccase/Mediator Systems: Their Reactivity toward Phenolic Lignin Structures. *ACS Sustain. Chem. Eng.* **2018**, *6*, 2037–2046. [CrossRef] [PubMed]
25. Zerva, A.; Koutroufini, E.; Kostopoulou, I.; Detsi, A.; Topakas, E. A novel thermophilic laccase-like multicopper oxidase from Thermothelomyces thermophila and its application in the oxidative cyclization of 2,3,4-trihydroxychalcone. *New Biotechnol.* **2019**, *49*, 10–18. [CrossRef]
26. Johannes, C.; Majcherczyk, A. Natural Mediators in the Oxidation of Polycyclic Aromatic Hydrocarbons by Laccase Mediator Systems. *J. Appl. Environ. Microbiol.* **2000**, *66*, 524–528. [CrossRef]

27. Galli, C.; Gentili, P.; Jolivalt, C.; Madzak, C.; Vadalà, R. How is the reactivity of laccase affected by single-point mutations? *Engineering laccase for improved activity towards sterically demanding substrates. Appl. Microbiol. Biotechnol.* **2011**, *91*, 123–131. [CrossRef]
28. Sitarz, A.K.; Mikkelsen, J.D.; Meyer, A.S. Structure, functionality and tuning up of laccases for lignocellulose and other industrial applications. *Crit. Rev. Biotechnol.* **2016**, *36*, 70–86. [CrossRef]
29. Subrahmanyam, V.V.; Kolachana, P.; Smith, M.T. Metabolism of hydroquinone by human myeloperoxidase: Mechanisms of stimulation by other phenolic compounds. *Arch. Biochem. Biophys.* **1991**, *286*, 76–84. [CrossRef]
30. Xu, F.; Shin, W.; Brown, S.H.; Wahleithner, J.A.; Sundaram, U.M.; Solomon, E.I. A study of a series of recombinant fungal laccases and bilirubin oxidase that exhibit significant differences in redox potential, substrate specificity, and stability. *Biochim. Biophys. Acta Protein Struct. Mol. Enzymol.* **1996**, *1292*, 303–311. [CrossRef]
31. Mehra, R.; Muschiol, J.; Meyer, A.S.; Kepp, K.P. A structural-chemical explanation of fungal laccase activity. *Sci. Rep.* **2018**, *8*, 17285. [CrossRef] [PubMed]
32. Glazunova, O.A.; Trushkin, N.A.; Moiseenko, K.V.; Filimonov, I.S.; Fedorova, T.V. Catalytic Efficiency of Basidiomycete Laccases: Redox Potential versus Substrate-Binding Pocket Structure. *Catalysts* **2018**, *8*, 152. [CrossRef]
33. Wang, T.-N.; Zhao, M. A simple strategy for extracellular production of CotA laccase in Escherichia coli and decolorization of simulated textile effluent by recombinant laccase. *Appl. Microbiol. Biotechnol.* **2017**, *101*, 685–696. [CrossRef]
34. Chen, Y.C. Beware of docking! *Trends Pharmacol. Sci.* **2015**, *36*, 78–95. [CrossRef] [PubMed]

catalysts

Review

Catalyst-Based Biomolecular Logic Gates

Dennis S. Winston and David D. Boehr *

Department of Chemistry, The Pennsylvania State University, University Park, PA 16802, USA; dxw343@psu.edu
* Correspondence: ddb12@psu.edu; Tel.: +1-814-863-8605

Abstract: Regulatory processes in biology can be re-conceptualized in terms of logic gates, analogous to those in computer science. Frequently, biological systems need to respond to multiple, sometimes conflicting, inputs to provide the correct output. The language of logic gates can then be used to model complex signal transduction and metabolic processes. Advances in synthetic biology in turn can be used to construct new logic gates, which find a variety of biotechnology applications including in the production of high value chemicals, biosensing, and drug delivery. In this review, we focus on advances in the construction of logic gates that take advantage of biological catalysts, including both protein-based and nucleic acid-based enzymes. These catalyst-based biomolecular logic gates can read a variety of molecular inputs and provide chemical, optical, and electrical outputs, allowing them to interface with other types of biomolecular logic gates or even extend to inorganic systems. Continued advances in molecular modeling and engineering will facilitate the construction of new logic gates, further expanding the utility of biomolecular computing.

Keywords: logic gate; allostery; enzyme computing; DNA computing; protein engineering; metabolic engineering; enzyme cascade

check for updates

Citation: Winston, D.S.; Boehr, D.D. Catalyst-Based Biomolecular Logic Gates. *Catalysts* **2022**, *12*, 712. http://10.3390/catal12070712

Academic Editor: Emmanuel M. Papamichael

Received: 2 June 2022
Accepted: 23 June 2022
Published: 29 June 2022

1. Introduction

Organisms have evolved a variety of mechanisms to respond, communicate, and adapt to their environments. For example, fundamental cellular processes including transcription and translation are highly regulated. Protein function, especially enzyme function, is often modulated by the concentrations of metabolites and signaling molecules. The activity of many enzymes depends not just on the concentration of the substrate, but on other metabolites that bind to allosteric sites, allowing for more complex regulation. The interconnected web of the reactions and regulation involved in metabolism can be conceptualized as a circuit of logic gates, analogous to a computer, where each individual reaction is a logic gate that produces output molecules in response to input molecules. In the case of an enzyme allosterically regulated by metabolites, the "decision" to produce more or less of a given metabolite depends on the concentration of the metabolites that control the activity of the enzyme. As such, biological regulation can be re-conceptualized in terms of logic gates, which then have a wide variety of biotechnology applications, including in synthetic biology for the optimized production of high value chemicals (e.g., [1,2]), for engineering of biosensors that sense multiple analytes (e.g., [3–5]), and for the programmed delivery of therapeutics based on biomarkers (e.g., [6,7]). In these applications, synthetic biological logic gates are often designed to interface with the existing biochemical logic gates in the cell. Here we provide some historical background for understanding biomolecular logic gates, outline recent work conceptualizing allosteric enzymes as logic gates, describe how enzymes (and their associated chemical reactions) can be engineered to behave as logic gates in response to biological inputs, and finally expand the discussion to nucleic acid-based logic gates, where allosterically regulated DNA-based enzymes (DNAzymes) can play important roles.

1.1. Fundamentals of Logic Gates

Logic gates are the basic building blocks of computers. In electronic Boolean logic gates, a low voltage/current is 0 and a high voltage/current is 1. This concept can be expanded to chemical reactions, where a low signal is 0 and a high signal is 1. Although the inputs and outputs of a molecular logic gate are not truly binary, molecular circuits have been designed to behave closer to binary logic gates with a sigmoidal response to inputs [8–12], though such behavior is difficult to apply generally. In living cells, biomolecules sense and transmit signals, behaving as complex circuits made of logic gates [13,14]. It has been shown theoretically that a series of chemical reactions occurring in solution without compartmentalization can be used to create logic circuits of arbitrary complexity [15] enabling mathematical operations to be carried out [16], although for multi-component systems noise reduction techniques may be necessary [17,18]. While our focus in this review is on idealized binary biomolecular logic gates, logic circuits in synthetic biology can be more generally understood using the theoretical framework of analog circuits to take into account the noise and leakiness that occur in real molecular systems [19,20].

A logic gate returns an output based on one or more inputs (Table 1). The simplest logic gate is the YES (or identity (ID)) gate, where the output is the same as the input. The YES gate is trivial to implement on its own but can be used as part of more complex logic operations. The NOT gate is the inverse of the YES gate, where the output is the opposite of the input. The YES and NOT gates can be implemented in biomolecular systems when a signal based on the product of a chemical reaction depends on the presence or absence of a reactant. The AND gate results in an output of 1 only if both inputs are 1. The OR gate results in an output of 1 if one or both inputs are 1. The AND and OR gates are the most commonly used biomolecular logic gates because of their relative simplicity. The NAND and NOR gates are the same as the AND and OR gates but with the output inverted. The XOR gate has an output of 1 only if one of the inputs, but not both, is 1. The NXOR gate is an XOR gate with inverted output. The NAND and NOR logic gates, while difficult to implement using biomolecules, are functionally complete operations, meaning they can be combined to produce any other logic operation. The inhibit gate (INH) describes a system in which the output is 1 when input A is 1, but is 0 when input B is 1. For example, if input A is the substrate for an enzyme and the output is the reaction product, input B is an inhibitor of the enzyme. The imply gate (IMP) is an INH gate with the output reversed. A summary of the fundamental logic gates and biomolecular examples is provided in [21].

Table 1. Fundamental Boolean logic gates.

Input Values		Output Values							
A	B	AND	OR	INH	XOR	NAND	NOR	IMP	XNOR
0	0	0	0	0	0	1	1	1	1
0	1	0	1	0	1	1	0	1	0
1	0	0	1	1	1	1	0	0	0
1	1	1	1	0	0	0	0	1	1

1.2. Early History of Biomolecular Logic Gates

A formal conception of chemical reactions as logic gates was published in 1961 [22]. Molecular logic gates have been developed for small molecules, but we will not focus on that work here and refer readers to other reviews in this area (e.g., [23–27]). In a biological context, the concept of molecular logic gates was initially applied to transcriptional regulation of enzyme activity [28,29]. In the 1980s, Okamoto and co-workers [30,31] applied the previously developed mathematical framework [32] directly to enzyme-catalyzed reactions. A decade later, the switching between metabolic pathways involving glucose was conceptualized by logic gates, and the dependence of the sharpness of transition between logical states on enzyme kinetic parameters was analyzed [33]. In one of the first experimental demonstrations of an enzyme-based logic gate, an AND gate was created by conformational switching of enzyme-conjugated azobenzene derivatives by light and oxidation/reduction of an inhibitor to control α-chymotrypsin activity [27,34]. In another experimental example, an XOR gate was realized via the Ca^{2+} and Mg^{2+} dependence of malate dehydrogenase activity [35]. Willner and colleagues [36,37] developed a half-adder and half-subtractor from parallel XOR and AND gates using horseradish peroxidase, glucose dehydrogenase, glucose oxidase, and catalase in the presence of reduced (NADH) and oxidized (NAD^+) forms of nicotinamide adenine dinucleotide. The first experimental demonstration of consecutive enzyme-based logic gates was in 2006, where a system of acetylcholine esterase, choline oxidase, microperoxidase-11, and glucose dehydrogenase was used to create logic circuits composed of concatenated AND, OR, and XOR gates [38]. In most examples of enzyme-based logic gates, output signals have been observed by spectrophotometry, although other forms of output are possible, including detecting changes in pH as an electrical signal [39,40] and interfacing to DNA computers [41]. Other methods for sensing outputs and additional examples are provided in [42]. We note that there are many strategies for designing biomolecular logic gates and circuits, including modulation of protein–protein interactions or dimerization [43,44] and regulation of transcription [45,46] including distributed computing across populations of cells [47,48]; however, in this review, we will focus primarily on catalyst-based logic gates that feature protein enzymes, nucleic acid enzymes, and catalytic DNA strands.

2. Protein-Based Logic Gates

2.1. Allosteric Enzymes as Logic Gates

In this review, we focus on recent work using biological catalysts to create biomolecular logic gates. Our own work has focused on understanding allostery and conformational dynamics of enzymes (e.g., [49–52]), which has implications in the development of biomolecular logic gates. Allosteric regulation can allow enzymes to behave as biomolecular logic gates [33]. For example, pyruvate kinase from *Mycobacterium tuberculosis* that uses adenosine monophosphate (AMP) and glucose-6-phosphate (G6P) as synergistic allosteric activators has been conceptualized as acting as an OR gate to regulate energy and glucose metabolism [53]. Binding of AMP and G6P induces similar allosteric pathways that likely help to regulate the active site. The sigmoidal response of enzymes that are allosterically activated by substrate binding can be useful for reducing noise [54]. Allosteric enzymes that follow the Monod–Wyman–Changeux (MWC) model of allostery, in which concerted conformational transitions occur across enzyme subunits, have been analyzed as logic gates and single-output molecules with two inputs have been found to be capable of producing YES, NOT, AND, OR, NOR, or NAND gates depending on the system [55,56] (Figure 1).

2.2. Engineered Protein Logic Gates

Chimeric proteins have been created to act as logic gates. Ostermeier's group has designed fusions of β-lactamase (BLA) and maltose-binding protein (MBP) that rely on a large conformational change induced by the binding of maltodextrins to activate β-lactamase [57–59]. The sequence for BLA was randomly divided and these two parts were inserted around a MBP sequence to mimic genetic recombination and proteins with reversible maltose-dependent BLA activity were identified [57]. This MBP-BLA fusion protein can be conceptualized as a variety of logic gates depending on the input. It functions as a YES gate for maltose, as BLA activity occurs in the presence of maltose but not in its absence. It is an OR gate for maltose and maltotriose, as either maltodextrin activates BLA activity. It is an IMP gate for maltose and β-cyclodextrin because β-cyclodextrin binding results in only a small conformational change that does not confer BLA activity [60] (Figure 2). They subsequently used directed evolution to select for sucrose-dependent BLA activity, creating an OR gate with sucrose and maltose as inputs [58]. They then engineered disulfide bonds into the MBP-BLA fusion to lock it in either the active/closed or inactive/open state in the absence of a reducing agent. With disulfide bonds to lock MBP-BLA in the closed/active state, the protein behaves as an IMP gate with maltose and the reducing agent as inputs; that is, in the absence of a reducing agent this variant is active, but in the presence of a reducing agent maltose is required for activity. With disulfide bonds to lock the protein in the open/inactive state, the protein acts as an AND gate for maltose and the reducing agent i.e., protein only has a chance to be in the active state in the presence of a reducing agent and needs maltose to be activated. With disulfide bonds in the hinge region and mutations that destabilize the open/inactive state, the BLA activity is regulated by a reducing agent but not maltose, converting it from a YES gate for maltose to a YES gate for the reducing agent. If an oxidizing agent is used as an input instead of a reducing agent, the IMP gate becomes an OR gate and the AND gate becomes an INH gate [60]. In general, using a reducing agent as an input for enzymatic logic gates may be useful for therapeutics by activating or inactivating enzymes upon entry to the cytosol.

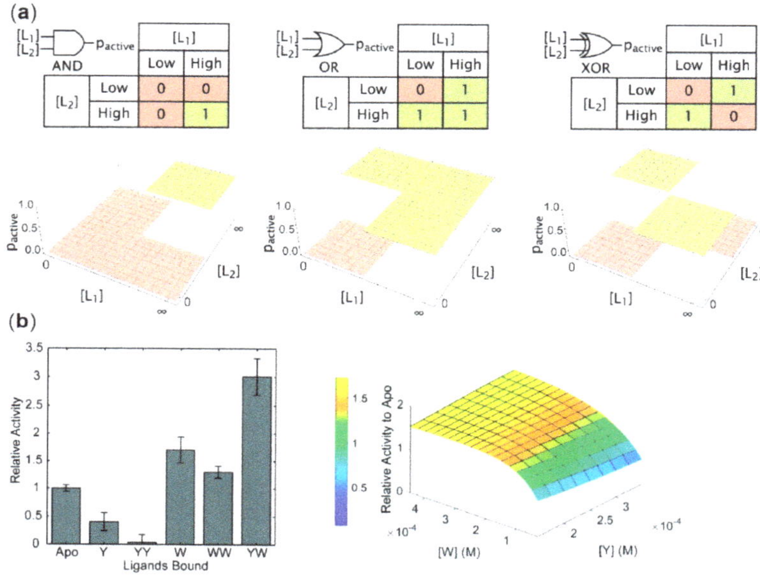

Figure 1. Allosteric enzymes as logic gates. (**a**) Binding of different allosteric modulators (L_1 and L_2) can result in different functional output (i.e., 0 for no or low enzyme activity; 1 for high enzyme activity). This panel is reprinted with permission from Galstyan, V., Funk, L., Einav, T., and Phillips, R. 2019. Combinatorial control through allostery. J. Phys. Chem. B., 123, 2792–2800. Copyright 2019 American Chemical Society [56]. (**b**) The functional landscape of *Saccharomyces cerevisiae* chorismate mutase (ScCM) is dependent on the concentration of the allosteric modulators tryptophan (W) and tyrosine (Y). ScCM is a homodimer in which the allosteric site binds W or Y (i.e., binding is mutually exclusive at each site). When ScCM binds Y at one site and W at the other site, ScCM displays the highest activity, which could be conceptualized as an AND gate. However, binding of Y alone leads to allosteric inhibition, providing complexity beyond the binary logic gates. This panel is reprinted with permission from Gorman, S.D. and Boehr, D.D. 2019. Energy and enzyme activity landscapes of yeast chorismate mutase at cellular concentrations of allosteric effectors. Biochemistry, 58, 4058–4069. Copyright 2019 American Chemical Society [49].

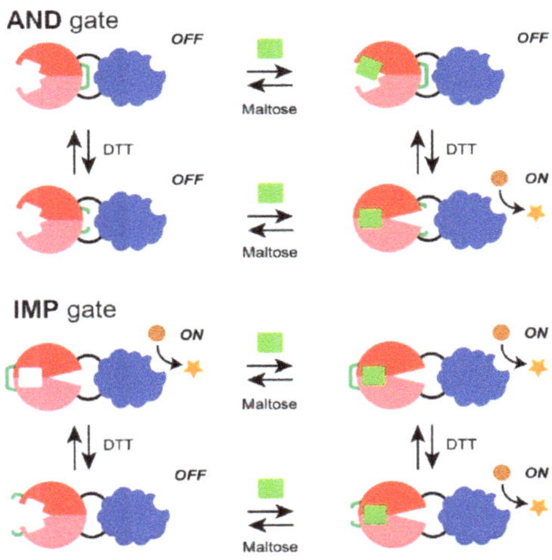

Figure 2. Redesign of logic gates using engineered disulfide bridges. The activity of BLA (blue) is dependent on the ability of MBP (red) to bind to maltose. In the AND gate design, a disulfide bond (green) holds the MBP domain in an open conformation, preventing the conformational change in MBP necessary to activate BLA. Addition of the DTT reducing agent breaks the disulfide bond to allow the MBP conformational change and BLA activation to occur. In the IMP gate design, the engineered disulfide bond now holds MBP in the closed conformation, such that BLA is always activated unless DTT is present and maltose is absent. This figure is adapted and reprinted with permission from Choi, J.H. and Ostermeier, M. 2015. Rational design of a fusion protein to exhibit disulfide-mediated logic gate behavior. ACS Synthetic Biology, 4, 400–406. Copyright 2015 American Chemical Society.

Another clever system that has been engineered to behave as a logic gate is pyrroloquinoline quinone-dependent glucose dehydrogenase (PQQ-GDH) fusion proteins, in which PQQ-GDH is fused to a reporter domain that binds a given ligand. GDH activity is allosterically regulated by ligand binding to the reporter domain. An advantage of using PQQ-GDH is that PQQ-GDH catalyzes a redox reaction, allowing the signal to be easily incorporated into an electronic circuit. Guo and colleagues [61] converted PQQ-GDH into an allosteric enzyme regulated by peptide binding by inserting calmodulin (CaM) at a loop by the glucose binding site, reducing GDH activity in the absence of peptide (Figure 3). To prevent activation at low peptide concentrations, they engineered calmodulin binding peptide (CaM-BP) to have a lower affinity for CaM and fused it with the FRB protein. They then fused the GDH-CaM to FK506 binding protein (FKBP), which binds FRB in the presence of rapamycin. By increasing rapamycin concentration, FKBP binds FRB, which then causes the local concentration of CaM-BP to increase, activating GDH. In this way, they were able to regulate the activity of GDH by adjusting the rapamycin concentration. They also showed that a variety of both ligands and reporters can be used. For example, they fused CaM-BP with different antibody domains to detect the proteins α-amylase, thrombin-activatable fibrinolysis inhibitor, clostridium TcdaA toxin, and interleukin-23. To demonstrate that different reporters can be used, instead of GDH activity they also used dihydrofolate reductase, green fluorescent protein (EGFP), and Nanoluciferase as reporters. Using the synthetic peptide receptor Clamp instead of CaM resulted in ligand binding turning off enzyme activity instead of turning it on. A further layer of regulation can be added to the CaM fusion system by adding a calmodulin domain to the CaM-BP to sequester CaM-BP that can be removed by protease cleavage to activate calmodulin [62]. The CaM fusion systems described here can be considered as four-input logic gates with analyte, substrate, cofactor, and Ca^{2+} as inputs [63], although for biosensing applications the substrate, cofactor, and Ca^{2+} are always present so it effectively behaves as a YES gate for the analyte.

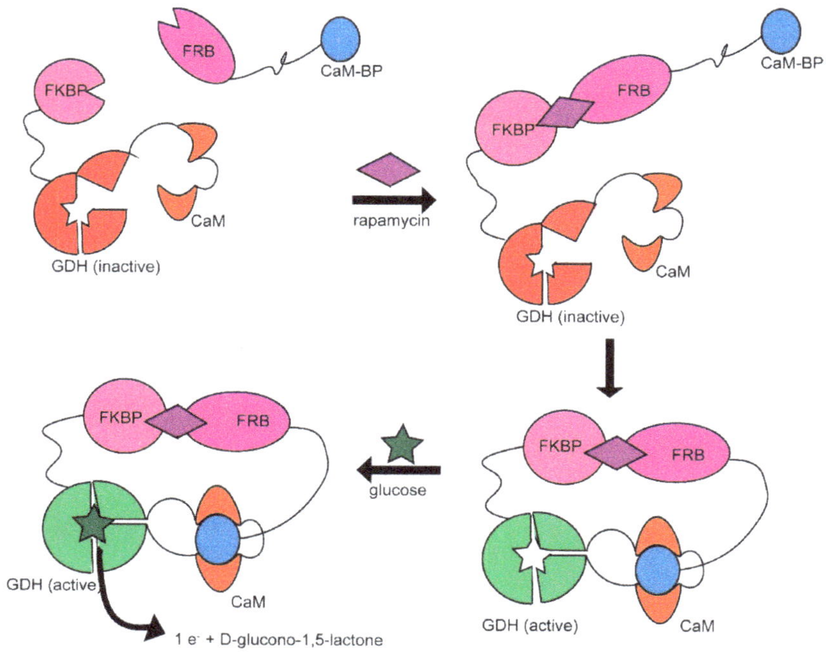

Figure 3. Calmodulin-glucose dehydrogenase (CaM-GDH) fusion protein that acts as an AND gate with rapamycin and glucose as inputs. Upon rapamycin binding, calmodulin binds calmodulin-binding peptide (CaM-BP). Binding of CaM-BP results in a conformational change in GDH, activating the enzyme and allowing it to catalyze the oxidation of glucose and produce an electrical signal as output. The electrical signal is produced only in the presence of both rapamycin and glucose.

Subsequent work by the Katz group [64] used the electrochemical signal generated by the PQQ-GDH-Clamp enzyme's reduction of glucose to release molecules into solution. They set up an IMP logic gate using Clamp's peptide and a PQQ-GDH without Clamp as input. In this system, PQQ-GDH-Clamp, glucose substrate, and cofactor were always present in the solution. Because GDH-Clamp is inhibited by peptide, but the GDH without Clamp can still reduce glucose in the presence of substrate, the reaction occurred for all cases except when both peptide and PQQ-GDH were absent. A current was produced by either conjugating the two PQQ-GDH enzymes to carbon nanotube buckypaper electrodes via an ester linker that both reacts with protein lysine residues and binds to the nanotubes or mediating electron transfer from the solution via 2,6-dichlorophenolindophenol (DCPIP) and phenazine methosulfate (PMS) [65]. This current was then used to drive the release of a molecule into the solution. The current generated by the reduction of glucose was coupled with the reduction of water catalyzed by bilirubin oxidase, which resulted in a local pH increase leading to the hydrolysis of an ester. The cleavage of the ester released the molecule into the solution. While they released a fluorescent dye, in principle any molecule that can be conjugated to the carbon nanotube could be released. Combined with the customizable GDH-CaM fusion proteins, the ability to release molecules via an enzyme-catalyzed reduction reaction can be used to release molecules or detect electric current in response to a wide variety of analytes.

Dokholyan's group constructed an OR gate in live cells by engineering focal adhesion kinase (FAK) to respond to chemical and optical inputs [66]. FAK is a kinase involved in cytoskeletal regulation and is allosterically regulated by the binding of regulatory proteins in its FERM domain. They inserted a previously constructed regulatory domain that allows for control of FAK activity via the binding of rapamycin [67], with the enzyme in an open active conformation in the presence of rapamycin. To regulate the enzyme by optical irradiation, they inserted a Light Oxygen Voltage 2 (LOV2) domain, which undergoes a conformational change in response to blue light, into a loop in the regulatory FERM domain determined to allosterically couple to the active site by computational analysis [68]. Upon irradiation, the kinase switches to an inactive closed conformation. First, constructs with the rapamycin-binding domain and LOV2 domain inserted separately were tested in HeLa cells and found to undergo regulation as expected. Because FAK activity results in the formation of large focal adhesions, the size of focal adhesions was used as the output for the logic gate. Then, light- and dark-stabilized variants of FAK with both allosteric regulatory domains inserted were used to test the logic gate behavior (due to phototoxicity, it was not possible to apply different combinations of rapamycin and light inputs at the same time.). In the dark, FAK was always active independent of rapamycin concentration. In the light, FAK was activated by addition of rapamycin. Therefore, when the rapamycin and absence of blue light are considered inputs and FAK activity is the output, the system behaves as an OR gate.

Ostermeier's group [69] used a pH-sensitive variant of the membrane protein listeriolysin O, which forms pores at low pH, to selectively release molecules from liposomes. By conjugating a designed ankyrin repeat protein (DARPin)-based inhibitor to the membrane, they controlled pore formation by releasing the inhibitor in response to either a reducing agent or protease cleavage. With pH, protease, and reducing agent as inputs, the system can be conceptualized as an AND-OR gate, where low pH in combination with either a protease or reducing agent results in pore formation. Under these conditions, the fluorescent probe calcein was released from the vesicles. It is noted that the enzyme serves as an input here, rather than as a gate itself.

2.3. Biosensors Based on Enzymatic Logic Gates

The chemical reactions catalyzed by a series of different enzymes can also be constructed to act as logic gates (e.g., see Figure 4). For example, Katz and Privman [70] used the controlled release of hydrogen peroxide (H_2O_2) in response to a logic gate made of a four-level enzyme cascade to release molecules from an Fe^{3+} cross-linked alginate hydrogel. Prior to this work, only two enzymatic reactions were able to be carried out in sequence with the enzymes immobilized on the hydrogel [71]. Their system consisted of four enzymes deposited on the interface of the alginate hydrogel film and the solution, which could be conceptualized as a series of interconnected AND gates where the inputs are the reactants in different steps of the cascade that are not products of a previous step. The first enzyme in the cascade was amyloglucosidase (AMG), which catalyzes the hydrolysis of maltose to two glucose molecules. The next enzyme was glucose dehydrogenase (GDH), which oxidizes glucose and reduces NAD^+ to NADH. The next enzyme was lactate dehydrogenase (LDH), which catalyzes the reduction of pyruvate into lactate with NADH as a cofactor. The final enzyme was lactate oxidase (LOx), which catalyzes the oxidation of lactate by O_2 to produce pyruvate and H_2O_2. The hydrogen peroxide then diffuses into the hydrogel and destroys it via a free radical mechanism, releasing the reporter molecule. The reporter molecule used was a fluorescent DNA, which in principle could be used as an input to DNA logic gates (e.g., see below) or could be a small molecule drug or input for downstream enzymatic reactions.

Katz's group also created a reversible CNOT logic gate using two enzymatic reactions, resulting in pH changes in opposite directions [72]. First, a XOR gate was created using two enzymes, urease and esterase, conjugated to the surface of a sensitive pH detector. Urease catalyzes the hydrolysis of urea, resulting in the production of ammonia and an increase in pH, while esterase catalyzes the hydrolysis of ethyl butyrate, resulting in the production of butyric acid and a decrease in pH. In the presence of either urea or ethyl butyrate, the pH at the surface of the detector changes drastically and a large absolute value of pH change is considered an output of 1. However, when urea and ethyl butyrate are present together, a weak acid and weak base are produced simultaneously, resulting in only a small pH change and an output of 0. In this way, an XOR gate was created. A similar XOR gate was also constructed using urease and penicillinase immobilized on a pH-sensing device, with urea and penicillin as inputs [73]. This XOR gate was run in parallel with a separate detector with only esterase conjugated to it. When the esterase-only detector senses a large pH change, ethyl butyrate is present, resulting in a YES gate. When considered in parallel with the XOR gate, the combined logic circuit has two inputs and two outputs, with each combination of outputs corresponding to a unique combination of inputs; hence, the logic gate is reversible. They also constructed a XNOR gate, where the output is 1 only when both inputs have the same value, by using pH as an input [74]. They used glucose oxidation by PQQ-GDH as the gate, with the enzymatic activity depending on the pH. After mapping the enzyme activity as a function of pH and pH as a function of acetic acid and ammonia concentration, they were able to define thresholds to convert the enzyme activity into a binary output. By controlling the pH via concentration of the input molecules, the XNOR gate was made. Together, logic gates that use pH as an output and input can be combined to construct multi-layered enzyme logic circuits.

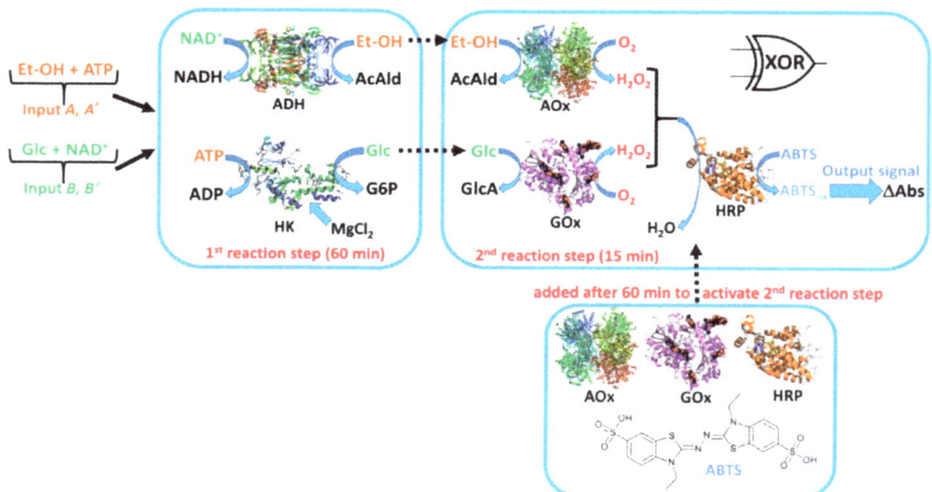

Figure 4. An enzyme cascade to model an XOR gate. When both inputs are provided (EtOH/ATP and Glc/NAD$^+$), the required substrate and cofactor are available for the ADH and HK enzymes, resulting in the removal of the required substrate for the second series of enzymes, AOx and GOx, and no production of H_2O_2. However, if only one of the inputs is provided, the EtOH or Glc substrates for AOx or GOx, respectively, will be available for H_2O_2 production. The figure was adapted with permission from [75].

Katz's group created two other enzyme-based XOR gates. The first system consisted of two sets of enzymes, with the second set of enzymes added after the first set of reactions was given enough time to reach completion [75] (Figure 4). The first set of enzymes consisted of alcohol dehydrogenase (ADH) and hexokinase (HK). ADH consumes ethanol (EtOH) only in the presence of NAD$^+$ and HK consumes glucose only in the presence of ATP. The second set of enzymes consisted of alcohol oxidase (AOx) and glucose oxidase (GOx). AOx uses EtOH to produce H_2O_2 and GOx uses glucose to produce H_2O_2. The production of H_2O_2 is considered the output, as it is used by horseradish peroxidase to produce a spectrophotometric signal. The system used two inputs: (1) EtOH and ATP and (2) glucose and NAD+. The key concept is that in the presence of both inputs, EtOH and glucose are both fully consumed by the first enzyme set so that no H_2O_2 is produced after addition of the second enzyme set. When only one input is added, the enzyme that would consume either EtOH or glucose in the first set does not have its cofactor and so the EtOH or glucose will be present in solution for the second enzyme set to use to produce H_2O_2. In the absence of both inputs, there is no glucose or EtOH so no H_2O_2 is produced.

In a later work, Katz' group constructed a different XOR gate to release nanoparticles from an alginate hydrogel by the production of H_2O_2 [76]. Two enzymes, GOx and L-glutamate oxidase (GluOx), were immobilized on the hydrogel surface. The solution contained two more enzymes: PQQ-GDH and glutamate dehydrogenase (GluDH). The two inputs to the XOR gate were (1) PMS (cofactor for GDH) and glutamate (substrate for GluDH and GOx) and (2) NAD+ (cofactor for GluDH) and glucose (substrate for GDH and GluDH). If H_2O_2 is produced at the surface of the hydrogel i.e., by GOx or GluOx, the H_2O_2 decomposes the hydrogel to release a nanoparticle. In the absence of any input, the nanoparticle is not released from the hydrogel because no H_2O_2 is formed. When just NAD+ and glucose were present, the glucose could not be reduced by GDH without PMS, so it was oxidized by GOx, producing H_2O_2 and releasing the nanoparticle. When just PMS and glutamate were present, glutamate was oxidized by GluOx, again resulting in the production of H_2O_2 and nanoparticle release. When both inputs were present (all four molecules), glucose was reduced by the GDH in solution before it could be oxidized by GOx on the hydrogel surface, preventing the production of H_2O_2. Similarly, the glutamate was reduced by GluDH before it could be oxidized by GluOx, preventing the production of H_2O_2. This system is a notable advancement over previous work, as the time-dependent addition of enzyme is not required. The nanoparticle released from this system can in principle be used for colorimetric sensing, as an input for further logic circuits, or for targeted drug delivery.

2.4. Protease-Based Logic Gates

A number of biomolecular logic circuits have made use of proteases to control protein–protein interactions and transcription. As these systems have been extensively reviewed in recent articles [77–79], we will focus here on two strategies for constructing protease-based cellular logic gates that rely on protein–protein interactions. Elowitz' group developed an elegant system of logic gates with viral proteases as inputs, called circuits of hacked orthogonal modular proteases (CHOMP) [80]. The output comes from a target protein that is activated or inactivated upon cleavage by a protease. The target protein is regulated by being linked with a protease cleavage site to a degron so that in the absence of protease the target is degraded. They created an AND gate by linking the target protein and degron by two sequential protease cleavage sites for two separate proteases. They created an OR gate by using two degrons with separate cleavage sites at the N- and C-termini of the target protein so that either protease will lead to the degradation of the target. Negation was achieved by using protease cleavage to reveal the degrons instead of removing them. To achieve more complex regulation, the protease itself was split in half, constituted by two leucine zipper domains that can be cleaved by a different, secondary protease. Moving beyond binary logic gates, they created a bandpass circuit where target protein activity increased over a period of hours before returning to the baseline over a day. The bandpass circuit was constructed by using both activator and repressor proteases (cleaving one degron, revealing another) and setting a threshold for the repressor protease by inactivating it with a third protease. They applied the CHOMP system to develop a proof of concept for a programmable therapeutic device by activating caspase-3 in cells with high levels of active Ras. By fusing half of the split protease to Ras and the other half to a domain that binds to activated Ras, they were able to reconstitute active protease in the presence of active Ras. The protease then cleaved an engineered variant of caspase-3 that is localized to the plasma membrane so that it is near the now Ras-associated protease, resulting in cell death.

Similar to the CHOMP system, split-protease-cleavable orthogonal-coiled-coil (SPOC)-based logic circuits were developed, in which split proteases are reconstituted in human cells by the formation of coiled-coils regulated by orthogonal proteases. In the absence of protease, an autoinhibitory coil is covalently linked to a target coil that is linked to the protein of interest (either a reporter or half of a split protease). Adding protease to cleave off the autoinhibitory coil allows for a different coiled-coil to displace the autoinhibitory coil. When this displacing coil contains the other half of a split protease, the protease is reconstituted and can be used to cleave the linker of another coil, allowing for logic cascades. Previously, this framework was used to construct an AND gate in an in vitro translation system by using a protease to cleave between the target and autoinhibitory helices [81] (Figure 5). Fink et al. expanded on this work by using human cell lines, inducing dimerization of the coiled-coils by the addition of small molecules, and allowing for negation by introducing cleavage sites between the split protease and the target coil. They constructed a three-layer logic cascade using three orthogonal proteases, which in principle could be extended with the addition of more orthogonal proteases or different small molecules for inducing coiled-coil dimerization. Compared to other methods such as CHOMP (Gao 2018) and transcription-based systems [82], SPOC is much faster, producing the output within 5 min rather than hours.

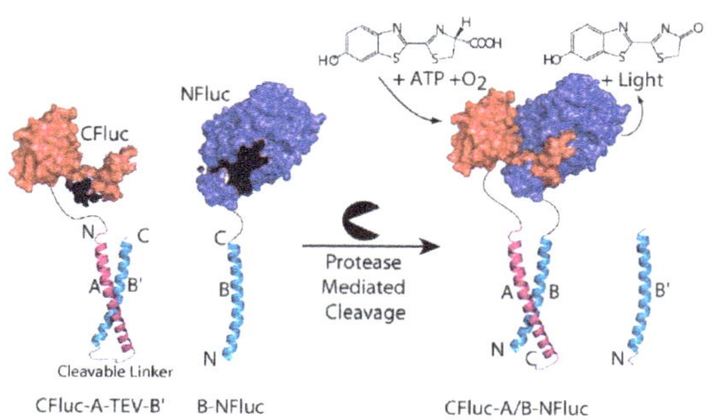

Figure 5. Split protein activation following protease addition. Here the firefly luciferase is separated into a C-terminal (CFluc) and N-terminal half (NFluc), both containing additional helices that can form a coiled-coil to bring both halves together to reconstitute the luciferase. This interaction can only occur when the autoinhibiting coil is released through addition of the tobacco etch virus (TEV) protease. Other reporter proteins can be similarly broken into two halves, including other proteases to generate extended logic cascades. Reprinted with permission from Shekhawat, S.S., Porter, J.R., Sriprasad A., and Ghosh, I. 2009. An autoinhibited coiled-coil design strategy for split-protein protease sensors. J. Am. Chem. Soc., 131, 15284–15290 [81].

3. Nucleic Acid-Based Logic Gates

3.1. *Catalytic (Entropy-Driven) Toehold-Mediated Strand Displacement*

While most catalysts in biological systems are protein enzymes, nucleic acids can also catalyze reactions and undergo allosteric regulation, allowing for the construction of logic gates. A common foundation for constructing logic gates with molecules of DNA in bulk solution is toehold-mediated strand displacement. In this method, signals are carried by single-stranded DNA with two recognition domains that can hybridize with the two strands of the gate that the signal is associated with. Each gate consists of double-stranded DNA, with one strand having a short overhang called a toehold in addition to the displacement domain. Both domains are complimentary to an input strand. The input strand is single-stranded DNA that is complimentary to the toehold and the displacement domain. It will preferentially displace one of the strands of the gate by first hybridizing with the toehold before proceeding by branch migration. After hybridizing with the toehold, the input strand hybridizes with the displacement domain, displacing the other strand that was initially part of the double-stranded gate. The displaced strand, which contains a sequence complementary to a different toehold than the displacing input strand, then proceeds as an input for the next strand displacement reaction at the next gate in the circuit. Two-input AND gates can be created by using the displacement of a strand by the first input to reveal a toehold that the second input can use to displace the output strand (Figure 6). OR gates can be created by having multiple single-input gates with the same output [83]. The outputs from strand displacement cascades can be reported by the displacement of a DNA strand with a fluorophore from its complement containing a quencher [84].

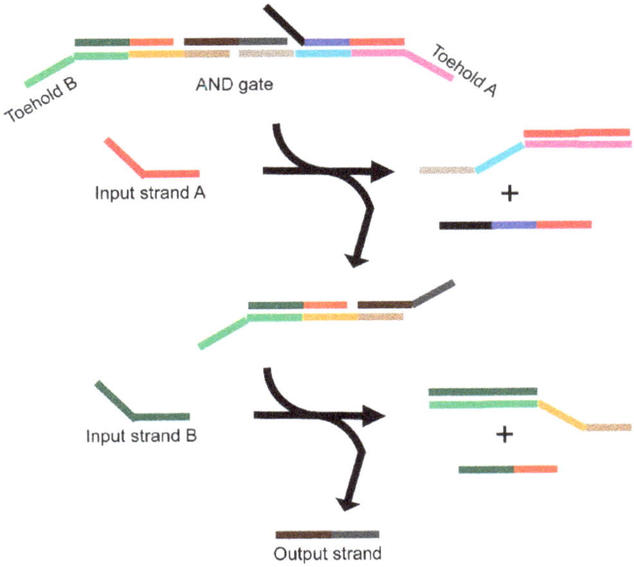

Figure 6. Schematic of a toehold-mediated strand displacement AND gate. Input strands A and B displace the complementary strands from the AND gate on either side of the output strand. The output strand is released only when both input strands hybridize with the gate.

An extension of the toehold-mediated strand displacement method is the use of toehold exchange and "Seesaw" gates (Figure 7). "Seesaw" gates rely on a "fuel" strand that allows the input strand to be regenerated at the end of each cycle so that only a low concentration of input strand is required to release all of the output strands from a gate. After the input strand hybridizes with the gate, the "fuel" strand displaces the input strand, allowing it to react with another gate and amplifying the input. Because the overall reaction for toehold exchange does not include the input strand, the input strand can be thought of as a catalyst for the displacement of the output strand by the fuel strand [85,86]. Double-stranded DNA that is complimentary to the input, called a threshold, can prevent hybridization of the input with the gate until a given concentration of input has been reached. Using thresholds and fuel together ensures that no output strands will be released (i.e., the output is 0) until a minimum number of input strands have been displaced. For any number of input strands above the threshold, the maximum number of output strands will be released (i.e., the output is 1), allowing for binary behavior. Without any thresholds or fuel, an integrating gate can be created where the output is simply the sum of any number of inputs, where each input has a sequence complementary to the integrating gate. AND and OR gates can then be constructed by placing an amplifying gate after an integrating gate and setting the threshold appropriately, with an AND gate having the threshold set above the maximum possible amount of one input. These logic gates can be chained together to form circuits of arbitrary complexity using dual-rail logic to construct complex logic functions from only AND and OR gates, with NOT gates realized by swapping the two wires of an input and output signal [87–89]. DNA logic gates can be constructed in the solution phase or with the gates immobilized on a surface [90–93], and can even output signals on a timer [94,95]. Huang and colleagues [96] expanded the catalytic strand displacement mechanism described above to create complex logic circuits, including three-input AND-AND and AND-AND-NOT gates as well as the first DNA-only two-output comparator and Feynman gates.

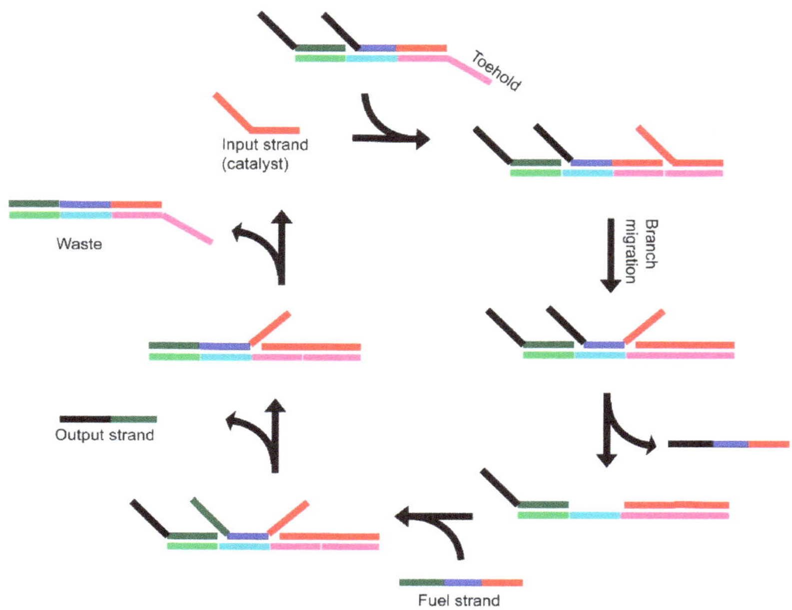

Figure 7. Schematic of toehold-mediated strand displacement using a seesaw gate. Different DNA sequences are shown in different colors, with black representing regions that do not form base pairs. Shades of the same color represent complementary sequences. First, the input strand hybridizes with the toehold (pink). The input strand further hybridizes to the gate via branch migration and displaces the initial complementary strand (black-blue-red). The fuel strand displaces both the output strand and input strand. The input strand can then continue the cycle with another molecule of gate DNA. The output strand can act as an input strand for a different gate.

3.2. Using DNAzymes as Part of Logic Gates

Instead of the toehold-mediated strand displacement mechanism described above, logic gates can also be constructed by using strand displacement to allosterically regulate the catalytic activity of DNAzymes [97–99] (Figure 8). Early work with DNAzymes used the E6 DNAzyme, which cleaves a fluorogenic RNA substrate. The E6 DNAzyme can have its catalytic site either blocked or revealed upon the binding of complimentary DNA to a regulatory stem loop, with logic gates created by using multiple stem loops around each catalytic site [83,100,101]. Multiple inputs and different fluorophores can be used, allowing for more complicated circuits [101–110]. By using allosterically regulated DNA-based ligases, the output of a logic gate can be oligonucleotides that can act as an input for another logic gate, allowing circuits to be constructed from layers of logic gates [111]. Similar systems using RNA-based inputs or catalysts and DNA substrates have also been developed [112–114], including systems that use substrate release to activate a downstream DNAzyme by strand displacement [115,116] and systems that release small molecules [117].

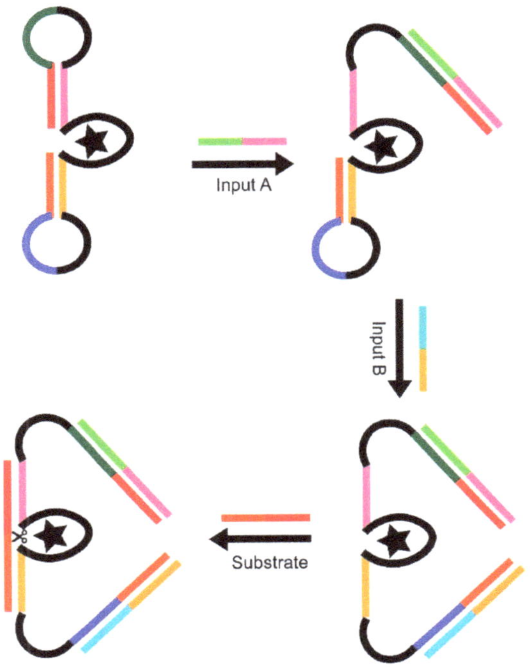

Figure 8. Schematic of a DNAzyme-based AND gate. Different DNA sequences are shown in different colors, with black representing regions that do not form base pairs. Shades of the same color represent complementary sequences. In the absence of input strands, the regulatory loops prevent the substrate strand from binding to the active site (represented by a star) and being cleaved. Binding of both input strands allows the substrate to bind at the active site, resulting in cleavage of the substrate strand.

3.3. Catalytic G-Quadruplexes for Visualizing the Output of DNA Logic Gates

G-quadruplexes, stacks of four planar guanine bases associated by Hoogsteen base pairing stabilized by a metal ion that are found in G-rich sequences of nucleic acids, have also been used as the basis for DNA-based logic gates. There is a large body of work on using G-quadruplexes that respond to input DNA oligonucleotides [110,118–121] or chemical conditions [122,123] by detecting the G-quadruplex via fluorescence, but our focus in this review is on the use of catalytically active G-quadruplexes that form in response to aptamer binding as part of logic gates. In these systems, a G-quadruplex intercalates hemin to catalyze the reduction of H_2O_2, oxidizing a substrate to produce a colored product or chemiluminescence analogous to horseradish peroxidase [124]. These catalytic G-quadruplexes have been used for the detection of metal ions [125–127], small molecules [128], and specific single-stranded DNA sequences. Catalytic G-quadruplexes are particularly well-suited to determine the output of a DNA logic circuit because the output can be detected visually without the need for potentially expensive equipment and G-quadruplexes can be formed in multiple ways. G-quadruplexes can be released from a duplex or triplex by strand displacement [129] or formed from two separate strands in response to strand displacement [110,118] or input oligonucleotides [130]. A notable example of a DNA-based logic gate for detection can be found in [131] (Figure 9). Yu et al. used DNA aptamers that recognize protein biomarkers at the surface of extracellular vesicles. The aptamers had an exposed toehold for hybridization with another strand of DNA upon binding the proteins. After hybridizing, DNA is cleaved by an endonuclease, releasing the DNA and allowing for signal amplification. One of the released DNA strands hybridized with DNA conjugated to a gold electrode and the other hybridized with the first strand and formed a G-quadruplex with peroxidase activity, resulting in an electrical signal only when both protein analytes were present.

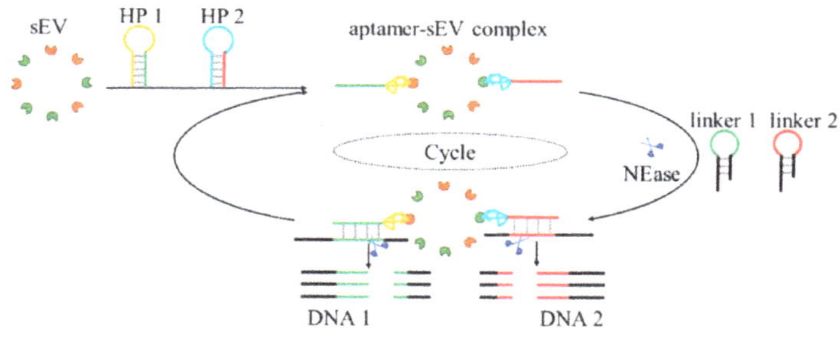

Figure 9. Dual-aptamer-assisted enzyme signal amplification. DNA hairpin structures 1 and 2 (HP1, HP2) can bind protein biomarkers on the surface of a small extracellular vesicles (sEV) leading to the release of the hybridization region (red and green parts of strands) that are complementary to regions in linkers 1 and 2 (color-coded the same). The restriction endonuclease (NEase) then recognizes and cleaves the new duplex. The resulting DNA1 and DNA2 can then have further downstream effects. Reprinted with permission from Yu, Y., Guo, Q., Jiang, W., Zhang, H., and Cai. C. 2021. Dual-aptamer-assisted AND logic gate for cyclic enzymatic signal amplification electrochemical detection of tumor-derived small extracellular vesicles. Anal. Chem., 93, 11298–11304.

4. Discussion

The logic gates presented here that depend on protein-level regulation can be compared against those that regulate at the level of transcription. In general, logic gates that depend on protein-level regulation are faster than transcription-based logic gates, occurring on the timescale of minutes rather than hours to days [46,132]. This is a major advantage because biological processes such as glucose metabolism occur too quickly for efficient regulation at the transcriptional level [132]. However, an advantage of transcription-based approaches is that when the final output of the logic operations is the regulation of transcription, a wider range of outputs are available. For example, in the cellular logic circuits reviewed in this article, the outputs of the circuits were limited to fluorescence, activation of a specific kinase that could be engineered to respond to light, and cleavage of an engineered caspase. In contrast, transcription-based approaches can use proteases to release transcription factors for virtually any gene [44,133].

Logic gates that rely on the detection of specific molecules, such as those presented in Section 2.3, will not be as useful for performing logic operations in cells because they are not based on generalizable principles and would experience interference from cellular metabolites. The GHD-CaM fusion system presented in Section 2.2, however, can potentially be very useful for biosensing applications because it can be used with antibodies for the detection of virtually any protein of interest. Similarly, the protease-based methods described in Section 2.4 are able to be modified to perform logic operators on a variety of protein targets. The detection of protein can in principle be coupled with drug release, although coupling the output of the logic gates with drug release in the manner described in Section 2.3 would be challenging to achieve in a biological system.

Some of the logic circuits in this review ultimately release molecules from hydrogels by the production of H_2O_2. These logic circuits could be combined with H_2O_2-releasing hydrogels that have been developed for therapeutic applications such as wound healing [134,135]. Currently, the H_2O_2-releasing hydrogels release H_2O_2 from the oxidation of glucose catalyzed by glucose oxidase. By using protein-based logic gates to regulate glucose oxidase activity, such as in the work described in Section 2.3, the formation and release of hydrogels can be regulated by logic gates. Similarly, catalytic G-quadruplexes that catalyze the production of H_2O_2 could be used as alternatives to glucose oxidase in these systems.

DNA-based logic gates have several advantages over protein enzyme-based logic gates for in vitro computations. One advantage is that they are easier to scale up due to the relative simplicity of Watson–Crick base pairing, as complimentary DNA sequences with well-understood displacement kinetics can be designed and synthesized much more easily than allosterically regulated proteins or protein–protein interaction interfaces [136]. An impressive example is reported in Cherry et al., where a DNA strand displacement-based logic circuit was used to distinguish between two handwritten numbers on a 10 × 10 pixel grid using 104 different DNA molecules [137]. A major disadvantage of using DNA for molecular logic circuits over proteins is speed—the image recognition circuit took hours to produce a result, on a similar timescale to transcription-based logic. Another disadvantage, not unique to DNA, is that scaling up to more complex logic circuits is challenging due to the large numbers of unique molecules needed and the extra time required to carry out the calculations.

While there are many examples of protein-based enzymatic logic gates that have been implemented in living cells [66,79,80,132], applications for DNA-based circuits integrated into cells have been more limited. DNAzymes have been introduced into cells for gene silencing and biosensing [138,139]. One of the major limitations is that DNA introduced into cells is susceptible to cleavage by nucleases. One strategy for improving the stability of DNAzymes, aptamers, and so on is to evolve the sequence with the desired activity in biological fluids [138], but this is not feasible for complex DNA logic circuits where the sequences must be carefully designed. There are many recent examples of DNA-based biosensors where fluorescence outputs are used to detect multiple types of molecules at the same time, including RNA and proteins at the cell surface [140–146]. Because the output is fluorescence, these biosensors have not been used for downstream logic processing. Chang et al. designed a DNA aptamer-based biosensor that forms a toehold when both analytes are present at the cell surface, allowing for downstream biomolecular logic processing [147]. Systems that use DNA logic circuits for delivery of therapeutics have been proposed or supported by in vitro experiments, but have not yet been implemented in cells. For example, a molecular device based on DNAzymes allosterically regulated by glucose has been proposed for the management of diabetes [148], but such a device has not been experimentally realized. In another example, an RNA-cleaving DNAzyme that can cleave RNA to prevent the transcription of cancer-associated mRNA in response to high concentrations of cancer marker RNA has been demonstrated to work in vitro [149] but not yet in living cells.

Biomolecular logic gates are likely most useful for diagnostics, while therapeutic applications are more challenging. For DNA-based biosensors, a logic circuit with inputs based on aptamer binding and an electronic output that relies on G-quadruplex formation as the output similar to that used by Yu et al. [131] is the most promising for diagnostics based on cell surface markers. While logic gates are not strictly necessary for the detection of multiple analytes, as different fluorescent signals can be used, DNA logic circuits can be useful for simplifying the output when different combinations of analytes are detected. We think the most promising protein-based logic gate biosensors are those that can be conjugated to antibodies for the detection of different proteins, such as the GDH-CaM system from Guo et al. described in Section 2.2 [61]. For protein enzyme therapeutics where a rapid response to changing conditions is required, the protease-based logic circuits that rely on the degradation or activation of protease are likely to be the most useful of the systems discussed in this article, as they have already been demonstrated to work inside cells.

5. Conclusions

The logic gate framework can be used to conceptualize, and in some cases simplify, complex regulatory networks in biology. In metabolic engineering, this framework can then be used to optimize the production of high value chemicals [1,2]. Advancements in biomolecular engineering will provide additional protein- and nucleic acid-based logic gates [43,150]. Protein- and nucleic acid-based logic gates can then be programmed to interact with each other [151], and catalyst-based logic gates can similarly interact with non-catalyst-based logic gates (e.g., transcription factors [152,153]). These logic gates have the potential to integrate multiple signals and thus make intelligent decisions in therapeutic and diagnostic applications. Several applications have already been realized, including in the detection and treatment of cancer (e.g., [140]) and the detection of various metabolic diseases (e.g., [154]).

Author Contributions: Conceptualization, D.S.W. and D.D.B.; writing—original draft preparation, D.S.W.; writing—review and editing, D.S.W. and D.D.B.; visualization, D.S.W. and D.D.B.; funding acquisition, D.D.B. All authors have read and agreed to the published version of the manuscript.

Funding: This research was funded by the National Science Foundation (NSF) grant number MCB-1615032 and National Institutes of Health (NIH) grant number AI104878.

Data Availability Statement: Not applicate.

Conflicts of Interest: The authors declare no conflict of interest.

References

1. Cui, S.; Lv, X.; Xu, X.; Chen, T.; Zhang, H.; Liu, Y.; Li, J.; Du, G.; Ledesma-Amaro, R.; Liu, L. Multilayer Genetic Circuits for Dynamic Regulation of Metabolic Pathways. *ACS Synth. Biol.* **2021**, *10*, 1587–1597. [CrossRef]
2. Hoynes-O'Connor, A.; Moon, T.S. Programmable genetic circuits for pathway engineering. *Curr. Opin. Biotechnol.* **2015**, *36*, 115–121. [CrossRef]
3. Katz, E.; Wang, J.; Privman, M.; Halámek, J. Multianalyte digital enzyme biosensors with built-in boolean logic. *Anal. Chem.* **2012**, *84*, 5463–5469. [CrossRef]
4. Halámek, J.; Windmiller, J.R.; Zhou, J.; Chuang, M.C.; Santhosh, P.; Strack, G.; Arugula, M.A.; Chinnapareddy, S.; Bocharova, V.; Wang, J.; et al. Multiplexing of injury codes for the parallel operation of enzyme logic gates. *Analyst* **2010**, *135*, 2249–2259. [CrossRef]
5. Privman, M.; Tam, T.K.; Bocharova, V.; Halámek, J.; Wang, J.; Katz, E. Responsive interface switchable by logically processed physiological signals: Toward "smart" actuators for signal amplification and drug delivery. *ACS Appl. Mater. Interfaces* **2011**, *3*, 1620–1623. [CrossRef]
6. Zhou, J.; Halámek, J.; Bocharova, V.; Wang, J.; Katz, E. Bio-logic analysis of injury biomarker patterns in human serum samples. *Talanta* **2011**, *83*, 955–959. [CrossRef]
7. Evans, A.C.; Thadani, N.N.; Suh, J. Biocomputing nanoplatforms as therapeutics and diagnostics. *J. Control. Release* **2016**, *240*, 387–393. [CrossRef]
8. Privman, V.; Arugula, M.A.; Halámek, J.; Pita, M.; Katz, E. Network analysis of biochemical logic for noise reduction and stability: A system of three coupled enzymatic and gates. *J. Phys. Chem. B* **2009**, *113*, 5301–5310. [CrossRef]
9. Rafael, S.P.; Vallée-Bélisle, A.; Fabregas, E.; Plaxco, K.; Palleschi, G.; Ricci, F. Employing the metabolic "branch point effect" to generate an all-or-none, digital-like response in enzymatic outputs and enzyme-based sensors. *Anal. Chem.* **2012**, *84*, 1076–1082. [CrossRef]
10. Halámek, J.; Zavalov, O.; Halámková, L.; Korkmaz, S.; Privman, V.; Katz, E. Enzyme-based logic analysis of biomarkers at physiological concentrations: And gate with double-sigmoid "filter" response. *J. Phys. Chem. B* **2012**, *116*, 4457–4464. [CrossRef]
11. Zavalov, O.; Bocharova, V.; Halámek, J.; Halámková, L.; Korkmaz, S.; Arugula, M.A.; Chinnapareddy, S.; Katz, E.; Privman, V. Two-Input Enzymatic Logic Gates Made Sigmoid by Modifications of the Biocatalytic Reaction Cascades. *Int. J. Unconv. Comput.* **2012**, *8*, 347–365.
12. Bakshi, S.; Zavalov, O.; Halámek, J.; Privman, V.; Katz, E. Modularity of biochemical filtering for inducing sigmoid response in both inputs in an enzymatic and gate. *J. Phys. Chem. B* **2013**, *117*, 9857–9865. [CrossRef]
13. Benenson, Y. Biomolecular computing systems: Principles, progress and potential. *Nat. Rev. Genet.* **2012**, *13*, 455–468. [CrossRef]
14. Bray, D. Protein molecules as computational elements in living cells. *Nature* **1995**, *376*, 307–312. [CrossRef]
15. Magnasco, M.O. Chemical kinetics is turing universal. *Phys. Rev. Lett.* **1997**, *78*, 1190–1193. [CrossRef]
16. Buisman, H.J.; ten Eikelder, H.M.M.; Hilbers, P.A.J.; Liekens, A.M.L. Computing algebraic functions with biochemical reaction networks. *Artif. Life* **2009**, *15*, 5–19. [CrossRef]
17. Privman, V.; Strack, G.; Solenov, D.; Pita, M.; Katz, E. Optimization of enzymatic biochemical logic for noise reduction and scalability: How many biocomputing gates can be interconnected in a circuit? *J. Phys. Chem. B* **2008**, *112*, 11777–11784. [CrossRef]
18. Privman, V. Control of noise in chemical and biochemical information processing. *Isr. J. Chem.* **2011**, *51*, 118–131. [CrossRef]
19. Teo, J.J.Y.; Woo, S.S.; Sarpeshkar, R. Synthetic Biology: A Unifying View and Review Using Analog Circuits. *IEEE Trans. Biomed. Circuits Syst.* **2015**, *9*, 453–474. [CrossRef]
20. Grozinger, L.; Amos, M.; Gorochowski, T.E.; Carbonell, P.; Oyarzún, D.A.; Stoof, R.; Fellermann, H.; Zuliani, P.; Tas, H.; Goñi-Moreno, A. Pathways to cellular supremacy in biocomputing. *Nat. Commun.* **2019**, *10*, 5250. [CrossRef]

21. Katz, E. Boolean Logic Gates Realized with Enzyme-catalyzed Reactions—Unusual Look at Usual Chemical Reactions. *ChemPhysChem* **2019**, *20*, 9–22. [CrossRef]
22. Sugita, M. The Switching Circuit Logically or Functionally Equivalent to a System of Biochemical Reactions. *J. Phys. Soc. Japan* **1961**, *16*, 737–740. [CrossRef]
23. Credi, A. Molecules that make decisions. *Angew. Chem.-Int. Ed.* **2007**, *46*, 5472–5475. [CrossRef]
24. Szaciłowski, K. Digital information processing in molecular systems. *Chem. Rev.* **2008**, *108*, 3481–3548. [CrossRef]
25. Ratner, M.A.; Jortner, J. Molecular Electronics: Some Directions. In *Molecular Electronics*; Blackwell Science Ltd.: Hoboken, NJ, USA, 1997; pp. 5–72.
26. De Silva, P.A.; Gunaratne, N.H.Q.; McCoy, C.P. A molecular photoionic and gate based on fluorescent signalling. *Nature* **1993**, *364*, 42–44. [CrossRef]
27. Sivan, S.; Tuchman, S.; Lotan, N. A biochemical logic gate using an enzyme and its inhibitor. Part II: The logic gate. *BioSystems* **2003**, *70*, 21–33. [CrossRef]
28. Sugita, M. Functional analysis of chemical systems in vivo using a logical circuit equivalent. V. Molecular biological interpretation of the self-reproducing automata theory and chemico-physical interpretation of information in biological systems. *J. Theor. Biol.* **1975**, *53*, 223–237. [CrossRef]
29. Jacob, F.; Monod, J. Genetic regulatory mechanisms in the synthesis of proteins. *J. Mol. Biol.* **1961**, *3*, 318–356. [CrossRef]
30. Okamoto, M.; Katsurayama, A.; Tsukiji, M.; Aso, Y.; Hayashi, K. Dynamic behavior of enzymatic system realizing two-factor model. *J. Theor. Biol.* **1980**, *83*, 1–16. [CrossRef]
31. Okamoto, M.; Sakai, T.; Hayashi, K. Biochemical switching device: Monocyclic enzyme system. *Biotechnol. Bioeng.* **1988**, *32*, 527–537. [CrossRef]
32. Rosen, R. Two-factor models, neural nets and biochemical automata. *J. Theor. Biol.* **1967**, *15*, 282–297. [CrossRef]
33. Arkin, A.; Ross, J. Computational functions in biochemical reaction networks. *Biophys. J.* **1994**, *67*, 560–578. [CrossRef]
34. Sivan, S.; Lotan, N. A biochemical logic gate using an enzyme and its inhibitor. 1. The inhibitor as switching element. *Biotechnol. Prog.* **1999**, *15*, 964–970. [CrossRef]
35. Zauner, K.P.; Conrad, M. Enzymatic computing. *Biotechnol. Prog.* **2001**, *17*, 553–559. [CrossRef]
36. Baron, R.; Lioubashevski, O.; Katz, E.; Niazov, T.; Willner, I. Logic gates and elementary computing by enzymes. *J. Phys. Chem. A* **2006**, *110*, 8548–8553. [CrossRef]
37. Baron, R.; Lioubashevski, O.; Katz, E.; Niazov, T.; Willner, I. Elementary arithmetic operations by enzymes: A model for metabolic pathway based computing. *Angew. Chem.-Int. Ed.* **2006**, *45*, 1572–1576. [CrossRef]
38. Niazov, T.; Baron, R.; Katz, E.; Lioubashevski, O.; Willner, I. Concatenated logic gates using four coupled biocatalysts operating in series. *Proc. Natl. Acad. Sci. USA* **2006**, *103*, 17160–17163. [CrossRef]
39. Vlasov, Y.G.; Tarantov, Y.A.; Bobrov, P.V. Analytical characteristics and sensitivity mechanisms of electrolyte-insulator-semiconductor system-based chemical sensors-a critical review. *Anal. Bioanal. Chem.* **2003**, *376*, 788–796. [CrossRef]
40. Privman, V.; Pedrosa, V.; Melnikov, D.; Pita, M.; Simonian, A.; Katz, E. Enzymatic AND-gate based on electrode-immobilized glucose-6-phosphate dehydrogenase: Towards digital biosensors and biochemical logic systems with low noise. *Biosens. Bioelectron.* **2009**, *25*, 695–701. [CrossRef]
41. Mailloux, S.; Gerasimova, Y.V.; Guz, N.; Kolpashchikov, D.M.; Katz, E. Bridging the Two Worlds: A Universal Interface between Enzymatic and DNA Computing Systems. *Angew. Chem.-Int. Ed.* **2015**, *54*, 6562–6566. [CrossRef]
42. Katz, E. Transduction of Signals Generated by Enzyme Logic Gates. In *Enzyme-Based Computing Systems*; Wiley-VCH: Weinheim, Germany, 2019; pp. 113–149. ISBN 9783527819997.
43. Chen, Z.; Kibler, R.D.; Hunt, A.; Busch, F.; Pearl, J.; Jia, M.; VanAernum, Z.L.; Wicky, B.I.M.M.; Dods, G.; Liao, H.; et al. De novo design of protein logic gates. *Science* **2020**, *368*, 78–84. [CrossRef]
44. Daringer, N.M.; Dudek, R.M.; Schwarz, K.A.; Leonard, J.N. Modular Extracellular sensor architecture for engineering mammalian cell-based devices. *ACS Synth. Biol.* **2014**, *3*, 892–902. [CrossRef] [PubMed]
45. Nielsen, A.A.K.; Der, B.S.; Shin, J.; Vaidyanathan, P.; Paralanov, V.; Strychalski, E.A.; Ross, D.; Densmore, D.; Voigt, C.A. Genetic circuit design automation. *Science* **2016**, *352*. [CrossRef] [PubMed]
46. Ding, N.; Zhou, S.; Deng, Y. Transcription-Factor-based Biosensor Engineering for Applications in Synthetic Biology. *ACS Synth. Biol.* **2021**, *10*, 911–922. [CrossRef]
47. Moškon, M.; Komac, R.; Zimic, N.; Mraz, M. Distributed biological computation: From oscillators, logic gates and switches to a multicellular processor and neural computing applications. *Neural Comput. Appl.* **2021**, *33*, 8923–8938. [CrossRef]
48. Guiziou, S.; Ulliana, F.; Moreau, V.; Leclere, M.; Bonnet, J. An Automated Design Framework for Multicellular Recombinase Logic. *ACS Synth. Biol.* **2018**, *7*, 1406–1412. [CrossRef]
49. Gorman, S.D.; Boehr, D.D. Energy and Enzyme Activity Landscapes of Yeast Chorismate Mutase at Cellular Concentrations of Allosteric Effectors. *Biochemistry* **2019**, *58*, 4058–4069. [CrossRef]
50. O'Rourke, K.F.; Sahu, D.; Bosken, Y.K.; D'Amico, R.N.; Chang, C.-e.A.; Boehr, D.D. Coordinated Network Changes across the Catalytic Cycle of Alpha Tryptophan Synthase. *Structure* **2019**, *27*, 1405–1415.e5. [CrossRef]
51. Boehr, D.D.; D'Amico, R.N.; O'Rourke, K.F. Engineered control of enzyme structural dynamics and function. *Protein Sci.* **2018**, *27*, 825–838. [CrossRef]

52. Gorman, S.D.; D'Amico, R.N.; Winston, D.S.; Boehr, D.D. Engineering allostery into proteins. *Adv. Exp. Med. Biol.* **2019**, *1163*, 359–384. [CrossRef]
53. Zhong, W.; Cui, L.; Goh, B.C.; Cai, Q.; Ho, P.; Chionh, Y.H.; Yuan, M.; Sahili, A.E.; Fothergill-Gilmore, L.A.; Walkinshaw, M.D.; et al. Allosteric pyruvate kinase-based "logic gate" synergistically senses energy and sugar levels in Mycobacterium tuberculosis. *Nat. Commun.* **2017**, *8*, 1986. [CrossRef] [PubMed]
54. Katz, E. Conclusions and Perspectives: Where Are We Going? In *Enzyme-Based Computing Systems*; Wiley-VCH: Weinheim, Germany, 2019; pp. 371–382. ISBN 9783527819997.
55. Agliari, E.; Altavilla, M.; Barra, A.; Dello Schiavo, L.; Katz, E. Notes on stochastic (bio)-logic gates: Computing with allosteric cooperativity. *Sci. Rep.* **2015**, *5*, 9415. [CrossRef] [PubMed]
56. Galstyan, V.; Funk, L.; Einav, T.; Phillips, R. Combinatorial Control through Allostery. *J. Phys. Chem. B* **2019**, *123*, 2792–2800. [CrossRef] [PubMed]
57. Guntas, G.; Mitchell, S.F.; Ostermeier, M. A Molecular Switch Created by In Vitro Recombination of Nonhomologous Genes. *Chem. Biol.* **2004**, *11*, 1483–1487. [CrossRef] [PubMed]
58. Guntas, G.; Mansell, T.J.; Kim, J.R.; Ostermeier, M. Directed evolution of protein switches and their application to the creation of ligand-binding proteins. *Proc. Natl. Acad. Sci. USA* **2005**, *102*, 11224–11229. [CrossRef]
59. Wright, C.M.; Majumdar, A.; Tolman, J.R.; Ostermeier, M. NMR chracterization of an engineered domain fusion between maltose binding protein and TEM β-lactamase provides Insight into Its structure and allosteric mechanism. *Proteins Struct. Funct. Bioinform.* **2010**, *78*, 1423–1430. [CrossRef]
60. Choi, J.H.; Ostermeier, M. Rational design of a fusion protein to exhibit disulfide-mediated logic gate behavior. *ACS Synth. Biol.* **2015**, *4*, 400–406. [CrossRef]
61. Guo, Z.; Johnston, W.A.; Whitfield, J.; Walden, P.; Cui, Z.; Wijker, E.; Edwardraja, S.; Lantadilla, I.R.; Ely, F.; Vickers, C.; et al. Generalizable Protein Biosensors Based on Synthetic Switch Modules. *J. Am. Chem. Soc.* **2019**, *141*, 8128–8135. [CrossRef]
62. Bollella, P.; Edwardraja, S.; Guo, Z.; Alexandrov, K.; Katz, E. Control of Allosteric Protein Electrochemical Switches with Biomolecular and Electronic Signals. *J. Phys. Chem. Lett.* **2020**, *11*, 5549–5554. [CrossRef]
63. Bollella, P.; Bellare, M.; Kadambar, V.K.; Guo, Z.; Alexandrov, K.; Melman, A.; Katz, E. Boolean Logic Networks Mimicked with Chimeric Enzymes Activated/Inhibited by Several Input Signals. *ChemPhysChem* **2020**, *21*, 589–593. [CrossRef]
64. Bollella, P.; Guo, Z.; Edwardraja, S.; Krishna Kadambar, V.; Alexandrov, K.; Melman, A.; Katz, E. Self-powered molecule release systems activated with chemical signals processed through reconfigurable Implication or Inhibition Boolean logic gates. *Bioelectrochemistry* **2021**, *138*, 107735. [CrossRef] [PubMed]
65. Lopez, R.J.; Babanova, S.; Artyushkova, K.; Atanassov, P. Surface modifications for enhanced enzyme immobilization and improved electron transfer of PQQ-dependent glucose dehydrogenase anodes. *Bioelectrochemistry* **2015**, *105*, 78–87. [CrossRef] [PubMed]
66. Vishweshwaraiah, Y.L.; Chen, J.; Chirasani, V.R.; Tabdanov, E.D.; Dokholyan, N.V. Two-input protein logic gate for computation in living cells. *Nat. Commun.* **2021**, *12*, 6615. [CrossRef] [PubMed]
67. Karginov, A.V.; Ding, F.; Kota, P.; Dokholyan, N.V.; Hahn, K.M. Engineered allosteric activation of kinases in living cells. *Nat. Biotechnol.* **2010**, *28*, 743–747. [CrossRef]
68. Wang, J.; Jain, A.; McDonald, L.R.; Gambogi, C.; Lee, A.L.; Dokholyan, N.V. Mapping allosteric communications within individual proteins. *Nat. Commun.* **2020**, *11*, 3862. [CrossRef]
69. Omersa, N.; Aden, S.; Kisovec, M.; Podobnik, M.; Anderluh, G. Design of Protein Logic Gate System Operating on Lipid Membranes. *ACS Synth. Biol.* **2020**, *9*, 316–328. [CrossRef]
70. Okhokhonin, A.V.; Domanskyi, S.; Filipov, Y.; Gamella, M.; Kozitsina, A.N.; Privman, V.; Katz, E. Biomolecular Release from Alginate-modified Electrode Triggered by Chemical Inputs Processed through a Biocatalytic Cascade—Integration of Biomolecular Computing and Actuation. *Electroanalysis* **2018**, *30*, 426–435. [CrossRef]
71. Gamella, M.; Privman, M.; Bakshi, S.; Melman, A.; Katz, E. DNA Release from Fe^{3+}-Cross-Linked Alginate Films Triggered by Logically Processed Biomolecular Signals: Integration of Biomolecular Computing and Actuation. *ChemPhysChem* **2017**, *18*, 1811–1821. [CrossRef]
72. Honarvarfard, E.; Gamella, M.; Poghossian, A.; Schöning, M.J.; Katz, E. An enzyme-based reversible Controlled NOT (CNOT) logic gate operating on a semiconductor transducer. *Appl. Mater. Today* **2017**, *9*, 266–270. [CrossRef]
73. Jablonski, M.; Poghossian, A.; Molinnus, D.; Keusgen, M.; Katz, E.; Schöning, M.; Notice, C. Enzyme-based XOR logic gate with electronic transduction of the output signal. *Int. J. Unconv. Comput.* **2019**, *14*, 375–383.
74. Karschuck, T.L.; Filipov, Y.; Bollella, P.; Schöning, M.; Katz, E. Not-XOR (NXOR) logic gate based on an enzyme-catalyzed reaction. *Int. J. Unconv. Comput.* **2019**, *14*, 235–242.
75. Filipov, Y.; Domanskyi, S.; Wood, M.L.; Gamella, M.; Privman, V.; Katz, E. Experimental Realization of a High-Quality Biochemical XOR Gate. *ChemPhysChem* **2017**, *18*, 2908–2915. [CrossRef] [PubMed]
76. Filipov, Y.; Gamella, M.; Katz, E. Nano-species Release System Activated by Enzyme-based XOR Logic Gate. *Electroanalysis* **2018**, *30*, 1281–1286. [CrossRef]
77. Chung, H.K.; Lin, M.Z. On the cutting edge: Protease-based methods for sensing and controlling cell biology. *Nat. Methods* **2020**, *17*, 885–896. [CrossRef] [PubMed]

78. Gräwe, A.; Ranglack, J.; Weber, W.; Stein, V. Engineering artificial signalling functions with proteases. *Curr. Opin. Biotechnol.* **2020**, *63*, 1–7. [CrossRef]
79. Chen, Z.; Elowitz, M.B. Programmable protein circuit design. *Cell* **2021**, *184*, 2284–2301. [CrossRef]
80. Gao, X.J.; Chong, L.S.; Kim, M.S.; Elowitz, M.B. Programmable protein circuits in living cells. *Science* **2018**, *361*, 1252–1258. [CrossRef]
81. Shekhawat, S.S.; Porter, J.R.; Sriprasad, A.; Ghosh, I. An autoinhibited coiled-coil design strategy for split-protein protease sensors. *J. Am. Chem. Soc.* **2009**, *131*, 15284–15290. [CrossRef]
82. Fernandez-Rodriguez, J.; Voigt, C.A. Post-translational control of genetic circuits using Potyvirus proteases. *Nucleic Acids Res.* **2016**, *44*, 6493–6502. [CrossRef]
83. Seelig, G.; Soloveichik, D.; Zhang, D.Y.; Winfree, E. Enzyme-Free Nucleic Acid Logic Circuits. *Science* **2006**, *314*, 1585–1589. [CrossRef]
84. Yang, J.; Wu, R.; Li, Y.; Wang, Z.; Pan, L.; Zhang, Q.; Lu, Z.; Zhang, C. Entropy-driven DNA logic circuits regulated by DNAzyme. *Nucleic Acids Res.* **2018**, *46*, 8532–8541. [CrossRef] [PubMed]
85. Zhang, D.Y.; Turberfield, A.J.; Yurke, B.; Winfree, E. Engineering Entropy-Driven Reactions and Networks Catalyzed by DNA. *Science* **2009**, *318*, 1121–1125. [CrossRef] [PubMed]
86. Zhang, D.Y.; Winfree, E. Control of DNA strand displacement kinetics using toehold exchange. *J. Am. Chem. Soc.* **2009**, *131*, 17303–17314. [CrossRef]
87. Thubagere, A.J.; Thachuk, C.; Berleant, J.; Johnson, R.F.; Ardelean, D.A.; Cherry, K.M.; Qian, L. Compiler-aided systematic construction of large-scale DNA strand displacement circuits using unpurified components. *Nat. Commun.* **2017**, *8*, 14373. [CrossRef]
88. Qian, L.; Winfree, E. Scaling Up Digital Circuit Computation with DNA Strand Displacement Cascades. *Science* **2011**, *99*, 1196–1202. [CrossRef] [PubMed]
89. Li, W.; Zhang, F.; Yan, H.; Liu, Y. DNA based arithmetic function: A half adder based on DNA strand displacement. *Nanoscale* **2016**, *8*, 3775–3784. [CrossRef]
90. Frezza, B.M.; Cockroft, S.L.; Ghadiri, M.R. Modular multi-level circuits from immobilized DNA-based logic gates. *J. Am. Chem. Soc.* **2007**, *129*, 14875–14879. [CrossRef]
91. Gerasimova, Y.V.; Kolpashchikov, D.M. Towards a DNA Nanoprocessor: Reusable Tile-Integrated DNA Circuits. *Angew. Chem.-Int. Ed.* **2016**, *55*, 10244–10247. [CrossRef]
92. Qian, L.; Winfree, E. Parallel and scalable computation and spatial dynamics with DNA-based chemical reaction networks on a surface. In *Lecture Notes in Computer Science*; Springer: Cham, Switzerland, 2014; Volume 8727, pp. 114–131.
93. Woods, D.; Doty, D.; Myhrvold, C.; Hui, J.; Zhou, F.; Yin, P.; Winfree, E. Diverse and robust molecular algorithms using reprogrammable DNA self-assembly. *Nature* **2019**, *567*, 366–372. [CrossRef]
94. Srinivas, N.; Parkin, J.; Seelig, G.; Winfree, E.; Soloveichik, D. Enzyme-free nucleic acid dynamical systems. *Science* **2017**, *358*. [CrossRef]
95. Fern, J.; Scalise, D.; Cangialosi, A.; Howie, D.; Potters, L.; Schulman, R. DNA Strand-Displacement Timer Circuits. *ACS Synth. Biol.* **2017**, *6*, 190–193. [CrossRef] [PubMed]
96. Huang, D.; Han, H.; Guo, C.; Lin, X.; Chen, D.; Yang, S.; Yang, Q.; Li, F. Information processing using an integrated DNA reaction network. *Nanoscale* **2021**, *13*, 5706–5713. [CrossRef] [PubMed]
97. Orbach, R.; Willner, B.; Willner, I. Catalytic nucleic acids (DNAzymes) as functional units for logic gates and computing circuits: From basic principles to practical applications. *Chem. Commun.* **2015**, *51*, 4144–4160. [CrossRef] [PubMed]
98. Stojanovic, M.N.; Stefanovic, D.; Rudchenko, S. Exercises in molecular computing. *Acc. Chem. Res.* **2014**, *47*, 1845–1852. [CrossRef] [PubMed]
99. Lakin, M.R.; Stojanovic, M.N.; Stefanovic, D. Implementing Molecular Logic Gates, Circuits Stefanovic, and Cascades Using DNAzymes. *Adv. Unconv. Comput.* **2017**, 1–28. [CrossRef]
100. Stojanovic, M.N.; Stefanovic, D. A deoxyribozyme-based molecular automaton. *Nat. Biotechnol.* **2003**, *21*, 1069–1074. [CrossRef]
101. Stojanovic, M.N.; Stefanovic, D. Deoxyribozyme-Based Half-Adder. *J. Am. Chem. Soc.* **2003**, *125*, 6673–6676. [CrossRef]
102. Lederman, H.; Macdonald, J.; Stefanovic, D.; Stojanovic, M.N. Deoxyribozyme-based three-input logic gates and construction of a molecular full adder. *Biochemistry* **2006**, *45*, 1194–1199. [CrossRef]
103. Macdonald, J.; Li, Y.; Sutovic, M.; Lederman, H.; Pendri, K.; Lu, W.; Andrews, B.L.; Stefanovic, D.; Stojanovic, M.N. Medium scale integration of molecular logic gates in an automaton. *Nano Lett.* **2006**, *6*, 2598–2603. [CrossRef]
104. Pei, R.; Matamoros, E.; Liu, M.; Stefanovic, D.; Stojanovic, M.N. Training a molecular automaton to play a game. *Nat. Nanotechnol.* **2010**, *5*, 773–777. [CrossRef]
105. Elbaz, J.; Wang, F.; Remacle, F.; Willner, I. PH-programmable DNA logic arrays powered by modular DNAzyme libraries. *Nano Lett.* **2012**, *12*, 6049–6054. [CrossRef] [PubMed]
106. Prokup, A.; Hemphill, J.; Deiters, A. DNA computation: A photochemically controlled and gate. *J. Am. Chem. Soc.* **2012**, *134*, 3810–3815. [CrossRef] [PubMed]
107. Orbach, R.; Remacle, F.; Levine, R.D.; Willner, I. DNAzyme-based 2:1 and 4:1 multiplexers and 1:2 demultiplexer. *Chem. Sci.* **2014**, *5*, 1074–1081. [CrossRef]

108. Orbach, R.; Wang, F.; Lioubashevski, O.; Levine, R.D.; Remacle, F.; Willner, I. A full-adder based on reconfigurable DNA-hairpin inputs and DNAzyme computing modules. *Chem. Sci.* **2014**, *5*, 3381–3387. [CrossRef]
109. Li, W.; Yang, Y.; Yan, H.; Liu, Y. Three-input majority logic gate and multiple input logic circuit based on DNA strand displacement. *Nano Lett.* **2013**, *13*, 2980–2988. [CrossRef]
110. Zhu, J.; Zhang, L.; Li, T.; Dong, S.; Wang, E. Enzyme-free unlabeled DNA logic circuits based on toehold-mediated strand displacement and split G-quadruplex enhanced fluorescence. *Adv. Mater.* **2013**, *25*, 2440–2444. [CrossRef]
111. Stojanovic, M.N.; Semova, S.; Kolpashchikov, D.; Macdonald, J.; Morgan, C.; Stefanovic, D. Deoxyribosyme-based ligase logic gates and their initial circuits. *J. Am. Chem. Soc.* **2005**, *127*, 6914–6915. [CrossRef]
112. Penchovsky, R.; Breaker, R.R. Computational design and experimental validation of oligonucleotide-sensing allosteric ribozymes. *Nat. Biotechnol.* **2005**, *23*, 1424–1433. [CrossRef]
113. Penchovsky, R. Engineering integrated digital circuits with allosteric ribozymes for scaling up molecular computation and diagnostics. *ACS Synth. Biol.* **2012**, *1*, 471–482. [CrossRef]
114. Kahan-Hanum, M.; Douek, Y.; Adar, R.; Shapiro, E. A library of programmable DNAzymes that operate in a cellular environment. *Sci. Rep.* **2013**, *3*, 1535. [CrossRef]
115. Brown, C.W.; Lakin, M.R.; Stefanovic, D.; Graves, S.W. Catalytic molecular logic devices by DNAzyme displacement. *ChemBioChem* **2014**, *15*, 950–954. [CrossRef] [PubMed]
116. Lakin, M.R.; Brown, C.W.; Horwitz, E.K.; Fanning, M.L.; West, H.E.; Stefanovic, D.; Graves, S.W. Biophysically inspired rational design of structured chimeric substrates for DNAzyme Cascade Engineering. *PLoS ONE* **2014**, *9*, e110986. [CrossRef] [PubMed]
117. Morihiro, K.; Ankenbruck, N.; Lukasak, B.; Deiters, A. Small molecule release and activation through DNA computing. *J. Am. Chem. Soc.* **2017**, *139*, 13909–13915. [CrossRef] [PubMed]
118. Zhu, J.; Li, T.; Zhang, L.; Dong, S.; Wang, E. G-quadruplex DNAzyme based molecular catalytic beacon for label-free colorimetric logic gates. *Biomaterials* **2011**, *32*, 7318–7324. [CrossRef] [PubMed]
119. Orbach, R.; Remacle, F.; Levine, R.D.; Willner, I. Logic reversibility and thermodynamic irreversibility demonstrated by DNAzyme-based Toffoli and Fredkin logic gates. *Proc. Natl. Acad. Sci. USA* **2012**, *109*, 21228–21233. [CrossRef]
120. Li, H.; Guo, S.; Liu, Q.; Qin, L.; Dong, S.; Liu, Y.; Wang, E. Implementation of Arithmetic Functions on a Simple and Universal Molecular Beacon Platform. *Adv. Sci.* **2015**, *2*, 1500054. [CrossRef]
121. Zhu, J.; Zhang, L.; Zhou, Z.; Dong, S.; Wang, E. Molecular aptamer beacon tuned DNA strand displacement to transform small molecules into DNA logic outputs. *Chem. Commun.* **2014**, *50*, 3321–3323. [CrossRef]
122. Miyoshi, D.; Inoue, M.; Sugimoto, N. DNA logic gates based on structural polymorphism of telomere DNA molecules responding to chemical input signals. *Angew. Chem.-Int. Ed.* **2006**, *45*, 7716–7719. [CrossRef]
123. Li, T.; Wang, E.; Dong, S. Potassium-lead-switched G-quadruplexes: A new class of DNA logic gates. *J. Am. Chem. Soc.* **2009**, *131*, 15082–15083. [CrossRef]
124. Travascio, P.; Li, Y.; Sen, D. DNA-enhanced peroxidase activity of a DNA aptamer-hemin complex. *Chem. Biol.* **1998**, *5*, 505–517. [CrossRef]
125. Moshe, M.; Elbaz, J.; Willner, I. Sensing of UO22+ and design of logic gates by the application of supramolecular constructs of ion-dependent DNAzymes. *Nano Lett.* **2009**, *9*, 1196–1200. [CrossRef] [PubMed]
126. Tang, W.; Wang, H.; Wang, D.; Zhao, Y.; Li, N.; Liu, F. DNA tetraplexes-based toehold activation for controllable DNA strand displacement reactions. *J. Am. Chem. Soc.* **2013**, *135*, 13628–13631. [CrossRef] [PubMed]
127. Zhang, L.; Zhang, Y.M.; Liang, R.P.; Qiu, J.D. Colorimetric logic gates based on ion-dependent DNAzymes. *J. Phys. Chem. C* **2013**, *117*, 12352–12357. [CrossRef]
128. Shlyahovsky, B.; Li, Y.; Lioubashevski, O.; Elbaz, J.; Willner, I. Logic gates and antisense DNA devices operating on a translator nucleic acid scaffold. *ACS Nano* **2009**, *3*, 1831–1843. [CrossRef]
129. Gao, W.; Zhang, L.; Zhang, Y.M.; Liang, R.P.; Qiu, J.D. DNA colorimetric logic gates based on triplex-helix molecular switch. *J. Phys. Chem. C* **2014**, *118*, 14410–14417. [CrossRef]
130. Fan, D.; Wang, K.; Zhu, J.; Xia, Y.; Han, Y.; Liu, Y.; Wang, E. DNA-based visual majority logic gate with one-vote veto function. *Chem. Sci.* **2015**, *6*, 1973–1978. [CrossRef] [PubMed]
131. Yu, Y.; Guo, Q.; Jiang, W.; Zhang, H.; Cai, C. Dual-Aptamer-Assisted and Logic Gate for Cyclic Enzymatic Signal Amplification Electrochemical Detection of Tumor-Derived Small Extracellular Vesicles. *Anal. Chem.* **2021**, *93*, 11298–11304. [CrossRef]
132. Fink, T.; Lonzarić, J.; Praznik, A.; Plaper, T.; Merljak, E.; Leben, K.; Jerala, N.; Lebar, T.; Strmšek, Ž.; Lapenta, F.; et al. Design of fast proteolysis-based signaling and logic circuits in mammalian cells. *Nat. Chem. Biol.* **2019**, *15*, 115–122. [CrossRef]
133. Barnea, G.; Strapps, W.; Herrada, G.; Berman, Y.; Ong, J.; Kloss, B.; Axel, R.; Lee, K.J. The genetic design of signaling cascades to record receptor activation. *Proc. Natl. Acad. Sci. USA* **2008**, *105*, 64–69. [CrossRef]
134. Lee, Y.; Son, J.Y.; Kang, J.I.; Park, K.M.; Park, K.D. Hydrogen Peroxide-Releasing Hydrogels for Enhanced Endothelial Cell Activities and Neovascularization. *ACS Appl. Mater. Interfaces* **2018**, *10*, 18372–18379. [CrossRef]
135. Kang, J.I.; Park, K.M. Advances in gelatin-based hydrogels for wound management. *J. Mater. Chem. B* **2021**, *9*, 1503–1520. [CrossRef]
136. Badelt, S.; Grun, C.; Sarma, K.V.; Wolfe, B.; Shin, S.W.; Winfree, E. A domain-level DNA strand displacement reaction enumerator allowing arbitrary non-pseudoknotted secondary structures. *J. R. Soc. Interface* **2020**, *17*, 20190866. [CrossRef] [PubMed]

137. Cherry, K.M.; Qian, L. Scaling up molecular pattern recognition with DNA-based winner-take-all neural networks. *Nature* **2018**, *559*, 370–388. [CrossRef] [PubMed]
138. Zhang, J. RNA-cleaving DNAzymes: Old catalysts with new tricks for intracellular and in vivo applications. *Catalysts* **2018**, *8*, 550. [CrossRef]
139. Li, X.; Yang, F.; Zhou, W.; Yuan, R.; Xiang, Y. Targeted and direct intracellular delivery of native DNAzymes enables highly specific gene silencing. *Chem. Sci.* **2020**, *11*, 8966–8972. [CrossRef] [PubMed]
140. You, M.; Zhu, G.; Chen, T.; Donovan, M.J.; Tan, W. Programmable and multiparameter DNA-based logic platform for cancer recognition and targeted therapy. *J. Am. Chem. Soc.* **2015**, *137*, 667–674. [CrossRef] [PubMed]
141. Gao, Q.; Zhao, Y.; Xu, K.; Zhang, C.; Ma, Q.; Qi, L.; Chao, D.; Zheng, T.; Yang, L.; Miao, Y.; et al. Highly Specific, Single-step Cancer Cell Isolation with Multi-Aptamer-Mediated Proximity Ligation on Live Cell Membranes. *Angew. Chem.-Int. Ed.* **2020**, *59*, 23564–23568. [CrossRef] [PubMed]
142. Chai, X.; Fan, Z.; Yu, M.M.; Zhao, J.; Li, L. A Redox-Activatable DNA Nanodevice for Spatially-Selective, AND-Gated Imaging of ATP and Glutathione in Mitochondria. *Nano Lett.* **2021**, *21*, 10047–10053. [CrossRef]
143. Wang, D.; Li, S.; Zhao, Z.; Zhang, X.; Tan, W. Engineering a Second-Order DNA Logic-Gated Nanorobot to Sense and Release on Live Cell Membranes for Multiplexed Diagnosis and Synergistic Therapy. *Angew. Chem.-Int. Ed.* **2021**, *60*, 15816–15820. [CrossRef]
144. Li, Y.; Deng, J.; Han, Z.; Liu, C.; Tian, F.; Xu, R.; Han, D.; Zhang, S.; Sun, J. Molecular Identification of Tumor-Derived Extracellular Vesicles Using Thermophoresis-Mediated DNA Computation. *J. Am. Chem. Soc.* **2021**, *143*, 1290–1295. [CrossRef]
145. Kamar, O.; Sun, S.C.; Lin, C.H.; Chung, W.Y.; Lee, M.S.; Liao, Y.C.; Kolpashchikov, D.M.; Chuang, M.C. A mutation-resistant deoxyribozyme or gate for highly selective detection of viral nucleic acids. *Chem. Commun.* **2017**, *53*, 10592–10595. [CrossRef] [PubMed]
146. Hemphill, J.; Deiters, A. DNA computation in mammalian cells: MicroRNA logic operations. *J. Am. Chem. Soc.* **2013**, *135*, 10512–10518. [CrossRef] [PubMed]
147. Chang, X.; Zhang, C.; Lv, C.; Sun, Y.; Zhang, M.; Zhao, Y.; Yang, L.; Han, D.; Tan, W. Construction of a Multiple-Aptamer-Based DNA Logic Device on Live Cell Membranes via Associative Toehold Activation for Accurate Cancer Cell Identification. *J. Am. Chem. Soc.* **2019**, *141*, 12738–12743. [CrossRef] [PubMed]
148. Taylor, S.; Stojanovic, M.N. Is there a future for DNA-based molecular devices in diabetes management? *J. Diabetes Sci. Technol.* **2007**, *1*, 440–444. [CrossRef]
149. Gomes de Oliveira, A.G.; Dubovichenko, M.V.; ElDeeb, A.A.; Wanjohi, J.; Zablotskaya, S.; Kolpashchikov, D.M. RNA-Cleaving DNA Thresholder Controlled by Concentrations of miRNA Cancer Marker. *ChemBioChem* **2021**, *22*, 1750–1754. [CrossRef]
150. Singh, A. Designing protein logic gates. *Nat. Methods* **2020**, *17*, 565. [CrossRef]
151. Katz, E. Bioelectronic Interface between Enzyme-Based and DNA-Based Computing Systems. In *Enzyme-Based Computing Systems*; Wiley-VCH: Weinheim, Germany, 2019; pp. 335–355. ISBN 9783527819997.
152. Groseclose, T.M.; Rondon, R.E.; Herde, Z.D.; Aldrete, C.A.; Wilson, C.J. Engineered systems of inducible anti-repressors for the next generation of biological programming. *Nat. Commun.* **2020**, *11*, 4440. [CrossRef]
153. Groseclose, T.M.; Rondon, R.E.; Hersey, A.N.; Milner, P.T.; Kim, D.; Zhang, F.; Realff, M.J.; Wilson, C.J. Biomolecular Systems Engineering: Unlocking the Potential of Engineered Allostery via the Lactose Repressor Topology. *Annu. Rev. Biophys.* **2021**, *50*, 303–321. [CrossRef]
154. Zhang, J.; Lu, Y. Biocomputing for Portable, Resettable, and Quantitative Point-of-Care Diagnostics: Making the Glucose Meter a Logic-Gate Responsive Device for Measuring Many Clinically Relevant Targets. *Angew. Chem.-Int. Ed.* **2018**, *57*, 9702–9706. [CrossRef]

catalysts

Review

A Review on Biological Synthesis of the Biodegradable Polymers Polyhydroxyalkanoates and the Development of Multiple Applications

Bryan Dalton [1], Purabi Bhagabati [2], Jessica De Micco [2], Ramesh Babu Padamati [2,3,*] and Kevin O'Connor [1,3,*]

1 School of Biomolecular and Biomedical Science, College of Science, University College Dublin, Stillorgan Road, Belfield, D04 V1W Dublin, Ireland; bryan.dalton@ucd.ie

2 The Centre for Research on Adaptive Nanostructures and Nanodevices, School of Chemistry, Trinity College Dublin, The University of Dublin, D02 W272 Dublin, Ireland; bhagabap@tcd.ie (P.B.); demiccoj@tcd.ie (J.D.M.)

3 BiOrbic Bioeconomy SFI Research Centre, University College Dublin, Stillorgan Road, Belfield, D04 V1W Dublin, Ireland

* Correspondence: babup@tcd.ie (R.B.P.); kevin.oconnor@ucd.ie (K.O.)

Abstract: Polyhydroxyalkanoates, or PHAs, belong to a class of biopolyesters where the biodegradable PHA polymer is accumulated by microorganisms as intracellular granules known as carbonosomes. Microorganisms can accumulate PHA using a wide variety of substrates under specific inorganic nutrient limiting conditions, with many of the carbon-containing substrates coming from waste or low-value sources. PHAs are universally thermoplastic, with PHB and PHB copolymers having similar characteristics to conventional fossil-based polymers such as polypropylene. PHA properties are dependent on the composition of its monomers, meaning PHAs can have a diverse range of properties and, thus, functionalities within this biopolyester family. This diversity in functionality results in a wide array of applications in sectors such as food-packaging and biomedical industries. In order for PHAs to compete with the conventional plastic industry in terms of applications and economics, the scale of PHA production needs to grow from its current low base. Similar to all new polymers, PHAs need continuous technological developments in their production and material science developments to grow their market opportunities. The setup of end-of-life management (biodegradability, recyclability) system infrastructure is also critical to ensure that PHA and other biobased biodegradable polymers can be marketed with maximum benefits to society. The biobased nature and the biodegradability of PHAs mean they can be a key polymer in the materials sector of the future. The worldwide scale of plastic waste pollution demands a reformation of the current polymer industry, or humankind will face the consequences of having plastic in every step of the food chain and beyond. This review will discuss the aforementioned points in more detail, hoping to provide information that sheds light on how PHAs can be polymers of the future.

Keywords: biodegradable polymers; food-packaging; biodegradation; recycling; bioeconomy; circular economy; sustainability

Citation: Dalton, B.; Bhagabati, P.; De Micco, J.; Padamati, R.B.; O'Connor, K. A Review on Biological Synthesis of the Biodegradable Polymers Polyhydroxyalkanoates and the Development of Multiple Applications. *Catalysts* **2022**, *12*, 319. https://doi.org/10.3390/catal12030319

Academic Editors: Evangelos Topakas, Roland Wohlgemuth and David D. Boehr

Received: 21 January 2022
Accepted: 16 February 2022
Published: 11 March 2022

1. Introduction

1.1. What Are PHAs

Polyhydroxyalkanoates (PHAs) belong to a class of biopolyesters, first discovered by Lemoigne in 1925 [1]. PHA is accumulated as intracellular granules (carbonosomes, Figure 1), in various Gram-positive and Gram-negative bacteria [2]. These bacteria can accumulate PHA using a wide range of carbon rich growth substrates [3–7]. Generally limitation of an inorganic nutrient (e.g., nitrogen) in the growth medium is needed to stimulate PHA accumulation from the carbon source which should be in excess [8]. However, Zinn et al., have reported PHA accumulation in continuous cultures grown under both carbon and nitrogen limitation [9].

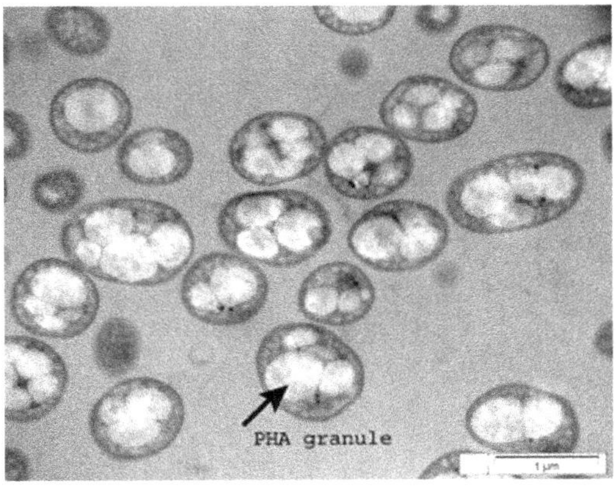

Figure 1. Transmission electron micrograph showing granules of PHA accumulated in the bacterium *P. putida* CA-3 from styrene [10].

Several enzyme driven pathways are used by microorganisms to metabolise carbon-based substrates and channel the carbon in those substrates towards PHA accumulation. These pathways will be described later in this review. Microorganisms can use a variety of carbon sources for growth and PHA accumulation [4] many of these carbon sources originate from waste or low value biomass [11]. The use of such carbon sources instead of higher cost starting materials makes PHA production via this route cheaper. The use of waste resources to make PHAs presents possibilities in a circular economy and an alternative to fossil-based resources to make conventional plastics. However, the heterogenous nature of waste materials as well as the security of supply provide challenges for industrial processes.

The most common PHAs are branched polymers with the carbon chains branching or extending out away from the 2-carbon repeating unit backbone of the polymer, as seen in Figure 2 [6].

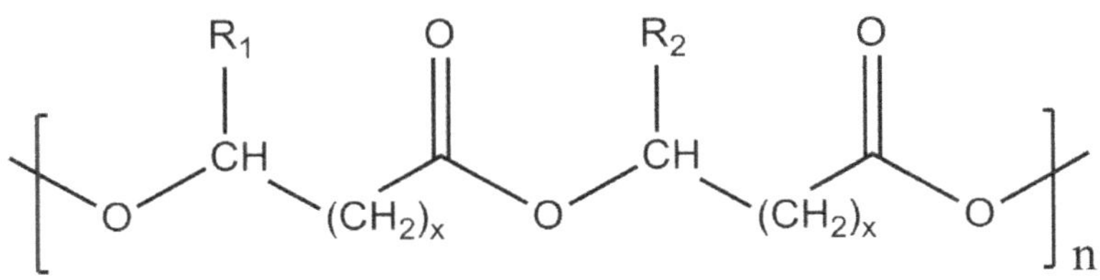

R_1/R_2 = alkyl groups C1–C16
X = 1,2,3,4
n = 100–30,000

Figure 2. General chemical structure of poly-3-hydroxyalkanoates (PHAs) [6].

Bacteria are capable of synthesising different subgroups of PHAs namely short-chain-length (scl), medium-chain-length (mcl) and long-chain-length (lcl) PHA [12]. scl and mcl-PHA groups are the most studied polymer chain lengths, with 3–5 carbons for scl-PHAs and 6–14 carbons for mcl-PHAs [13]. lcl-PHAs (>14 carbon atoms per monomer unit) are the least common occurring chain length of PHAs, with limited studies on their production and applications [14]. The molecular weight and type of PHA monomers determine the physical and thermal properties of the biopolymers; the branched nature of the polymer carbon means that the side chain interferes with crystallinity. Furthermore, the interaction of the side chains affects other properties such as tensile strength, elongation to break ratio, glass transition temperature and melting points [15]. Currently, PHAs and their blends are used in various applications. It is important to assess these composites' mechanical, barrier, and thermal properties for optimal application. PHAs and blends/composites with other polymers have been spun into fibres but also processed into trays and films with high gas and liquid barrier capabilities [13]. PHA properties are often comparable to conventional fossil fuel plastics, with authors suggesting PHAs as a sustainable alternative to plastics originating from fossil-based resources [16,17].

PHB (scl-PHA) is made via a three-step pathway once acetyl-CoA is made in the cell. While mcl-PHAs are made in microorganisms through two major pathway β-oxidation, starting from fatty acids, and de novo fatty acid synthesis using acetyl-CoA. Acetyl-CoA is a central metabolite produced by microorganisms when they metabolise sugars and other carbon substrates [18,19]. Within bacterial species, a small number of strains have been studied in depth and brought to an industrial scale for production. Various factors such as carbon source cost, efficiency of bacterial growth, PHA accumulation levels, properties of the PHA, and downstream processing costs, directly affects the scale-up potential of polymer production [12]. *Pseudomonas putida* (*P. putida*), *Pseudomonas oleovorans*, *Burkholderia sacchari* and *Cupriavidus necator* are some of these promising bacterial strains that have been studied extensively for their efficacy in PHA production.

PHAs are considered to be bioplastics, meaning they are biobased (originating from renewable sources) and biodegradable (can be broken down naturally through biological processes) (European Bioplastics, 2016). PHA's potentially offer more end of life options when compared to non-biodegradable biobased materials such as bio-polyethylene (bio-PE) and bio-terephthalate (bio-PET). Although PHA's look like a genuine green alternative to other plastics, much research still needs to be carried out to employ more control over the quality, quantity, and general economics for the production of PHA's [20]. The synthesis, properties, applications, and biodegradability of PHAs will be discussed in more detail in the following sections of this review.

1.2. The Synthesis of PHAs—Types of Microorganisms That Accumulate PHA and the Processes Within

PHAs are macromolecules, allowing for the storage of carbon and energy in PHA producing microorganisms [4]. In nature, PHA synthesis can occur through a number of different metabolic pathways, with these pathways determined by the microorganism producing it and the conditions in which this microorganism lives [20]. The synthesis of PHAs is regulated through three genes and three enzymes and is one of the simplest biosynthetic pathways [4]. One of the most common, and shorter pathways for PHA production is that of polyhydroxybutyrate (PHB) (Figure 3). Out of the 150 plus monomers that can be produced by PHA producing bacteria, only a small number of these, under natural conditions, can be formed into the homopolymers and copolymers required for the formation of PHAs [21]. Due to this natural bottleneck for inhibiting polymer formation, scl-PHAs such as PHB and P(3HB:HV) are more commonly found in more typical physiological conditions due to their shorter metabolic pathways [22]. With over 300 microorganisms capable of producing PHA, there are relatively few that can synthesize PHA at a level that is sufficient for scaling up to an industrial level. There are several bacterial species that can produce PHA at sufficient levels, though (Table 1). Although PHA can be formed by bacteria, fungi, and microalgae, there is significantly more research and promise for the efficiency of bacteria to produce PHA [23].

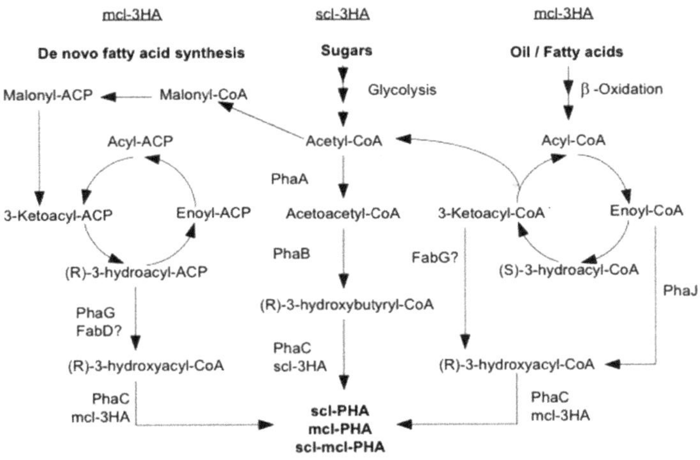

Figure 3. Biosynthesis pathways of short chain length (scl) PHA, medium chain length (mcl) PHA and short-medium-chain length (scl-mcl) PHA from sugars and oils. PhaA, β-ketothiolase; PhaB, NADPH-dependent acetoacetyl-CoA reductase; PhaC, PHA synthase; PhaG, 3-hydroxyacyl-ACP-CoA transferase; PhaJ, (R)-enoyl-CoA hydratase; FabD, malonyl-CoA-ACP transacylase; FabG, 3-ketoacyl-CoA reductase [19].

Table 1. Microorganisms extensively studied for their ability to accumulate PHA.

Microorganism	PHA Type	Reference
P. putida KT2440	mcl	[24]
P. oleovorans ATCC 29347	mcl	[25]
Burkholderia sacchari LFM 101	scl	[26]
Cupriavidus necator H16/DSM 428	scl	[27]

P. putida strain KT2440, is one of the most studied bacterial strains in the pseudomonas genus. *P. putida* KT2440 is a Gram-negative bacterium that has been extensively studied due to its ability to degrade aromatic compounds [28] and accumulate mcl-PHA. It is one of the most studied bacteria strains as it can synthesize high quantities of polymer per cell mass, amounts up to 75% w/w make it an attractive strain for production scaleup of PHA [8]. *P. putida* KT2440 also has advantageous features for biotechnology due to it metabolic diversity and genetic tractability [18]. It is capable of utilizing a wide range of carbon sources such as glucose, glycerol and fatty acids [29]. PHA accumulation in KT2440 from sugars requires nitrogen limitation but high levels of PHA can be accumulated, without nitrogen limitation while utilising fatty acids as the carbon source [29]. *P. putida* KT2440 has two main metabolic pathways that provide precursor molecules for the synthesis of mcl-PHA's (Figure 3) namely de novo fatty acid synthesis when using substrates such as glucose and glycerol and β-oxidation when using fatty acids. Another mcl-PHA producing bacterial strain that has been studied extensively is *Pseudomonas oleovorans* ATCC 29347. This strain type can grow on substrates such as n-alkane, n-alkene and long-chain fatty acids to accumulate copolymers of mcl-PHA [19]. Monomer units ranging from C6 to C14 are produced via β-oxidation of acyl-CoA [30], with monomers below C6 bypassing the PHA producing phase with these monomers converted into acetyl-CoA. These monomers below C6 are then harnessed via the TCA cycle for energy and as a carbon source for cell growth [19].

Burkholderia sacchari (stain IPT101 in the case of UCD), is a Gram-negative bacterium that was originally isolated from sugarcane crops in Brazil [31]. It is capable of utilizing the components of lignocellulose (cellulose, hemicellulose, lignin) to produce high-value chemicals and products such as poly-3-hydroxybutyrate [P(3HB)] [32]. The capability of *B. sacchari* to produce such high-value products from the most abundant renewable resource worldwide (lignocellulose) was the driving force for its use as a strain for the industrial scale-up of PHA production processes [32,33]. This is particularly apparent in studies that have shown this bacterium can accumulate P(3HB) up to 80% of its cell dry weight from sucrose [34]. It also grows efficiently on glucose, glycerol, organic acids and hydrolysed straw [35]. Another well studied PHB producing bacterial strain is *Cupriavidus necator* H16, previously known as *Ralstonia eutropha*. This bacterium is considered a model organism due to the availability of its complete genome sequence as well as its genetic tractability [36,37]. On top of its metabolic diversity, *C. necator* can achieve high cell densities aerobically under both heterotrophic and chemolithoautotrophic growth conditions. Under heterotrophic conditions *C. necator* can utilize substrates such as fructose, fatty acids and *N*-acetylglucosamine in tandem with oxygen. Under chemolithoautotrophic conditions *C. necator* can utilize carbon dioxide as it's substrate in tandem with hydrogen and oxygen [36]. When carbon dioxide is the single available source of carbon *C. necator* can assimilate via the reductive pentose phosphate cycle, otherwise known as the Calvin-Benson-Bassham cycle [37]. Due to its capability to produce a variety of valuable chemicals and polymer from an array of sources, *C. necator* is considered a good candidate for biotechnological processes [36,38].

1.3. Substrates Utilised by Microorganisms to Accumulate PHA

Microorganisms use a wide variety of carbon based substrates for growth and PHA accumulation. Monosaccharides such as glucose and fructose, disaccharides such as sucrose, and lactose have been used by microorganisms to accumulate PHA [31]. Sucrose can be generated from sugar beet and sugar cane while lactose is a major product and by-product of the dairy industry [33,39]. Complex carbohydrates such as cellulose and starch can be hydrolysed to produce glucose, arabinose, xylose and other monosaccharides which can subsequently be supplied to microorganisms for PHA accumulation [39,40]. Fatty acids can also be utilized as the sole carbon source for the production of PHA allowing for the use of waste, low-value or renewable oils for PHA production in place of other higher valued food-chain crops [41]. Examples such as palm kernel oil (PKO), where lauric acid is the pre-dominant 12-carbon saturated fatty acid of PKO, has shown promise a substrate for the production of mcl-PHA with strains of *P. putida* [41,42]. In order for PHA production to compete with that of petroleum based polymers in the world today, the substrates must originate from a biomass source that is considered either of low value or from a waste stream, to make the process as economically attractive as possible. The following sub-sections will cover the wide array of substrates that are utilised by microorganisms in the accumulation of PHA in more detail.

1.4. Sugar Containing Feedstocks for PHA Accumulation (Including Wastes)

1.4.1. Dextrose from Corn

Corn grain and corn stover (leaves, stalks) have been investigated as a carbon source for PHA production, focussing on the 1st generation substrates such as dextrose, corn oil and corn gluten [43]. The consistency and quantity in the production of the sugar dextrose from agricultural corn grain farming provide a reliable feedstock for PHA production, but with this comes the overlapping and competition with the demand for corn grain in the food chain. A past example of commercial-scale production of PHA from corn can be seen from the U.S. biotech company Metabolix, Inc. in Cambridge, MA, USA. Metabolix opened their first commercial-scale plant in Iowa in 2010, where they produced a corn-syrup based PHA up to 50,000 tons per year [44]. The commercial-scale production of PHA from 1st generation substrates of corn is possible, but it's viability is still uncertain due to the environmental burdens associated with corn cultivation [45]. For example, when compared to polystyrene production, PHA production does not improve overall environmental impacts regarding photochemical smog, acidification and eutrophication [43]. In order for corn to PHA production to be viable, PHA fermentation technologies must improve in tandem with increased integrated systems involved with corn cultivation.

1.4.2. Sucrose from Sugar Beet/Cane

After reviewing the viability of corn to PHA, a clear path to making commercial-scale production of PHA viable is to obtain the substrate from a plentiful low-value or waste carbon source stream. An example of this is low-grade mollases, which is a residual syrup by-product of sugar refining mills that is high in sucrose but unsuitable for food products [46]. Using such as low-value or waste stream has potential, but most studies show that cell production and polymer accumulation within are still not commercially viable [47]. Although studies by Chaudhry et al. [48] using *Pseudomonas* species growing on sugar wastes have shown little promise with maximum cell dry weights of of 12.53 g L^{-1} and PHA content of 35.63%, other studies by Kulpreecha et al. [49] with *Bacillus megaterium* on sugar wastes have shown more promise with cell dry weights of up to 72.7 g L^{-1} and PHA content of 42% achieved over a 24 h period. The more promising latter study indicates that variables such as microorganism strain, carbon source, carbon source production and pre-treatments as well as fermentation conditions all play a role in optimizing PHA production from low-value and waste streams for industrial scaleup.

1.4.3. Whey to PHA

Dairy whey is another promising food waste stream that has the potential for use as a commercial-scale carbon source for PHA production. Whey is a by-product of the cheese making industry comprised of lactose, proteins, fats, water-soluble vitamins, mineral salts and other key nutrients that microorganisms require to grow [47]. Up to 1.60×10^8 tons of whey produced worldwide, surpassing the quantities needed for whey powder production [50]. Whey is currently a problematic by-product for the dairy industry as only ~50% is utilised for the production of lactose, casein and protein powder with the remainder either disposed of or inefficiently treated and processed for animal feed [51]. For these reasons, whey has potential to be a reliable and cheap carbon source stream for PHA production via microbial fermentation. Whey has the additional benefit of not requiring any enzymatic or chemical pre-treatments prior to being used as a carbon source. Life cycle assessments also suggest that the whey to PHA production is comparable to the ecological footprint of existing petroleum based plastics [47,50]. However, there are some disadvantages involved with using whey for PHA production. The fermentation process energy requirements, minimal PHA accumulation yields per whey input as well as the inability of typical PHA accumulating microorganisms to directly metabolise whey have hindered the development for the commercial scale-up of the whey to PHA process [47]. Some studies have altered PHA producing bacterial strains to combat the metabolic issue, with the transformation of *Cupriavidus necator* DSM 545 to include genes *lacZ*, *lac* and *lacO* from of *E. coli* to produce *C. necator* mRePT that can directly metabolize whey to produce PHA [52]. Whey is a promising carbon source stream with many benefits for PHA production, but much development of the fermentation process and microorganism strain manipulation are still required to improve current energy inefficacies, output yields (per kg of input) and the quality of the polymer produced before whey to PHA is commercially viable across the world.

1.4.4. Grass to PHA

Many researchers are investigating the short-circuiting of the anaerobic digestion (AD) process to produce fatty acids such as acetic, propionic and butyric acid [53]. We have previously reported on the conversion of grass via AD generated fatty acids to PHA production by *Pseudomonas* strains which is comparable to the same strains on laboratory fatty acids [54]. The availability of glucose rich hydrolysates from pre-treatment and de-lignification of perennial grasses also allowed the production of PHA.

1.4.5. Waste Cooking Oils (WCO) as Substrates for PHA Accumulation

Waste cooking oils could be used as cheap substrates for microbial production of PHA. Waste cooking oil is comprised of animal and/or vegetable matter that's been previously used in the frying of foods both in the service industry and domestic households [55]. In the EU, up to 60% of 1.748 millions tonnes of WCO is improperly disposed of annually [56]. Some WCOs can be utilized as biofuels, but it is advantageous to apply this carbon to materials rather than for energy due to its carbon-rich composition. Deep fat frying food has become more common in households and the food industry, leaving up to 29 million tonnes of inedible oils needing disposal of per year [57]. This leaves a carbon rich waste product with a poor end of life management system in place. However, one can keep the cooking oil in the materials cycle by converting it into a material, burning it for energy or converting it to biofuels (biodiesel). The conversion of WCO to the biodegradable polymer PHA is a promising route given that WCO is rich in fatty acids (conjugated to glycerol) and fatty acids are well known substrates for PHA accumulation. The conversion of waste cooking oils into PHAs has been reported for a number of bacterial species (Table 2). Plant oils (triacylglycerols) can support growth and PHA production in some mcl-PHA producing strains and could lead to polymers with altered properties to those produced using pure fatty acids as substrates [58]. Few of the well characterized, commonly used, mcl-PHA accumulating bacterial strains have the ability to grow and produce PHA directly on triacylglycerols [59].

Table 2. Bacterial strains growing on different plant oils and waste substrates to produce PHA.

Substrate	Strain	mcl-PHA ($g\ L^{-1}$)	Biomass ($g\ L^{-1}$)	Time (h)	References
Lard, Butter Oil, Olive Oil, Coconut Oil & Soybean Oil	*P. saccharophila* NRLL B-628	2.1	1.6–2.8	48	[60]
Lard And Coconut Oil	*P. putida* KT2442	0.9–1.6	4	72	[61]
Waste cooking oil (1%)	*Pseudomonas* sp. D12	0.1	0.54	72	[62]
Waste cooking oil (2% *w/v*)	*P. aeruginosa* L2-1	3.43	6.8	24	[63]
Waste cooking oil (20 g/L)	*P. aeruginosa* 42A2	1–1.6	3.5–5.5	72	[64]
Waste Fat (20 g/L)	*P. aeruginosa* 42A2	2.3–3.6	3.5–5.5	72	[65]
Soy Molasses (2% *w/v*)	*P. corrugate* 388	0.07–0.6	1.5–3.6	72	[65]
Lard, Butter Oil, Olive Oil, Coconut Oil & Soybean Oil	*P. saccharophila* NRLL B-628	2.1	1.6–2.8	48	[59]

Lipase mediated hydrolysis of plant oils or WCO results in glycerol and fatty acid production, both of which have been shown to be converted to PHA via de novo fatty acid synthesis and β-oxidation respectively [66]. The general structure of triglycerides and their hydrolysis products can be seen in Figure 4. Mcl-PHA has been produced by *Pseudomonas aeruginosa* from palm oil [67], *Brassica carinta* oil [68] and waste frying oil [65]. However, this species is an opportunistic pathogen and not suitable for PHA production at an industrial level. *Pseudomonas chlororaphis* was previously used to produce mcl-PHA from palm oil [69]. *Comamonas testosteroni* produced mcl-PHA from caster seed oil, coconut oil, mustard oil, cottonseed oil, groundnut oil, olive oil and sesame oil [70]. *Pseudomonas resinovorans* used lard, butter oil, olive oil, sunflower oil, coconut oil and soybean oil to produce mcl-PHA [58] and *Pseudomonas saccharophilia* used soybean oil and sunflower oil to produce mcl-PHA [60]. Genetic engineering has also been used to express lipase genes in well-known PHA producing strains such as *P. putida*, *P. oleovorans* and *P. corrugata* to allow them to grow directly on oils [61,71]. Ruiz et al. (2019) demonstrated high cell density (100 g CDW L^{-1}) using WCO but with a modest PHA accumulation level (25–30% CDW) [56] compared to PHA accumulated from fatty acids where cells can accumulate between 60 and 70% PHA. In addition to this, Ruiz et al. (2019) demonstrated that hydrolysed waste cooking oil could be utilised by *P. putida* KT2440 to achieve biomass of 159.4 g L^{-1} with 36.4% of this CDW composed of PHA [72].

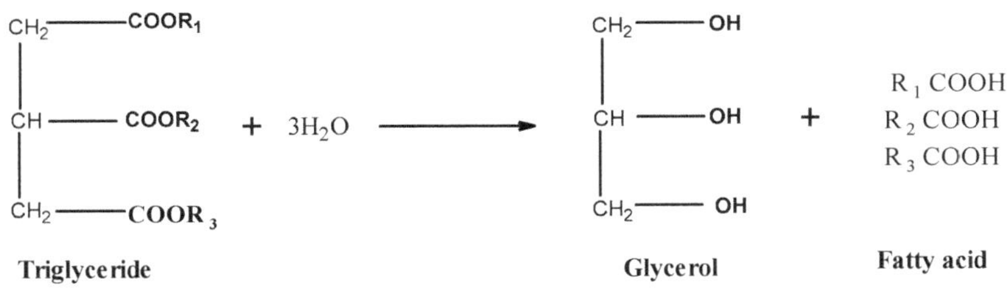

Figure 4. General schematic showing the general structure of triglycerides present in plant oils and the hydrolysis products of these triglycerides (glycerol and fatty acids).

1.4.6. PHA from Gases

Utilising gases, such as CH_4, as a carbon feedstock is one interesting option as it provides a cheap and accessible substrate, removes or prevents greenhouse gases from entering our atmosphere and at the same time produces biobased biodegradable polymer [73]. Industrial waste gases such as methane (CH_4) from biogas, carbon dioxide (CO_2) and syngas (CO_2 + H_2) can be utilized as substrates for PHA production [17]. Biogas is a renewable resource that is comprised mostly of CH_4 and CO_2 with trace amounts of nitrogen (N_2), oxygen (O_2) and hydrogen sulfide (H_2S) [74]. It's a renewable resource produced through the anaerobic digestion of organic substrates, this process is found at waste-water treatment plants where sewage (the organic substrate) is broken down to produce biogas [75]. There are also dedicated sites that use biomass such as agricultural crops and food waste to produce biogas. Due to the widespread production of biogas across the world, it is estimated a total of 58.7 billion Normal cubic metres (Nm^3) of biogas was produced in 2014 (World Bioenergy Association, 2017). These significant quantities of biogas can be harnessed for energy through combined heat and power production (CHP) gas engines that utilise the majority component CH_4 to produce electric and thermal energy [75]. Although CHP gas engines can harness more than 80% of the potential energy from biogas, the economic feasibility is challenged due to factors such as initial high capital investment, operation and maintenance costs, and depreciation/outdating of the CHP gas engines themselves [76]. For these reasons, an alternative method for harnessing this abundant, renewable and carbon rich biogas is highly sought after.

Through a bacterial fermentation process, it is possible for microorganisms to use gases such as CO_2, O_2 and CH_4 as feedstocks to produce PHA [77]. These are methanotrophic bacteria that have been used by biotech start-up Mango Materials, who use biogas as their sole carbon source in tandem with O_2 for the production of PHB (National Science Foundation, 2014). After a life cycle assessment, the production of PHB from methane originating from wastewater treatment plants and landfills requires >10% less energy when compared to other substrates such as corn-derived sugars [78]. Mango Materials currently use the excess biogas from a CHP gas engine system in a waste-water treatment plant for the production of PHB, that would otherwise be wasted through flare burning. On top of this, the use of this biogas to PHA technology closes the carbon cycle through sequestration of the carbon within the final polymer produced [77]. The abundance of wastewater treatment plants and landfills, current inefficient use of biogas, benefits of PHA production and carbon sequestering all make biogas to PHA a good candidate for commercial scale-up across much of the world.

2. PHA Fermentation Production Processes

There are three common fermentation strategies for the cultivation of bacteria to produce PHA: batch; fed-batch and continuous, all having shared goals for the control and optimization of factors such as carbon to nitrogen ratio, length of the fermentation run, temperature, pH and others [79]. The scale-up of PHA production processes and the downstream processing of fermentation harvests will be discussed in this section also.

2.1. Batch Cultivation

In a batch cultivation the media for the microbial culture (including the carbon source) is provided at the beginning of the fermentation, with no addition or removal of these throughout the process. Cells are harvested at the end of the batch run and the removal of the PHA cell mass at the end of the process [80]. This batch fermentation strategy can then be broken into two types based on the form of bacterial growth, a one stage process and two stage process [81]. The one-stage process involves the simultaneous growth of biomass and PHA in the closed fermentation system. The two-stage process is divided into a biomass growth phase with a subsequent PHA accumulation phase [82]. Although there is simplicity and cost effectiveness for the batch cultivation process, it does present issues regarding the optimisation of yields. Final PHA yields can be low when conditions such as the exhaustion of the carbon source occur, ultimately resulting in PHA within the cell mass being depolymerised to provide carbon and energy for the growth of bacteria [9]. The growth of the bacterial strain *Azohydromonas australica* DSM 1124 with sucrose as the carbon source in a batch fermentation resulted in a final biomass of 8.71 g L^{-1} CDW with PHB accumulation of 6.24 g L^{-1} over a 36 h period [83]. This means a growth rate of only 0.17 g L^{-1} h^{-1} was achieved and points to the limitations of batch fermentation processes which is not efficient and suited for industrial scale-up.

2.2. Fed-Batch Cultivation

Fed-batch cultivations are the most common biotechnological processes [84] including PHA production due to the ability to control nutrient and carbon input which allows for the avoidance of carbon limitations throughout the process. Controlling these conditions allows for a more efficient process resulting in an increased cell density and PHA accumulation within the process [81]. The real-time monitoring and regulation of substrate concentration and the adjustment of conditions within the process presents challenges in maintaining a standardised repeatable process but once the predicted growth rate and nutrient consumption are controlled/maintained then reproducible processes can be achieved. Some high cell density fed-batch operations involving *P. putida* and *B. sacchari* have shown growth of up to 159.4 g L^{-1} and 221 g L^{-1} respectively, with % PHA of this CDW at 51% and 45%, respectively [12,72,85,86]. *P. putida* KT2440 produced a biomass of 141 g L^{-1} CDW and PHA (mcl) accumulation of 72.6 g L^{-1} after 38 hrs when growing on oleic acid achieving a PHA accumulation rate of 1.91 g L^{-1} h^{-1} [12]. Ramsay and co-workers reported on a number of fed batch fermentation strategies to produce high cell density cultures with a high content of PHA achieving up to 71 g L^{-1} and 56% PHA and 1.44 g PHA L^{-1} h^{-1} [87]. Thus fed batch fermentations are far more productive compared to the batch cultivation mentioned above in Section 2.1—making fed-batch cultivations a far more suitable fermentation production process for scaling up.

2.3. Continuous Cultivation

Continuous cultivations are very different when compared to batch and fed-batch cultivations as once the desired PHA and biomass accumulation rates are reached, these conditions are kept constant making for a chemostat culture [9,88,89]. The harvesting of PHA accumulating cells is continuous. This fermentation strategy is considered to be beneficial in that it allows for a consistent production of PHA with near identical quantities and qualities from start to finish of the continuous process [90]. Even though this process aims to establish long-term genetic stability of the PHA producing strain, the very nature of the long lasting continuous process can result in the contamination of the culture and genetic changes over time [91]. Another challenge is the large volume of spent fermentation media that is produced on a daily basis. This has to be cleaned up and the water re-used. An example of continuous cultivation was demonstrated using *Pseudomonas oleovorans* ATCC 29347 on single carbon substrates such as citrate, hexanoate and octanoate. This involved a constant concentration of nitrogen with a step-wise increase in carbon concentrations which allowed for the production of mcl-PHA during the growth phase on fatty acids and the nitrogen limitation phase once carbon was present [92].

2.4. Scale up of PHA Fermentation Processes

Although PHAs are accepted as a promising biopolymer for a green economy compared to petrochemical polymers, their production at an industrial scale does present challenges. The production of PHA's at a commercial scale is made possible through the manipulation and optimization of the process with modern scientific and engineering practices to achieve high cell densities with a higher percentage of PHA within the cell [21]. The cost of producing PHA is dependent upon the cost of the fermentation substrate, which can account for up to 50% of the overall production costs, the volumetric productivity of the process ($g{\cdot}L^{-1}\ h^{-1}$) and the level of PHA in the cell (% of cell dry weight) as the efficiency of the downstream process is higher with higher PHA content in the bacterial cells [4]. Choi and Lee estimated that at 100 kt scale, PHA production costs between USD 2.6–6.7 $ kg^{-1}, depending on the PHA producing microorganism [93]. Compared to plastics such as polyethylene (USD 0.9–1.0 kg^{-1}), production costs are up to 7 times higher for PHA and some reports indicate as much as 10 times higher [4]. One method for reducing this ten-fold cost difference is the scale-up of the fermentation process. In Austria, PHB was produced at 1000 kg/week in a 15 m^3 fermenter using bacterial species *Alcaligenes latus* which grew rapidly on sucrose [94]. Up to 90% of the CDW from this process contained PHB which allowed for the high productivity, although the carbon source is still in direct competition with the food chain [95]. PHBHHx was produced in a 20 m^3 using glucose and lauric acid achieving 50 g L^{-1} and 50% of CDW was PHBHHx, using the strain *Aeromonas hydrophila* [96]. The major issue for this scale-up was the cost of DSP, specifically extraction, where the use of ethyl acetate and hexane increased overall production costs dramatically. Other research has looked at the production of PHAs by halophilic, salt requiring microorganisms. The use of such microorganisms is advantageous as they can grow optimally under conditions of high salt concentrations, with such high salt concentrations capable of reducing the chance of microbial contamination to a large extent, and the inherent low value of substrates that can be utilised from these high salinity environments [97,98]. However, it has been reported by the company BluePHA in China that the production of PHB and PHBV using *Halomonas* species can present problems when scaling up. The high salinity causes difficulties with the downstream processing effluent, corrosion among fermentation equipment, and the lack of well-defined genetic and system engineering, making PHA production using halophiles challenging [98,99]. Despite this BluePHA appear to be operating at an industrial scale and producing PHAs. The use of methanotrophic (methane consuming) microorganisms to produce PHB is also of high interest, as utilising a renewable and potent greenhouse gas and creating a strong carbon capture technology within the polymer itself is highly advantageous [100]. Even with this double-barrel benefit, the use of methane (often fed in the form of scrubbed biogas) as a substrate for PHA producing microorganisms does present significant challenges. The low solubility of methane in aqueous solutions under atmospheric pressure (22 mg L^{-1}) results in a low mass transfer, which in turn can result in low cell growth and density [100,101]. Methane can be sourced from anaerobic digesters or landfills, so scaling up fermentation systems using this carbon substrate is feasible [102]. The cost of PHB production using methane can range from $4.1–8.5/kg, depending on production capacity [103,104]. These are not economically competitive prices and the market may require incentives for methane to PHB production facilities in tandem with rising fossil fuel prices, in order to be a viable technology [100]. Even with these challenges, some biotech start-ups such as Mango Materials in the San Francisco Bay Area are scaling up their methane to PHB process. They have most recently implemented their 'Launch Facility' which includes a 5000 L fermentation tower system with complete DSP setup at the Silicon Valley Clean Water waste water treatment plant in California.

Another important issue for biopolymer production is the sourcing of a sustainable carbon feedstock, which should not compete with food-chain feedstocks [105,106] as this creates competition with food supply and land use consequently increasing the cost of that raw material affecting both food prices and PHA production costs [85]. Waste substrates are a good alternative to food based resources but the availability, heterogeneity and quality of the starting material are challenges that need to be addressed.

Although the upstream costs for the scale-up of PHA production is vitally important (e.g., carbon feed cost) to make it competitive with other fossil fuel polymer production processes, an important research and development focus for successful up-scaling is on the downstream processing that ensures a reproducible polymer with desired physical and chemical properties. The drying of bacterial cells, the method of extracting the polymer from those cells affects the molecular weight of the PHA. The properties of polymers are dependent on the monomeric composition and molecular weight of the polymer and this affects the types of applications the polymer can be utilised for, hence the biotechnological and downstream processes need to economically produce a polymer with a consistent and desirable molecular weight.

2.5. Downstream Processing to Harvest PHA

The downstream processing of PHA and the extraction methods involved are critical for the commercial scale-up of PHA production, with many of these technologies currently in the development stage [50]. Many downstream processing recovering methods have advantages or disadvantages in regard to economics, ecological impacts, safety, recovery yields, production purity/quality and difficult scalability [83]. Some of the more well-known recovery methods involve the use of solvents to obtain a good quality product, but solvents such as chloroform should have no place in the PHA production chain on a larger scale [107]. The extraction of PHA using the halogenated solvent chloroform is the performance benchmark for the extraction process of PHB and PHB co-polymers from its cell biomass, particularly at a laboratory scale [108]. It is the benchmark as studies by Rebocho et al. showed that their PHBHV/3HB polymer grown from halophilic yeast *Pichia kudriavzevi* VIT-NN02 from agricultural waste materials has a yield recovery of 99.99% using this method [109]. Solvent extraction using the likes of chloroform then require an antisolvent (usually at low freezing temperatures) to precipitate the polymer out of solution. This increases the resource input thus increasing waste potential that requires even further significant energy input to separate the solvent and antisolvent for reuse [108]. For these reasons the commercial scaling of such a process is not feasible. The extraction of mcl-PHA using acetone is the most common solvent extraction method for this type of polymer, with precipitation induced most prominently by the anti-solvents methanol and ethanol (typically at freezing temperatures) [110]. It is possible to obtain a pure polymer with a high molecular weight, while removing the bacterial endotoxin using this method [9,111]. Although a portion of the utilised acetone can be recycled, the environmental impacts and capital costs for using solvents in such extraction processes is not recommended for industrial scale-up [9]. Larger scale studies (medium scale @ ~200L) on cells produced by *P. Putida* involved heat pre-treatment at 121 °C, followed by digestion using Alcalase, EDTA and SDS with an optional final chloroform extraction to increase PHA purity from 95% to 99% [111,112]. This method is preferable as the quantity of chloroform used it significantly reduced. The used of SDS and enzymes will be discussed in more detail in the following paragraphs.

The use of sodium hypochlorite (halogenated) in tandem with anionic surfactant SDS (irritant) for the removal of PHA from cell mass goes against the theory of a greener plastic production process [83,113]. The use of sodium hypochlorite has also been reported to reduce the molecular weight of the polymer by half, even post optimization of the digestion time and parameter conditions [83,114].

Other reported methods for the extraction of PHA from biomass involves the use of enzymes to breakdown the outer non-PHA component of the biomass [83]. This non-PHA part of the biomass usually makes up <10% of the total cell dry weight, with the use of enzymatic catalysts for this step highly beneficial compared to chemicals such as sodium hypochlorite. Although enzymes have advantages in theory, the methods are still in their infancy and not cost effective due to the slow reaction rates of enzymes and low final product purities [34]. For these reasons, enzyme cocktails are often used in tandem with oxidants such as hydrogen peroxide to achieve the yields and purities required for commercial-scale up [115].

3. Properties of PHAs

Conventional fossil fuel derived plastics are used across every facet of life today, from the home to nearly all industries due to their properties allowing for their manipulation into convenient, durable and specialised products [4]. The formation of such products has paved the way for applications that would not otherwise be possible. However, in recent decades it has become apparent that this has come at a high cost with little or no sign of this cost subsiding. Common fossil derived plastics such as polyethylene, polypropylene and nylon are xenobiotic. Their release into the environment is resulting in environmental pollution that is further complicated by their mechanical degradation into microplastics [4,116]. The environmental damage and unsightly nature of non-biodegradable polymers must be halted but the dependency that the modern world has on these plastics is strong with little sign of subsiding. Plastics have many positive attributes that the world needs but a lack of waste management safety nets to capture the waste plastic contributes to a world where plastic pollution is taking place unabated. Society is calling for the reduction in plastic use and the use of a suitable replacement. A number of polymers such as polylactic acid and thermoplastic starch are in the market and their production is increasing. PHAs (Figure 5) show great promise but have limited market share [4].

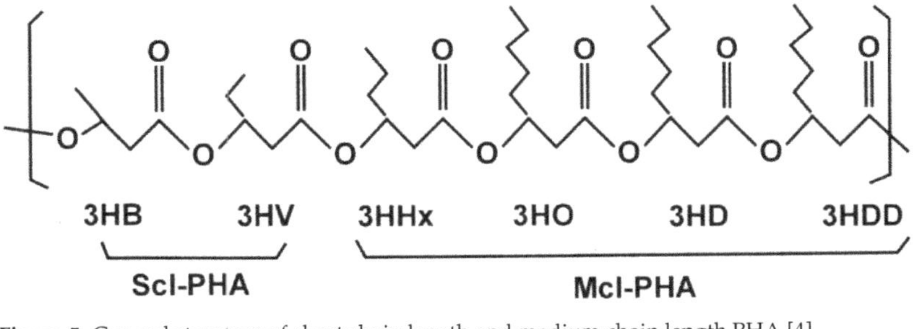

Figure 5. General structure of short chain length and medium chain length PHA [4].

The replacement of fossil plastics with biobased plastics addresses the issue of renewable carbon but not end of life management. The biodegradability of biobased plastics can help to address end of life management by offering more options. Biodegradable plastics should not be produced so that they can be released into the environment but rather that they increase the end of life management options for plastic and increase the chances of the plastic being collected and managed. Plastic products could be designed with emergency biodegradation in the event that they are accidentally released into the environment but this should not be the driver for plastic resource management as it could promote bad behaviour and continue to perpetuate plastic pollution.

3.1. General Properties of PHA

The properties of PHAs, such as processability, mechanical properties, UV resistance, are similar to commercially available thermoplastic fossil-based polymers such as polypropylene (PP) and polystyrene (PS). Depending on the composition of its monomers, with over 150 types of PHA structures reported in the scientific literature, an array of different properties and functionalities exist within this biopolyester family [44]. One universal property of PHAs is that they are thermoplastic, with other properties varying depending upon their chemical structure [117]. The number of carbon atoms present in the monomer dictates their molecular structure, along with the chain length of these associated monomers, which results in the diverse and multifaceted characteristics of PHAs [118]. The thermal, mechanical, and barrier properties are the most important and studied features of PHAs. Thermal properties such as glass transition temperature (T_g) and melting temperature (T_m), along with mechanical properties such as tensile strength (MPa) and extension to break (%) are key for determining the appropriate use of PHA polymers for various practical applications [119]. The thermal properties of PHAs depend on the polymer's chain length, and these attributes directly affect the polymer's mechanical properties. As mentioned earlier, based on the length of side hydrocarbon chain, PHAs can be broadly classified into three categories; scl-PHA, mcl-PHA and lcl-PHA. The subsequent section of this review will cover mechanical, thermal, and other functional properties like PHA's gas barrier properties and its structure-property relationships. The discussion will primarily focus on scl-PHA and mcl-PHA; however, lcl-PHA is less common for applications due to its inferior properties, so it will not be discussed.

3.2. Properties of Scl-PHAs

The scl-PHAs are a class of PHAs with a number of side chain carbon atoms ranging from 3 to 5. For example, poly(3-hydroxybutyrate) (PHB), poly(3-hydroxyvalerate) (PHV) and their copolymer poly(3-hydroxybutyrate-co-3-hydroxyvalerate) (PHBV) are typical examples of short-chain-length PHAs [120]. The properties of these polymers vary depending on their substrate type, length of side chain, co-monomer and co-polymer composition. In general, PHA polymers produced by bacteria are semicrystalline, and crystallinity can influence the polymer properties. PHB homopolymer is known to develop high crystallinity (>60%), making it stiff and brittle [121]. Despite PHB having mechanical properties close to commercial polypropylene (PP), the challenge in melt processing PHB into products is one of the major factors restricting its broad adoption in commodity applications. The thermal properties, mechanical properties and other physical properties of PHB are discussed here in this sub-section.

3.2.1. Thermal Properties

The scl-PHA polymers are thermoplastics and differ in their thermal properties depending on their chemical composition. PHB homopolymer has a melting temperature (T_m) around 180 °C, close to its decomposition temperature [122]. Incorporating co-monomers such as 3-hydroxyvalerate (3HV) into the polymer chain decreases T_m and crystallinity of the polymer [123]. The glass transition temperature (T_g) is linked to the segmental mobility of polymer chains, which play a critical role in dictating the thermal and mechanical properties of the polymer. Previous studies have indicated that the number of carbons has a direct effect on the glass transition (T_g) and melting temperature (T_m), and crystallization temperature (T_c), of PHAs [119]. The T_g and T_m are essential thermal properties for various applications of PHAs. PHB is one scl-PHA that is highly explored for its properties. The linear chain structure of PHB leads to a predominantly crystalline phase, which is interconnected through the amorphous sections creating a semicrystalline polymer. A large number of reports suggest the range of T_g is −15 °C to 9 °C, depending upon the structure of PHB and its copolymers [124–127]. Similarly, the melting point (T_m) of PHB is reported to be between 160 °C to 180 °C, which depends upon the type of its crystalline phase [128]. Isostatic PHB displays a high melting temperature close to 175 °C and the T_g is close to room temperature (21 °C), resulting in a very narrow processing temperature window for PHB [129].

The thermal stability and degradation mechanism are important as they affect processability and mechanical properties. Thermal decomposition of PHB takes place just above melting temperature, making it vulnerable to degradation during processing at 180 °C. A degradation temperature of 210–220 °C of PHB limits the melt processability of PHB due to a narrow temperature window between melt and degradation temperatures. Temperature-induced chain scission in PHB can result in a rapid decrease in molecular weight [130]. Thermal analysis of melt-processed PHB indicates a double melting peak due to bimodal distribution of crystallite size resulting from changes in molecular weight due to random scission of PHB during the melt processing [131]. Various approaches have been implemented to improve the melt processibility of the PHB polymers. The addition of nucleating agents, plasticizers, blending with other polymers, co-polymerisation, nano additives etc. have been evaluated to improve the processability of PHB [132–135]. The addition of plasticizers reduces the interchain entanglement within the polymer and promotes PHB processing without thermal degradation [136]. Also, the inclusion of nucleating agents increases the spherulite counts due to the formation of numerous small and imperfect crystallites, leading to the lowering of T_g and the overall decrease in crystallinity of the polymer [137–139]. However, it is important to mention that some of the approaches to improve thermal stability and processability of PHAs can affect biodegradation [140].

3.2.2. Mechanical Properties

PHB polymers are traditionally stiff and brittle due to their high crystallinity. The mechanical properties of PHBs are similar to polypropylene; however, the lower elongation and high stiffness prevent PHB polymers from replacing fossil-based polymers such as polypropylene. Tensile strength measures the amount of force required to pull a material until it breaks and is typically within a range of 8.8 to 50 MPa for PHB synthesized from different raw materials. Similarly, Young's modulus presents the stiffness characteristics, and scl-PHAs show a modulus of 3.5×10^3 MPa [141]. Elongation at break is the measure of the material's ability to be stretched until it breaks, and it is expressed in the percentage of the original length of the material. The scl-PHAs show percentage elongation at break ranging very low to 15% to >200% based on its chemical and physical structure [142]. Pure isotactic PHB is brittle that show a tensile strength of 30–35 MPa, elongation at break around 15% and modulus of 1.2 GPa [117,143]. The primary reason for the high brittleness of PHB is due to slow secondary crystallization that occurs within the amorphous phase of PHB [117].

Though PHB has poor ductile properties, the addition of co-monomers, plasticizers, various polymers and fillers significantly enhance its elongation and tensile strength by reducing the crystallinity of the polymer [144,145]. The mechanical properties of the PHB polymers are altered by incorporating co-monomer units, side chains, and bulky functional groups by controlling the overall crystallinity of the polymers. Mangeon et al., reported a 6.5 fold improvement in percentage elongation at break values of natural terpene based plasticized PHB [145]. Table 3 below presents the effect of copolymerization on the mechanical properties of PHB.

Table 3. Mechanical properties of various types of PHB.

PHB Type	Tensile Strength (MPa)	Modulus (MPa)	Elongation to Break (%)	Reference
P(3HB)	40	3500	6	[146]
P(4HB)	104	149	1000	[147]
P(3HB-co-17%3HHx)	20	173	850	[148]
P(3HB-co-20 mol% 3HV) [a]	20	800	5	[149]
P(3HB-co-6 mol% 3HA) [b]	17	200	680	[149]
Polypropylene	38	1325	400	[150]

[a] 3-hydroxyvalerate [3HV]. [b] 3-hydroxydecanoate (3 mol%), 3-hydroxydodecanoate (3 mol%), 3-hydroxyoctanoate (<1 mol%), 3-hydroxy- cis-5-dodecenoate (<1 mol%) [3HA].

Thus, the rigidity of PHB can be well-tuned by introducing some co-monomers into the backbone, which increases the flexibility to a significant extent. For example, the introduction of 3-hydroxyhexanoate (3HHx) comonomer to PHB by 17 mol% increased the ductility in poly-3-hydroxybutyrate-co-3-hydroxyhexanoate, P(3HB-co-3HHx) [148]. The physical blending of PHB with P(3HB-co-3HHx) through melt or solvent mixing lead to a similar improvement in flexibility of the blend system [151].

The ageing process largely impacts the mechanical properties of PHA overtime during the shelf-life of the polymer. The secondary crystallization phenomenon in the amorphous phase of PHB reduces flexibility and makes the material brittle. As a consequence of ageing in PHB, the tensile moduli increase while stress at break remains stable [152]. Further discussion on the ageing process for PHB is discussed in the next section.

3.2.3. Ageing of PHB

One major drawback with PHB materials is their slow ageing phenomenon over time, decreasing mechanical properties. Ageing of PHB occurs by slow changes in its amorphous and crystalline phases, resulting in either hardening or softening of the material. The ageing phenomenon in PHB primarily occurs through secondary crystallization as well as physical ageing [153]. At the first stage, after melt processing, PHB cools down without undergoing any crystallization; this is followed by the second stage of autocatalytic crystallization that occurs at a high rate. In the final stage, secondary crystallization occurs at a very slow rate within the amorphous regions of PHB [154]. The secondary crystallization that usually occurs within the amorphous phase leads to the formation of imperfect crystals along with the interlamellar spherulitic spaces. As a result, the polymer chain mobility in the amorphous phase becomes restricted, making the polymer more brittle [129]. The physical factors of ageing in polymers are related to the relaxation of polymer chains below its T_g due to the residual mobility of polymer chains in its glass phase [155]. Overall, the ageing of PHB leads to deterioration of mechanical and physical performances. Biddlestone et al., reported that the embrittlement effect of ageing in PHB can be reduced satisfactorily by annealing the PHB samples at 77 °C prior to storage. Another method to control the secondary crystallization of PHB is the addition of nucleating agents and then anneal is at 146 °C. Such annealed PHB samples are reported to retain their ductility during complete storage period [156].

3.2.4. Gas Barrier Properties

PHAs have an excellent barrier to air and moisture, making them suitable for packaging applications. The gas barrier property of polymers has been considered an essential property in packaging applications due to its associated advantages of lightweight, easy processing, and forming characteristics. Most biodegradable polymers have similar oxygen barrier properties to conventional petroleum based polymers, however, the barrier properties of biodegradable polymers decreases with the increase in humidity. PHB show better moisture and oxygen gas barrier properties than polypropylene (PP) and polyethylene terephthalate (PET), respectively [157]. The lamellar structure in the PHB crystalline phase contributes to its superior aroma barrier properties and other gas and moisture barrier properties [158]. Unlike amorphous polymers, the crystalline phase of PHB restricts the passage of gas molecules, making it more relevant to food packaging applications. Given the advantage of the high barrier properties of PHB, it can be blended with other biodegradable polymers and obtain a polymer blend with improved gas barrier properties suitable for food packaging applications. Thellen et al., reported extruded PHB and PHBV films with OTR (0% relative humidity (RH) and 23 °C) and WVTR (100% RH, 23 °C) values ranging from 193–410 cc-mil × m^{-2} × day^{-1} and 114–217 g-mil × m^{-2} × day^{-1}, respectively [159]. However, further developments are needed to improve PHB barrier properties to replace fossil-based polymers with PHB for packaging applications.

3.3. Properties of Mcl-PHA

The mcl-PHA polymers typically have more than 5 side-chain carbon atoms and the physical nature varies from semicrystalline to an amorphous nature. The long side-chains of PHA affect the crystallisation phenomenon and lead to a reduction in thermal and mechanical properties. Also, the variation in monomer composition and bulky functional groups influence the elastomeric properties of mcl-PHA [160]. The typical examples of mcl-PHA are poly(3-hydroxyoctanoate) (PHO) and poly(3-hydroxynonanoate) (PHN), which are primarily formed as copolymers with 3-hydroxyhexanoate (HHx), 3-hydroxyheptanoate (HH) and/or 3-hydroxydecanoate (HD). Wide variation in properties of mcl-PHAs has been reported in the literature due to the lack of commercial polymer availability, variation in the production methods, type of carbon source that is used for production of mcl-PHA polymers [161]. A discussion on the thermal and mechanical properties of mcl-PHA are detailed below.

3.3.1. Thermal Properties of Mcl-PHA

The glass transition temperature (T_g) is closely associated with the polymer chains' segmental mobility, which dictates the toughness and other physical properties of the PHA polymers. The mcl-PHAs have a T_g value ranging between −65 °C and −25 °C with a melting temperature between 40–70 °C [140]. An increase of 4 to 7 carbons in the side chain of PHAs, increases the melting temperature (T_m) from 45 °C to 69 °C due to increased crystallisation upon the participation of both the main chain and side chain carbons in the smectic structure [162]. The T_g of mcl-PHAs decreases with an increase in the carbon chain length and presence of pendant groups. The mcl-PHA and its co-polymers have lower thermal properties due to large and irregular side chains that inhibit polymer chains' close packaging to crystalline structures [163]. Several other literatures support the tendency of side chains of mcl-PHA to crystallize and thus alter the thermal properties [164]. In the case of mcl-PHA with a large fraction of 3-hydroxydodecanoate (3HDD) and 3-hydroxytetradecanoate (3HTD), these show two melting peaks separated by a cold crystallization peak [165]. This dual-mode crystallization has been attributed to the presence of two distinct crystalline phases, which have different crystallization kinetics due to the variation in side chains lengths. PHA co-polymers containing 12–15 mol% aromatic side chains have shown increased T_g due to the rigidness of the side groups in the copolymers [166]. However, some co-polymers of mcl-PHA polymers do not crystalise due to the disorder introduced by the side chains present in the polymers [56,167]. The crystallization phenomenon of PHO is prolonged and requires several days to weeks to complete at room temperature; however, lower temperatures are more favourable to achieve maximum degree of crystallisation [168]. PHO polymers can attain 30% crystallinity (max) and having a T_g close to −35 °C and T_m nearly at 61 °C [30]. The crystalline regions act as physical crosslinks between amorphous phases of PHO and hence the polymer behaves like thermoplastic elastomer [167]. In isothermal crystallization studies using DSC, it is understood that the melting peak of PHO becomes more defined and increases at higher crystallization temperatures. It refers to the ability of the polymer chains to reorganize in more ordered crystalline domains at higher crystallization temperatures [169]. Copolymerization of mcl-PHA alters thermal properties like T_g, T_m, T_c and heat of fusion

(ΔH_m). In particular, the side chain carbon length of co-monomers has a greater influence on the crystallization mechanism and thermal properties of mcl-PHA [170]. For example, as the number of side-chain carbon atoms increases from 3 to 4 or higher, the layered crystal structure of packing with both main and side chains is developed, increasing T_g and T_m [164]. As the number of -side chain carbon becomes 7 or more, smectic liquid crystalline phases form at low temperatures and leads to cold crystallization [162]. Hence, considering its thermal characteristics are more relevant to elastomers, PHO has gained significant importance as an additive in blending with low ductile biodegradable polymers.

3.3.2. Mechanical Properties of mcl-PHA

Mechanical properties are critical when selecting PHA polymers for a specific application. In general, mcl-PHAs are more flexible and suitable for food packaging and tissue engineering, where biodegradability and biocompatibility are required [141]. The mcl-PHA co-polymers containing 3-hydroxyalkonate are much more flexible due to their lower crystallinity and close to properties that of LLDPE polymer [171]. The crystalline parts of mcl-PHA act as the physical crosslinks, contributing to its mechanical strength. Marchessault et al., reported the modulus and elongation at break of PHO is 17 MPa and 250–320%, respectively [170]. In another report (Gagnon et al.) mcl-PHA with 86% of 3HO presents modulus and elongation to break of 9.3 ± 1.4 MPa and 380 ± 40%, respectively [168]. The elongation at break for mcl-PHA is also reported to be as high as 1000%, which varies with the side chain's length and chemical nature [88]. Table 4 presents the mechanical properties of various mcl-PHA polymers.

Table 4. Mechanical properties of various types of mcl-PHA.

mcl-PHA Type	Tensile Strength (MPa)	Modulus (MPa)	Elongation to Break (%)	Reference
P(HO)	NA	17	250–350	[170,172]
P(3HO-co-12%3HHx-co-2%3HD *)	9.3	147.69	380	[168]
P(3HO-co-4.6%3HHx)	2022.9	173599.9	6.5	[173]
P(3HO-co-5.4%3HHx)	23.9	493.7	17.6	[141]
P(3HO-co-7%HHx)	17.3	288.9	23.6	[141]
P(3HO-co-8.5%HHx)	15.6	232.3	34.3	[141]

* 3-hydroxydecanoate (3HD).

From Table 4, it is evident that there is a change in the mechanical properties of the PHA with the addition of one or more co-monomers into the PHA main chain backbone. For example, the copolymer containing 4.6 mol% of 3HHx improves tensile strength and modulus to a greater extent, while the elongation to break is reduced by 97–99% relative to PHO. In general, it can be understood that the high elasticity of mcl-PHA is relevant to flexible packaging applications. At the same time, the poor mechanical strength and low melt temperature limit its widespread acceptance. Further, an emphasis has been made to modify mcl-PHA adopting different techniques discussed later sections in the review.

3.4. PHA Modification

PHAs are a type of polymer that is derived from bacteria and are degraded by bacteria. Hence, this polymer holds great potential for the circular bioeconomy. Though PHAs possess a wide variety of properties, they are difficult to use on their own in certain applications [117]. In this regard, modification of PHAs is an essential prerequisite to make them suitable for various applications. In this section, a detailed discussion on PHA modification will be covered.

In the broad sense, PHA modification is classified into three primary sections as presented in Figure 6:

1. Chemical modification
2. Physical modification
3. Biological modification

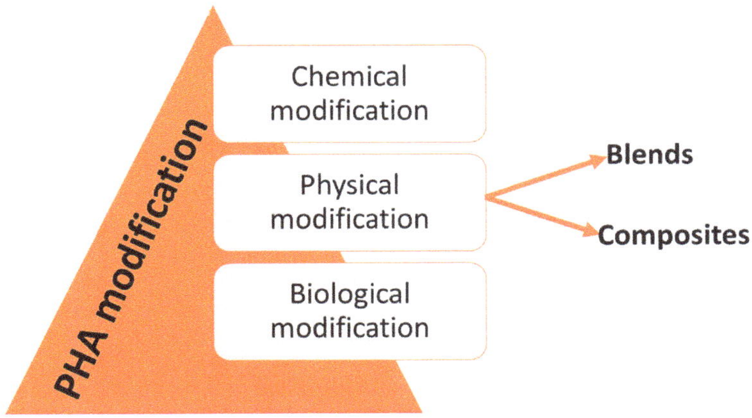

Figure 6. Different modifications of PHA.

Each section contains multiple subsections that will be discussed individually, along with the underlying mechanism associated with the modification process.

3.4.1. Chemical Modification of PHA

The primary reason for chemical modification is introducing functional groups to PHA that add valuable attributes that biotechnological conversion processes cannot easily achieve. These chemically modified PHAs, possessing improved properties, can be utilized as multifunctional materials [174]. The actual processing temperatures of PHAs are essentially determined by their crystalline melting points (T_m) because of their poor thermal stability during melt conditions. Larsson et al., performed a stability study on both the PHAs after Soxhlet extraction with $CHCl_3$ followed by washing with aq. HCl solution, respectively. An increase in thermal decomposition temperature of 50 °C was reported for PHA washed with an aqueous solution of HCl acid [175]. In another attempt of mcl-PHA modification, Nerkar et al. used Lauroyl peroxide (L-231) as a crosslink agent of PHO, before melt blending with PHB [176]. Due to the variation in the melt viscosity between PHB and PHO there is poor compatibility between the two polymers during the melt mixing. However, the chain extension of PHO using peroxide prior to melt mixing with PHB enhances the viscosity of PHO, which improves both modulus and elongation to break. In similar context, Xiang et al., modified PHBV using DCP as crosslinking agent and the resultant long-chain branched copolymeric poly(3-hydroxybutyrate-co-3-hydroxyvalerate) (LCB-PHBV) showed an accelerated crystallization rate [177]. Gopi et al., reported improvement in crystallization characteristics of PHO upon chemical modification with DCP and triallyl trimesate coagent [178]. Bhatia et al., modified PHBV co-polymer with ascorbic acid using *Candida antarctica* lipase B mediated esterification [179]. The ascorbic-modified PHBV showed a lower degree of crystallinity (99.6%), making it more ductile and easy to process. Also, the thermal degradation temperature increases to 294.9 °C and biodegradability enhanced by 1.6 fold compared to unmodified PHBV. PHA can be modified by chemical treatment with para-toluene sulfonic acid monohydrate (APTS) catalysed by acid, which leads to PHA linear chains forming with one end hydroxyl group and the other end carboxyl group [180]. These monohydroxyl terminated PHA act as an initiator for ring-opening polymerization of lactones. Polyurethanes based on different PHAs like P(3HB), P(4HB) etc. were synthesized by melt polymerization of PHA with Hexamethylene diisocyanate (HDI) under relatively mild processing condition [181]. Similarly, dihydroxyl terminated PHA is synthesized by the reaction of PHA with 1,4-butanediol in the presence of p-toluenesulfonic acid catalyst [182]. Diblock and triblock blends were prepared by reacting dehydroxylated PHA with polyethylene glycol (PEG), which show improvement in properties [183]. A range of literature is available on PHA chemical modification, and some of the methods can be adopted for large scale industrial applications.

3.4.2. Physical Modification of PHA

The blending approach and incorporation of additives are widely adopted methods to modify the individual polymers' deficiencies. Moreover, the blending approach is more economical than co-polymerisation and other chemical modification methods. The goal for developing biodegradable polymer blends and composites is to improve the adhesion between the individual polymers components, reduce the interfacial tension between these components, and generate desired phase morphology and improved performance [3]. However, three factors that limit the blending with various polymers is; morphology of the biphasic system, degree of miscibility, and extent of compatibility between the phases. At thermodynamic equilibrium, the mixture of two polymers in an amorphous state exists as a single-phase; hence, the blend is treated as compatible, improving the physical and mechanical properties of the resulting blend polymer. The blend systems reviewed below focus on PHA based blend composites suitable for packaging film applications.

The selection of plastic packaging materials for food packaging largely depends on the type of food and shelf life of the product. The plastic film materials targeted for food packaging need to be versatile enough to withstand the handling process and should be able to maintain the physical and chemical integrity while retaining the aroma and keeping the food products fresh [184]. For example, stereoregular poly(L-lactic acid) [PLLA] cannot withstand high-temperature beverages; however, stereo complex PLA can be high heat resistant and suitable for high-temperature beverage applications [185]. While PLA alone does not provide good gas and moisture barrier properties [186], specific grades of PHA with a high degree of crystallinity can provide better barrier properties than PLA [152]. Therefore, biodegradable polymer blends are desirable to produce products similar to fossil-based polymer products.

PHA–PLA Blends: PLA is one of the most extensively used biobased and biodegradable plastics to date and has already gained attention for use in food packaging applications. The difference in theoretically calculated solubility parameters of PLA and PHB is very low and this represents possibility for good miscibility between the two polymers [187]. However, it is not always the case since the miscibility between PLA and PHB is also dependent on the molecular weight of both polymers, processing temperature and composition of blends. One component with low molecular weight in these blend systems leads to better miscibility, whereas PLA and PHB are higher in molecular weight and distinctively phase-separated [188]. Melt mixing of PHB with PLA in different compositions while keeping PLA as the matrix phase leads to a blend system with a higher degree of crystallinity and improved thermal properties [189]. The better mechanical property of PLA/PHB at 75/25 ratio is attributed to the fact that the added PHB acted as a nucleating agent in the PLA matrix. It is speculated that the synergistic effect of PLA/PHB blend system within the range of 75/25 ratio refers to good compatibility between the two phases [190]. The major difficulty experienced during film extrusion of PLA/PHB blends is due to its high brittleness and poor processing characteristics [191]. To improve these blends' processability, blending the third component like a plasticizer or compatibilizer enhances the material flexibility and makes it suitable for film extrusion [188]. Numerous biocompatible and biobased plasticizers, used both for PHB and PLA, improve the flexibility of PLA and PHB [192]. Using acetyl tri-n-butyl citrate (ATBC) and polyethylene glycol as plasticizers in PLA/PHB blend led to improved thermal stability, which was studied under isothermal TGA analysis at 180 °C for less than 6 min [187]. Also, adding additives like cellulose nanocrystals (CNC) in the PLA/PHB system retain thermal stability while enhancing the mechanical properties significantly [193]. Apart from cellulosic fillers, the mechanical properties PLA/PHB blends can be improved by the addition of fillers like catechin, and nanofillers like, organically modified nanoclay etc. [194].

PHA holds great promise to be used in blends due to its high elongation properties [195,196]. The blending of PLA with PHA can be beneficial as it improves both polymers' mechanical performance, in terms of reducing their brittle nature [196]. Most PHA/PLA blends are formed through conventional melt processing. Noda et al. (2004) reported melt-processed blends of poly(L-lactic acid) PLLA with Nodax™, a family of mcl-PHA, poly(3-hydroxybutyrate)-co-(3-hydroxyalkanote) bacterial copolyesters. It was found that a wide range of thermal and mechanical properties could be obtained by varying the type and quantity of mcl-PHA [197]. They reported that the elongation to break ratio was increased as well as improved energy at break was observed in solvent cast PLA-PHB and PLA-PHBV blends. Melt mixing of mcl-PHA with PLA leads to a reduction in brittleness of PLA while improving the ductility in the blend system. The addition of 5–15% of mcl-PHA into PLA reduces modulus of the blend compared to neat PLA, which is attributed to the plasticization effect of low molecular weight PHA and lubrication effect imparted by the high molecular weight mcl-PHA chains. Hence, the material's ductility, which is usually inverse of modulus, increases in the blend system upon increasing the mcl-PHA content. Subsequently, the formability of these blends into the desired shape also enhances [187]. There is evidence of improved miscibility between PLA and mcl-PHA phases in blends where PHA fraction is higher than 50% [198].

PHA-PCL Blends: To improve their properties, PHA polymers can be blended with fossil-based biodegradable polymers such as polycaprolactone (PCL). PHB blended with PCL to overcome the limitation of brittleness, though several studies reflect the that PHB is immiscible with PCL [199–201]. Another member of the PHA family is poly(3-hydroxybutyrate-co-3-hydroxyhexanoate) [P(3HB-co-3HH)], which shows better ductility compared to PHB; however, the brittleness induced by the slow ageing of P(3HB-co-3HH) limits its application [202].The inclusion of 2.5–20 wt% of PCL into P(3HB-co-3HH) and PHB leads to an appreciable improvement in the ductility of the resultant blend, which is attributed to the fine dispersion of PCL phase within the PHA matrix [203,204]. In the same study, it is observed that there is a severe issue with the embrittlement of PCL/PHA blends with time due to the development of cracks or voids in PHA phase due to secondary crystallisation over a period of time [203]. Garcia et al., studied PHB-PCL blends along the entire composition with 25 wt% increment and found the blends are immiscible at all compositions [205]. Further, in the study, PCL was found to act as an impact modifier for PHB as the elongation to break changed from 11.2% for neat PHB to 1000% for blend with 75% PCL [205]. Nonetheless, the selective solvent etching of one phase of the these blends readily shows immiscibility in scanning electron micrographs, which is supported by two T_gs Przybysz et al., reported improved compatibilization between PCL and PHB through reactive extrusion using di-(2-tert-butyl-peroxyisopropyl)-benzene (BIB) as peroxide crosslinking agent. Compatibilized PCL-PHB (75/25) blend system with 0.5 wt% BIB show elongation to break 305 ± 14%; whereas the un-crosslinked PCL-PHB (75/25) blend show 125 ± 5%. Such improvement in properties is attributed to partial cross-linking/branching of studied blends confirmed by the melt flow rate and gel fraction measurements [206].

PHA-PBAT Blends: PBAT has good melt processability and shows ductile behaviour suitable for film extrusion and high thermal stability. It is chosen as one of the desired biodegradable polymers to blend with PHA. However, PBAT is not miscible or compatible with any of the PHAs, and requires compatibilizers to produce miscible blends with improved properties [207]. The addition of compatibilizers, such as organically modified natural Fibres, clay etc. are some of the preffered approaches to improve compatibility. Though scl-PHA like PHB or PHBV shows high barrier properties due to its high crystallinity, blending of PBAT diminishes its mechanical strength and gas barrier property due to the amorphous nature of PBAT [208]. However, co-polymers such as, poly(3-hydroxybutyrate-co-3-hydroxyvalerate) (PHBV)/poly(butylene adipate-co-terephthalate) (PBAT) blend (weight ratio of PHBV:PBAT is 30:70) modified with silane treated 10 wt% of recycled wood Fibre (RWF) improved mechanical morphological and thermal properties of

PHBV-PBAT blends [209]. Nagarajan et al., observed improved interphase compatibility in PHBV-PBAT blends upon adding 0.75 phr poly diphenylmethane diisocyanate (pMDI) as compatibilizer in association with switchgrass Fibre loading of up to 30 wt% [210,211]. Larsson et al., carrier out reactive extrusion study of PHB (ENMAT Y3000P)/PBAT (Ecoflex) blend across the complete range of composition using dicumyl peroxide (DCP) as crosslinking agent. The peroxide free radical initiator DCP provided compatibility between the two constituent polymers and this is evident from the improved dynamic shear modulus, an apparent increase in interfacial adhesion, an increase in tensile storage modulus [175].

PHA-PBS Blends: Polybutylene succinate (PBS) is a thermoplastic biodegradable polymer synthesized by polycondensation reaction of 1,4-butanediol with succinic acid [212]. The advantages of PBS over PHB are its high flexibility and impact strength, thermal degradation stability, and chemical resistance makes it a good candidate for blending with scl-PHA [213]. At the first instance, it is reported that PHBV/PBS blends are immiscible, as evidenced by the decreasing crystallization rate of PHBV upon the addition of PBS [214]. In-situ compatibilization of PHBV-PBS blends using DCP as crosslinking agent improved the interfacial adhesion between PBHV and PBS phases due to formation of PHBV-g-PBS copolymers which subsequently acted as compatibilizer and partially cross-linked networks in the blends. An addition of 0.5 wt% DCP in PHBV/PBS (80:20) blend, the elongations at break increased from <10% (for neat blend) to 400% and the un-notched Izod impact toughness values increased from 10 $kJ \cdot m^{-2}$ (for neat blend) to 50 $kJ \cdot m^{-2}$ [215]. The increased rate of crystallization of PHBV in the presence of PBS indicated the developed miscibility induced by the crosslinking. Recently, Righetti et al., studied thermodynamically immiscible blends of poly(3-hydroxybutyrate) (PHB)/poly(butylene succinate) (PBS) and PBSA. The resultant blends showed improved ductility with minimal reduction in elastic modulus due to successive solidification of crystalline phases of the individual polymers [216].

PHA-Natural Polymer Blends: Natural polymers are attractive in designing biodegradable polymer blends based on PHA due to their degradability and abundant availability. Poly(3-hydroxybutyrate-co-3-hydroxy valerate) PHBV and cellulose acetate were blended with plasticiser and chain extending agent [217] to improve the ductility of the composites. However, the resultant blends had significantly reduced mechanical and thermal properties due to poor miscibility between the polymers. PHA blended with cellulose acetate butyrate (CAB) is reported to be miscible blends as evidenced by single T_g of PHB [218]. The improved compatibility was due to reduced spherulite growth of PHB in the presence of CAB, leading the amorphous phase of PHB. The morphology investigation of the blend through small-angle X-ray scattering (SAXS) indicates the presence of a homogeneous amorphous phase situated mainly in the interlamellar regions of crystalline PHB [219]. Chiulan et al., carried out a systematic cytocompatibility studies using L929 cell line of a triblend system comprising poly(3-hydroxybutyrate) (PHB), polyhydroxyalkanoate (PHA), predominantmy poly(3-hydroxyoctanoate) with high amorphous content and bacterial cellulose (BC) [220]. The blends show high surface roughness, medium hydrophobicity and enhanced cytocompatibility, making it suitable for biomedical applications. Blending natural rubber (NR) with mcl-PHA was reported to widen the application range of medium chain length polyhydroxyalkanoates by altering its thermal properties while reducing the overall cost of polymer. Also, it was observed that the degradation behaviour of the blend could be further tailored based on the blend composition [221]. The potential applications of these blends include flexible conventional and barrier packaging films, medical materials such as absorbable surgical sutures, matrices for drug delivery systems, and biodegradable moulded goods, paper coatings, non-woven fabrics, adhesives, films and performance additives [219,222,223].

Starch and thermoplastic starch are home compostable and mostly soluble in water, whereas PHAs are industrially compostable and hydrophobic. Hence, blending both the polymers not only eliminates the limitations related to starch films for packaging applications and aids flexible film processing of the blend system [224]. Lai et al., reported a compatible blend of modified corn starch-PHB with a single T_g at 37 °C [225]. However, the large difference in polarity between starch and PHA makes the two systems incompatible. Therefore, crosslinking or a compatibilizer are required to improve the compatibility between the two phases [226]. Sun et al. developed acid crosslinked blend of hydroxypropyl di-starchphosphate (HPDSP)/PHA. Different acids like citric acid, adipic acid, and boric acid were used as a crosslink agent. The superior thermal degradation stability of the crosslinked blends represented effectively is claimed to be due to intermolecular interactions of grafting or cross-linking between starch and PHA. Further, the citric acid crosslinked blend showed 45–72% improvement in Oxygen permeability (OP) and 45.6–270% improvement in water vapour permeability (WVP) in comparison to the control starch/PHA blend. Improved crystallinity and the compact and uniform crystal microstructure in crosslinked crystalline blend systems is the primary reason of increased gas barrier property and light transmission property of the blends.

PHA Bionanocomposites: Many researchers have studied composites of PHA, and a large variety of natural biofibres used as reinforcing fillers in PHA and the resultant composites are termed 'Bionanocomposites'. For example, plant-based natural fibres like flax fibre [227,228], jute fibre [229], wheat [230], rice straw fibre [220], cellulose fibre [221], wood fibre [222], pineapple fibre [223], and bamboo fibre [231], kenaf fibre [232] etc. are reported to have improved mechanical properties of PHAs. PHA biocomposites with 40% kenaf fibre lost their flexural modulus and strength under the condition of aqueous exposure for 60 days. However, incorporating a suitable compatibilizer to the composite system suggested improvement in the homogeneity between the two phases of polymer and kenaf fibre [233]. Short abaca fibres were melt mixed with PHA and subsequently, injection moulded to demonstrate products. These biofibres were surface modified with butyric anhydride, and the effect of surface treatment of these fibres, fibre length, fibre content were thoroughly investigated by Shibata et al. [234]. The surface treatment of abaca fibres improved the flexural properties of the PHA/abaca fibre biocomposites and 5 mm length is the appropriate length to obtain best mechanical properties. Sisal fibres were added to PHA under inert and oxidative condition resulting in improved thermal stability [235]. Hemp fibres were embedded in PHA polymer matrix and the crystallization studies were conducted to understand the behaviour of the fibres in the PHA matrix [236]. Transcrystallization took place at the surface of the hemp fibres indicating nucleation and improvement in the crystallization process. Wheat starch granules were melt mixed with PHBV at 160 °C and the tensile strength of the composite was found to reduce from 18 MPa to 8 MPa, whereas the Young's modulus has increased by ~63%, making the material useful within the range of flexible film packaging applications. As part of an interesting observation of this biocomposite, the degree of biodegradation significantly improved upon addition of starch into the PHA matrix. While 20 days were required to degrade 100% pure, 100 micron thick PHBV film, the biocomposite of PHBV with 50 wt% starch granules took only 8 days to completely disappear under aerobic conditions [237]. Wood celluloses are one of the cost-effective biofillers in polymer matrices and incorporation of these fillers into PHBV enhance the stiffness, brittleness and other mechanical properties. The reduced percentage elongation at break, due to the addition of the cellulose, was mitigated by the increased HV content of PHBV [238]. Posidonia oceanica (PO), an abundant Mediterranean seagrass, was used as biofiller for PHBV. The resultant biocomposites were analysed for their physical properties and biodegradability in the marine environment separate to its routine performance analysis [239]. The addition of 10 wt% of acetyl tributyl citrate (ATBC) as a plasticizer along with 30 wt% of PO fibres enhance the processibility of the composite. An increase in PO fibre content from 0 wt% to 30 wt% results increase in tensile strength and Charpy's impact energy from 2 GPa to 2.4 GPa and 3.6 kJ m^{-2} to 4.4 kJ m^{-2}, respectively at the cost of a moderate reduction in the elongation to break from 3.2% to 1.9%. Also, biocomposites with 20 wt% PO fibre caused 10% rise in the biodegradation rate of the polymeric matrix after 216 days. Under real-time marine environment, higher degradation of composites was observed compared to neat PHBV film. It was understood that the PO fibres were capable of developing biofilm consisting of bacteria and fungi on the surface after only 3 months of incubation in marine sediments [239].

Inorganic fillers are also used as additives in PHA primarily to enhance its functional properties while maintaining biodegradation characteristics. It is essential to use non-toxic and biocompatible minerals as additives and hence unlike biofibres, selection of mineral additives is challenging. Limited studies were made on layered silicate based PHA composites and Maiti et al., which reported melt extrusion technique to develop PHB/organically modified montmorillonite (MMT) clay composites [240]. These composites have shown improvement in mechanical properties like storage modulus by 40% at 2 wt% filler concentration compared to pristine PHB. Furthermore, the modified clay particles were found to act as strong nucleating agents of PHB and improved thermal stability. The time for 30% biodegradation at 60 °C is 9 and 6 weeks for pristine PHB and PHB with 2 wt% clay particles, respectively [240]. In a similar line, Choi et al., reported improvement in tensile properties like Young's modulus from 480 MPa for neat PHBV to >790 MPa of PHBV upon addition of intercalated MMT clays. Similarly, from thermogravimetric analysis of the composites, an increase of 11 °C in degradation temperature corresponding to 3% weight loss ($T_{3\%}$) of the composites were observed [241]. Bordes et al. reported to have improved the elongation at the break by 21% and 85% for PHB and PHBV while retaining modulus (>1884 MPa) and tensile strength (>24 MPa) upon adding an optimum of 1 wt% of organically modified montmorillonite (OMMT) due to high degree of nanolayer dispersion within PHB and PHBV matrices [242]. Xu et al., presented the work to develop high-performance PHA-based nanocomposites with long alkyl chain quaternary salt (LAQ) functionalized graphene oxide (GO-g-LAQ). The obtained nanocomposites show improvement in tensile strength and storage modulus from neat PHA by 60% and 140%, respectively. Incorporation of 5 wt% of GO-g-LAQ reduced oxygen permeability of the nanocomposite (0.21 $cm^3 \times m^{-2} \times d \times atm$) by 86% from neat PHA film (0.21 $cm^3 \times m^{-2} \times d \times atm$). These nanocomposites also provide inherent antibacterial performance against Gram-negative (*E. coli*) and Gram-positive (*S. aureus*) bacteria [243]. The tensile strength, storage modulus at room temperature improved by 60% and 140%, respectively, and it is suggested to have the potential for food packaging applications [243]. Other biobased inorganic fillers like silicon dioxide (SiO_2) [244], titanium dioxide (TiO_2) [245], calcium carbonate ($CaCO_3$) [246], polyhedral oligomeric silsesquioxane (POSS) [247] etc. have great potential for improvement in multiple properties of PHA based composites.

3.4.3. Biological Modification of PHA

PHAs can be biologically modified in a number ways; simultaneous addition of two substrates in the culture medium and; feeding of substrates containing functional groups [248].

Simultaneous addition of two substrates in culture medium: Many PHA formations can be modified through the co-feeding of substrates in the bacterial fermentation [248]. *Halomonas bluephagenesis* can produce a PHBV copolymer depending on the concentrations of glucose and propionic acid (or other propiongenic substrates) supplied within the fermentation media [98]. A study by Xu et al. showed that the co-metabolism of glycerol and lignin derivates simultaneously improved cell dry weight and biosynthesis of PHA for *P. putida* KT2440 [249].

Feeding of substrates containing functional groups: The feeding of different but specific substrates with functional groups is a method for biologically modifying PHA [248]. For example, *P. putida* can be grown with the presence of ω-phenoxyalkanoates in its culture medium, producing PHA with phenoxy groups in the side chains [250]. Another study found that the introduction of functional groups to PHA through fatty acids allowed for the formation of functional PHA for further grafting [251]. This opens the door to widening PHA diversity further through functional PHA site chains that can be modified through controllable homopolymerization, random copolymerization, block copolymerization and grafting [252].

4. Applications of PHAs

4.1. Food Packaging Applications

The use of plastics in food packaging is ubiquitous as plastics provide physical, mechanical, chemical and microbial protection from the external environment. Plastics are lightweight and have good transparency, making them highly convenient to use as display products. Conventional petroleum-based plastics have advantages like good gas and moisture barrier properties, thermal seal-ability, easy processing, non-toxic, low price, and readily available. Most plastics used in food packaging are non-polar in characteristics and are non-biodegradable e.g., polyethylene's (PE), polypropylenes (PP), polystyrenes (PS), polyvinyl chloride (PVC), polyethylene terephthalate (PET) etc. These non-biodegradable plastics are becoming the primary source of plastic waste that can and are causing severe environmental pollution. The single-use culture of these non-biodegradable plastics represents a severe global ecological problem, also commonly known as *"White pollution"* [253,254]. In recent years, biodegradable polymers produced from renewable resources are emerging as alternative packaging materials to mitigate the environmental consequences caused by packaging waste. Examples of such biodegradable polymers include starch-based plastics, polylactic acid, polyhydroxyalkanoates, Polybutylene Succinate, cellulose esters etc. However, the major drawback of biodegradable plastics used in food packaging is their poor mechanical, thermal, and barrier properties.

Among commercially available biodegradable polymers, PHA polymers are promising candidates for packaging applications due to their similar barrier and mechanical properties compared to fossil-based plastics such as LDPE and polypropylene [255]. Table 5 presents the barrier properties of some of the biodegradable polymers.

Table 5. Water vapor transmission rate [WVTR] and oxygen transmission rate [OTR] of certain biodegradable plastics [256–260].

Biodegradable Polymers	WVTR ($g/m^{-2}/24\ h^{-1}$) *	OTR ($cm^3\ m^{-2}/24\ h^{-1}$) **
PCL cast film	800	500
PLA cast films	350	410
PHB films	300	425
PHBV	138	218
PBS	84	737
Pea-based TPS	200	276
PBAT	540	1530
Cellulose acetate	1090	650

* 38 °C, 90% RH, ** 23 °C, 0% Relative Humidity (RH).

Comparing the moisture barrier property among the mentioned polymers in the Table 5, it is realized that PHBV and PBS exhibit better moisture barrier properties compared to PLA and other commercial bioplastics. Similarly, the PHBV shows the lowest (best) oxygen gas barrier properties, contributing to its valued addition for packaging applications. In a similar context, it is worth mentioning that Thellen et al., observed ten-times higher oxygen gas barrier property in PHA with high degree of crystallinity compared to PLA [159]. PHAs show good gas barrier properties, and it is the appealing point of this bioplastic in numerous packaging applications. Keskin et al., reported having better moisture and gas barrier properties for PHA than its commercial counterparts PLA and PCL [261]. Water vapour permeability (WVP) of commercial PHA like Enmat Y1000P from Tianan and Mirel F1006, Mirel F3002 from Telles show intermediate water vapour barrier properties compared to commercial PP and is much better than PLA [152]. However, PHA polymer containing processing aids such as plasticizers has inferior barrier properties due to decreased crystallinity. Crystallinity is the factor that dictates the gas barrier property of PHAs. scl-PHAs show better barrier performance compared to mcl-PHAs due to their high crystallinity. The side chain length naturally affects the barrier properties of PHA by affecting the crystallinity of the polymer. In the list of biodegradable plastics that are

100% biobased, PHA is rated at the top in terms of gas barrier properties, which makes it a suitable candidate to blend with other bioplastics to improve their mechanical properties.

PHAs are thermosensitive at temperatures generally considered not high in the polymer processing industry and have inferior mechanical properties to conventional fossil-based polymers. Also, widening the applications of PHAs as commodity plastics requires a reduction in the cost of production. Comparatively high production costs of PHA to conventional plastics present problems in its competitiveness for production in the global market [3,262]. Therefore, blending PHA with commercial biodegradable polymers could be an viable approach to PHA polymers for various applications.

4.2. Biomedical Applications

The biomedical applications of PHA and their blends have been widely reported [255]. The properties of these biodegradable, biocompatible and non-toxic blends makes their use in medical applications such as wound management, drug delivery carriers, biodegradable implants (vascular system, orthopaedy), tissue engineering and anticancer agents an attractive option. Some of these applications found in Table 6 will be reviewed in this section.

Table 6. Applications of PHA in medicine [9].

Type of Application	Products
Wound management	Sutures, skin substitutes, nerve cuffs, surgical meshes, staples, swabs
Vascular system	Heart valves, cardiovascular fabrics, pericardial patches, vascular grafts
Orthopaedy	Scaffolds for cartilage engineering, spinal cages, bone graft substitutes, meniscus regeneration, internal fixation devices (e.g., screws)
Drug delivery	Micro- and nanospheres for anticancer therapy
Urology	Urological stents
Dental	Barrier material for guided tissue regeneration in periodontitis

A high biocompatibility is necessary when exposing the human or animal body to an object that isn't naturally occurring to that organism [9]. This biocompatibility of a foreign object used in a human or animal is determined by factors such as shape, surface porosity, material chemistry and the environment in which the object is in-situ (tissue) [9]. This is why PHAs have high potential in biomedical applications, as tests have shown that components of PHA, such as (*R*)-3-hydroxybutyric acid from PHB, are already present in human blood [263]. This ultimately reduces the potential for the immune system to reject an implant that is comprised of natural occurring elements. The natural occurring properties of PHAs are a good starting point for their use in medical applications, but the potential and optimization of such applications requires improvement of properties through the blending of polymers. Raza et al. (2017) reported that mice fibroblast cell growth was increased when they were grown on a PHB/PHBHHx polymer blend when compared to the individual PHB and PHBHHx components. The monomeric units of PHAs can reduce bacterial infection like that from *Staphylocuccus aureus* with P3HB/P4HB known for aiding in the enhancement of angiogenic properties of the skin and wound healing [255]. Along with increased biocompatibility and regenerative benefits, there is also a significant increase in the tensile strength of blended films when the PHBHHx content of the blend is increased [264,265]. As PHA is so biocompatible and biodegradable it's also a promising candidate for use as a drug carrier. It was shown by Kassab et al. (1997) that microspheres of PHB can be formed in a range of sizes, these spheres being loaded with rifampicin, allowing for the release of the drug over a 24 h period. This allowed for the rate of the drug released to be increased or decreased depending on the width of the PHB sphere [266]. As well as a wound management and as a drug carrier tool, PHA can also be utilized as a scaffold material in tissue engineering [9]. This is a complex application as it requires the biopolymer scaffold to serve a wide array of functions: biocompatibility; support of cell growth and cell adhesion; guide and organize cells; allow ingrowth of cells and flow of nutrients/wastes and biodegrade in a non-toxic manner [267]. PHA production costs are currently high and so cannot compete with petroleum-based polymers in most applications today, so using PHAs in these complex biomedical applications can warrant the initial production cost.

5. Biodegradability of PHAs and Composites

The detrimental impact of the continuous accumulation and physical degradation of petroleum-based plastics to form microplastics is putting unprecedented pressure on our natural Environment [268]. The total cumulative global plastic production reached 7.82 billion metric tons by 2015 [269]. Increased efforts are required to prevent the release of plastics into the environment, to increase the rate of plastics recycling and to find alternatives to these fossil based recalcitrant plastics. The two major movements circular economy and bioeconomy seek to address resource efficiency (circular economy) and the switch to the use of biobased resources (bioeconomy). The latter makes up one half of the circular economy. Thus, society is shifting towards efficiency of resource use, increased recycling and the use of biobased and biodegradable polymers. Not all biobased polymers are biodegradable and not all biodegradable polymers are biobased. For example there are technologies emerging to produce bio-based equivalents to fossil based plastics such as poly(ethylene) and poly(ethylene terephthalate) [270]. There are also polymers such as Polybutylene adipate terephthalate (PBAT) and polycaprolactone (PCL) that are fossil based and biodegradable. Biobased polyethylene can be manufactured from sugar cane through a multi-step process involving fermentation, to produce ethanol, and chemical catalysis to produce ethylene and subsequently polyethylene [271], refer to 'Figure 7' below for more details on this multi-step process. PET has two monomers terephthalate and ethylene glycol. Terephthalate is made from oxidation of para-xylene (1,4 dimethylbenzene) which can be synthesised from 2,5-dimethylfuran (DMF) and ethylene or 4-methyl-3-cyclohexene-1-carboxaldehyde (4-MCHCA). Ethylene glycol, having a similar to structure to glycerol is normally produced from ethylene (fossil fuel based) but can also be produced from biomass such as sugars via microorganisms such as *Pseudomonas syringae* [272]. Ethylene glycol can then be synthesized from p-xylene (originating from biomass sugars), polymerized with terephthalic acid from biomass to form PET [273]. DMF can be synthesised from fructose [274] while 4-MCHCA can be produced from acrolein and isoprene. Acrolein can be produced through the dehydration of glycerol, a major component of plant oils and a by-product of biodiesel production [275]. Biobased Isoprene can be derived from plants and more recently genetically engineered microbes using glucose as the carbon substrate [276].

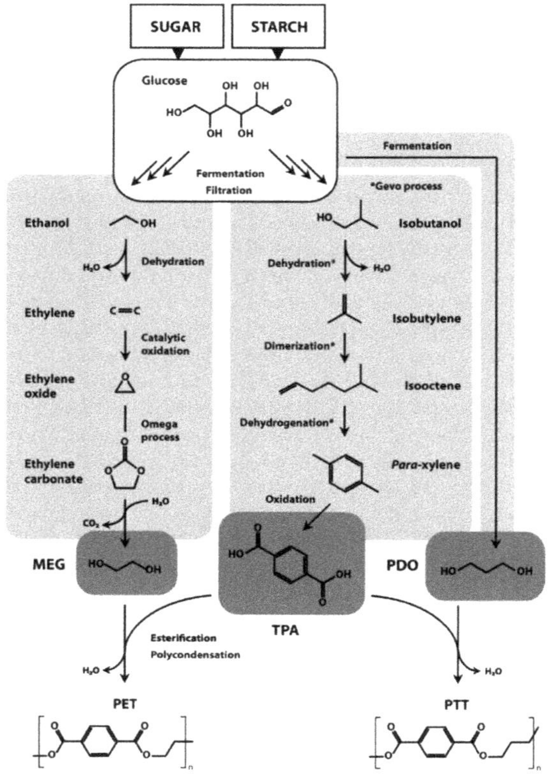

Figure 7. Production routes for fully biobased polyethylene terephthalate (PET) and polytrimethylene terephthalate (PTT). Biomass-derived substrates are marked by white boxes, biobased synthesis of monomers is highlighted in light gray, key monomers in dark gra y, key monomers in dark gray. Abbreviations: MEG, monoethylene glycol; PDO, 1,3-propanediol; TPA, terephthalic acid [277].

Biodegradable polymers such as PHAs are compostable in industrial composting facilities. PHB is also home compostable and a particularly good example of a highly degradable polymer in multiple environments i.e Marine pelagic, fresh water aerobic, anaerobic aquatic, and soil. However PHO, a medium chain length PHA is not home compostable and degraded slowly in aerobic marine and fresh water environments. Furthermore, it exhibited very poor biodegradation in anaerobic aquatic and soil environments [278]. The biodegradation profile of PHO in managed environments is similar in many ways to polylactic acid which is industrially compostable but not home compostable. However, PLA shows no biodegradation in marine fresh water, or soil. Thermoplastic starch (TPS) is the single biggest selling biobased polymer while poly(lactic acid) is the fastest growing. PHA production remains low but its growth potential remains high [279].

The use of PHAs in polymer composites can improve the mechanical and thermal properties of the final biobased products, while also improving the biodegradability properties of the overall biocomposite product [280]. This section will review PHA biodegrading microorganisms and their extracellular biodegradation of PHA, the properties of PHAs affecting biodegradation and the biodegradation of PHA blends and their additives.

Short and medium chain-length PHAs can be degraded by an array of bacteria and fungi under both aerobic and anaerobic conditions. Some of the most effective PHA biodegrading bacteria belong to genera's such as *Pseudomonas*, *Bacillus* and *Burkholderia* with fungi in the *Ascomycota* and *Zygomycotina* genera's proven to be effective and sometimes even more effective PHA degraders than bacteria [281,282]. Enzymes called PHA depolymerases are mediators of the extracellular biodegradation of PHA, where they breakdown the polymer into shorter chains through hydrolytic depolymerisation to form oligomers with further biodegradation into trimer and dimer units either through the action of PHA polymerase [280] or through lipases and hydrolases [283]. PHB is a common PHA that is broken down in this manner, as shown in Figure 8 below. This figure depicts the protein structure of the extracellular PHB depolymerases where there are three domains: binding, which is responsible for absorption and spreading across the polymer; linker, which is responsible for linking the binding domain to the catalytic domain and; catalytic, which cleaves the PHA and any available dimers and trimers [280].

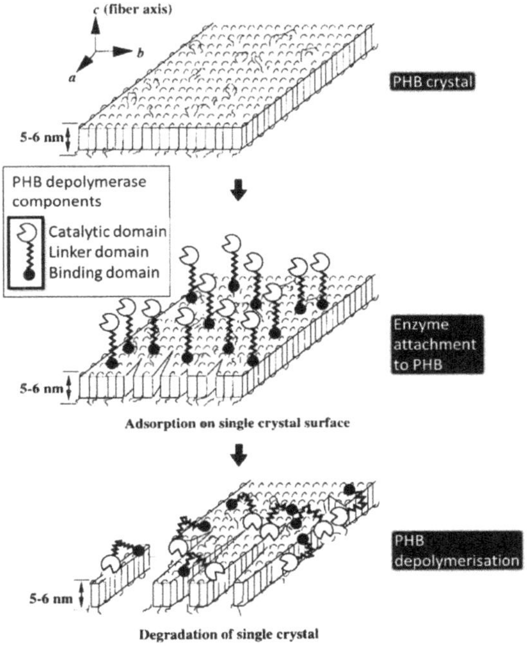

Figure 8. A Single PHB crystal enzymatic degradation by PHB depolymerase [280].

Although there are three domains involved with the enzyme attachment to the PHB surface, the enzymatic hydrolysis of PHA's is a two-step process. The first being the absorption of enzymes onto the polymer surface and the second being the hydrolytic cleavage of PHA bonds induced by the absorbed enzyme [284]. This process favours less crystalline polymers that have amorphous surface crystals as the hydrolytic enzymes can access the less ordered structure more easily [285]. The by-products and efficiency of this enzymatic degradation are dependent on the type of PHA. PHA biodegradation rate or efficiency depends on the physical and chemical properties of the polymer such as crystallinity, types of copolymers, and the copolymeric structure [280,286]. For example, PHBV has a greater amorphous region than PHB due to the presence of 3-hydroxyvalerate meaning it has a reduced water barrier, increasing absorption which allows for an easier access for the enzyme catalytic domain to cleave the PHA [287]. Other factors such as unknown enzymes, PHA side chain length and frequency, environmental conditions (e.g., water temperature) and molecular weight can affect the rate of a polymers biodegradation [280,288].

6. Recycling of PHAs

Recycling is the most desired option to manage plastic waste and associated environmental pollution and greenhouse gas emissions. There is increasing use of biobased and biodegradable plastics to replace fossil-based plastics in various applications to curb the ever-growing plastic pollution and terrestrial water littering. As a result, there is considerable growth in biodegradable plastics in packaging and other commodity products due to more comprehensive end-of-life options available with biodegradable plastics like PLA, PHA, PBS, TPS, and other biodegradable plastics. Figure 9 shows the end-of-life options available for biodegradable polymers. Different recycling processes recycle only 9% of total plastics produced [289].

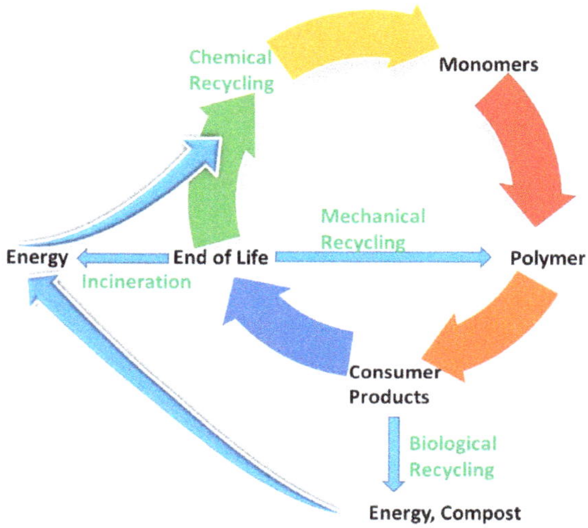

Figure 9. End-of-life options for biodegradable plastics.

Mechanical recycling is the most used recycling process, which involves collecting, sorting, washing, and grinding the plastics to obtain homogeneous material for reuse and further processing [149]. However, not all the biodegradable polymers can be recycled in this way as they could degrade due to the reprocessing (thermal-mechanical degradation) and factors during lifetime (exposure to environmental factors like heat, oxygen, light, moisture) [290]. Chemical recycling involves the degradation of polymer into their chemical constituents that can be polymerized to the original product or converted into other valuable products [291]. Even though mechanical recycling is the most preferred methodology, chemical recycling can produce value-added materials and fully represent the circular polymer production economy [292]. Indeed the use of chemical and biological technologies to both depolymerise plastics and convert the individual monomers to value added products including biodegradable plastics is being investigated in two joint EU-China projects i.e Mix-up (https://www.mix-up.eu accessed on 20 January 2022)and BIOICEP (https://www.bioicep.eu accessed on 20 January 2022) Overall, the lack of an ample quantity of biodegradable polymer waste makes recycling less economically attractive than conventional plastics. Chemical recycling is possible with biodegradable polymers; however, the current processes are expensive and not practised on a commercial scale. Apart from traditional recycling techniques, biodegradable polymers can be recycled by organic recycling under industrial composting conditions, which is a unique advantage with biodegradable polymers compared to fossil based plastics [293]. In polymers that cannot be recycled, energy recovery is the preferred option through an incineration process.

PHA polymers have the fastest market growth rate since 2014, and its production reached about 100,000 tonnes in 2020 [294]. A variety of PHA polymers produced on an industrial scale have shown similar fossil-based plastics performances for various applications. Mechanical recycling of PHA polymers is not recommended for PHA polymers due to their poor thermal properties. Generally, PHA polymers are recycled by the industrial composting process, with very few studies in terms of PHA recycling available [295]. Poly(hydroxybutyrate) (PHB) recycling via extrusion studied by [296], showed a significant reduction in the physical properties after two extrusion cycles and 50% decrease in the tensile strength at complete degradation after the third recycling cycle. However, the degraded products of PHB can be used as plasticizer for making PLA based composites [297].

Furthermore, PHA based polymers are often blended with other materials to reinforce their properties, and consequently, mechanical recycling system becomes more complex and expensive. However, it was reported that by blending a small amount of PLA with PHB, PLA could act as a stabiliser by improving the stability of the PHB up to 5 cycles of recycling with a slight decrease in mechanical properties [298,299]. Furthermore, chemical recycling of PHA is reported through the pyrolysis process, leading to the production of crotonic acid monomers [300]. These resultant monomers can be used to produce other polymers, plasticizers, herbicides and cosmetic products. Table 7 shows the recycling pathways of PHA polymer by different recycling and associated products. Currently, PHA polymers not separated from mixed plastic waste streams due to their low volume and non-existence of an eco-system for the recovery of PHA polymers.

Table 7. Recycling Methods for various PHAs and its blends.

Recycling Type	Methods/Technology Adopted	Type of PHAs	Final Product as an Outcome	Reference
Mechanical	Melt extrusion technique	PHBV	The process can be repeated up to five times; an 8% loss of tensile strength.	[301]
	Melt extrusion technique	PHBV/PLA blend	The process can be repeated up to six times.	[298]
	Melt extrusion technique	PHB	A grafted PLA/PHB blend with better physical properties.	[296]
	Melt extrusion technique	PHB	Significant reduction of physical properties after two cycles.	[149]
Chemical	Pyrolysis at 170–290 °C	PHB, PHBV	Crotonic acid, its diamer and trimer.	[302]
	Pyrolysis with magnesium oxide and magnesium hydroxide as catalyst at 160–280 °C	PHB, PHBV	Highly selective crotonic acid, PHB oligomers, and 2-pentenoic acid.	[303]
	Pyrolysis conducted at temperature range of 220–670 °C	P(3HD)	2-decnoic acid the primary product which is a carbon source for new P(3HD).	[304]
	Pyrolysis at 310 °C	PE/PHBV (70/30) blend	Higher composition of crotonic acid compared to neat PHBV, separation of two materials without contamination.	[258]
	Microwave-assisted degradation in green solvents (water, methanol, ethanol) under alkaline condition, mild condition of 20 min microwave heating at 110 °C	PHB	3-hydroxybutanoic acid, 3-methoxybutanoic acid, and crotonic acid.	[305]

Table 7. *Cont.*

Recycling Type	Methods/Technology Adopted	Type of PHAs	Final Product as an Outcome	Reference
Biological	Enzymatic degradation	PHB and PHBV	Range of microorganisms.	[306]
	Anaerobic sludge and aerobic seawater conditions	Poly(3-hydroxybutyrate-co-3-hydroxyhexanoate) (poly(3HB-co-3HHx))	Under anaerobic condition dominant bacteria *Cloacamonales* and *Thermotogales*, 77.1 ± 6.1% of the carbon converted to gas, methane yield ($p \leq 0.05$), 483.8 ± 35.2 mL·g^{-1} volatile solid (VS). Under aerobic seawater condition dominant bacteria *Clostridiales*, *Gemmatales*, *Phycisphaerales*, and *Chlamydiales*, 83% polymer weight loss in about 6 months' time.	[307]
	Marine environment	PHAs	Mean rate of biodegradation was 0.04–0.09 mg·day^{-1}·cm^{-2} (p = 0.05), the measurement was calculated in terms of respirometric methods and mass loss experiments.	[308]
	Laboratory compost condition, pH 7.5, at 37 °C for 8 weeks, with cultures of *Pseudomonas chlororaphis* B-561 and *Streptomyces* sp. BV315	mcl-PHA films	Moderate weight loss of maximum 13.1% observed in case of mcl-PHA compost with Streptomyces sp. BV315.	[309]
	Biodegradation in static composting pile at industrial condition	PHB/PLA blend films	Reduction in molecular weight by 47.6% in 70 days.	[310]
	Under controlled composting conditions according to ISO 14855-1	PHBV and PHB	As per ISO 14855-1 standard, a range of 79.7–90.5% biodegradation.	[311]

The introduction of biodegradable plastics into the market has given rise to the need for more efficient recycling techniques. Currently, the traditional recycling methods are not compatible with most of the biodegradable polymers. A further problem in the recycling of biodegradable plastics comes from the separation of different types of plastics. Therefore, further efforts and investments are required to improve the segregation of biodegradable plastics to re-use the biodegradable material most efficiently.

7. Conclusions

The current economic viability for PHA production is still challenging when competing with fossil-based plastics sold into high-volume, low-value applications. Given that starting materials to make PHAs is a significant cost factor in their production, waste or low-value substrates are of interest in this review and continue to be across much of the research and development on PHAs. The fermentation production processes, scale-up of these processes, and the downstream processing involved are key addressable pillars of the PHA production chain. An in-depth knowledge of the polymers' properties and how these relate to applications is critical in steering the production of PHAs in the right direction. While the biodegradability of PHAs and composites is a critical characteristic, the performance of these polymers in products will also be a major factor affecting their market uptake. Thus, research into substrates, production processes, properties, applications, modifications, biodegradability, and even recyclability of PHAs are areas that require further research and development.

In general, biodegradable plastic can reduce plastic littering and return clean organic matter to the soil in the form of high-quality compost by avoiding contamination with persistent plastics. Products based on biodegradable polymers are designed to safeguard our ecosystem and minimise plastic pollution while improving the circularity of resources used to produce the polymers. At the same time, creating circular products, such as biodegradable polymers, provides solutions to environmental issues and supports the creation of new jobs and infrastructures that unlock the potential for local agricultural and forestry value chains. Currently, the share of biodegradable plastic products is tiny (1–2%) and is expected to grow 5–8% in the next 5 years. However, the entry of biodegradable-based plastic products into the plastic market is slow due to high cost, nonavailability of polymers, and widespread confusion among the consumers regarding biodegradable plastic products. Promoting biodegradable-based plastics requires further investments and improving the scale of manufacturin and availability of these polymers to use in various applications. Also, appropriate regulatory policies are necessary to minimise barriers and create end-of-life infrastructure to promote products based on biodegradable polymers.

Further, an appropriate labelling system is needed to distinguish the products based on biodegradable plastic products to create proper recycling and disposal infrastructure to capture the valuable biodegradable plastics entering into mixed-plastic waste streams. In addition, to realise the true environmental benefits of biodegradable plastics, correctly verified biodegradation standards are essential to boost the use of biodegradable polymers. Therefore, biodegradable plastics such as PHA and their composites can contribute to a sustainable circular plastic economy and reduce unavoidable plastic littering.

Author Contributions: Conceptualization, K.O. and R.B.P.; investigation, B.D., P.B. and J.D.M.; writing—original draft preparation, B.D., P.B. and J.D.M.; writing—review and editing, K.O. and R.B.P.; funding acquisition, R.B.P. and K.O. All authors have read and agreed to the published version of the manuscript.

Funding: The researchers were funded by Environmental Protection Agency, grant number 2019-RE-LS-4 and Science Foundation Ireland (BiOrbic Bioeconomy SFI research centre 16/RC/3889). R.B. would like to acknowledge funding from the European Union's Horizon 2020 research innovation program under grant agreement No. 870292 (BIOICEP). KOC would like to acknowledge funding from the European Union's Horizon 2020 research innovation program under grant agreement No. 870292 (870294).

Data Availability Statement: Data can be shared with permission from the authors.

Conflicts of Interest: The authors declare no conflict of interest.

References

1. Alamgeer, M. Polyhydroxyalkanoates (PHA) genes database. *Bioinformation* **2019**, *15*, 36–39. [CrossRef] [PubMed]
2. Valappil, S.P.; Boccaccini, A.R.; Bucke, C.; Roy, I. Polyhydroxyalkanoates in gram-positive bacteria: Insights from the genera Bacillus and Streptomyces. *Antonie Leeuwenhoek* **2006**, *91*, 1–17. [CrossRef] [PubMed]
3. Muthuraj, R.; Misra, M.; Mohanty, A.K. Biodegradable compatibilized polymer blends for packaging applications: A literature review. *J. Appl. Polym. Sci.* **2018**, *135*, 45726. [CrossRef]
4. Raza, Z.A.; Abid, S.; Banat, I.M. Polyhydroxyalkanoates: Characteristics, production, recent developments and applications. *Int. Biodeterior. Biodegrad.* **2018**, *126*, 45–56. [CrossRef]
5. Steinbüchel, A.; Valentin, H.E. Diversity of bacterial polyhydroxyalkanoic acids. *FEMS Microbiol. Lett.* **1995**, *128*, 219–228. [CrossRef]
6. Madison, L.L.; Huisman, G.W. Metabolic engineering of poly(3-hydroxyalkanoates): From DNA to plastic. *Microbiol. Mol. Biol. Rev.* **1999**, *63*, 21–53. [CrossRef]
7. Nikodinovic-Runic, J.; Guzik, M.; Kenny, S.T.; Babu, R.; Werker, A.; Connor, K.E. Carbon-rich wastes as feedstocks for biodegradable polymer (polyhydroxyalkanoate) production using bacteria. *Adv. Appl. Microbiol.* **2013**, *84*, 139–200.
8. Le Meur, S.; Zinn, M.; Egli, T.; Thöny-Meyer, L.; Ren, Q. Production of medium-chain-length polyhydroxyalkanoates by sequential feeding of xylose and octanoic acid in engineered Pseudomonas putida KT2440. *BMC Biotechnol.* **2012**, *12*, 53. [CrossRef]

9. Zinn, M.; Witholt, B.; Egli, T. Occurrence, synthesis and medical application of bacterial polyhydroxyalkanoate. *Adv. Drug Deliv. Rev.* **2001**, *53*, 5–21. [CrossRef]
10. Ward, P.G.; de Roo, G.; O'Connor, K.E. Accumulation of polyhydroxyalkanoate from styrene and phenylacetic acid by Pseudomonas putida CA-3. *Appl. Environ. Microbiol.* **2005**, *71*, 2046–2052. [CrossRef]
11. Bhuwal, A.K.; Singh, G.; Aggarwal, N.K.; Goyal, V.; Yadav, A. Isolation and screening of polyhydroxyalkanoates producing bacteria from pulp, paper, and cardboard industry wastes. *Int. J. Biomater.* **2013**, *2013*, 752821. [CrossRef] [PubMed]
12. Lee, S.Y.; Wong, H.H.; Choi, J.; Han, C.S. Production of medium-chain-length polyhydroxyalkanoates by high-cell-density cultivation of Pseudomonas putida under phosphorus limitation. *Biotechnol. Bioeng.* **2000**, *68*, 466–470. [CrossRef]
13. Ryu, H.W.; Hahn, S.K.; Chang, Y.K.; Chang, H.N. Production of poly(3-hydroxybutyrate) by high cell density fed-batch culture of *Alcaligenes eutrophus* with phospate limitation. *Biotechnol. Bioeng.* **1997**, *55*, 28–32. [CrossRef]
14. Licciardello, G.; Catara, A.F.; Catara, V. Production of polyhydroxyalkanoates and extracellular products using Pseudomonas corrugata and P. mediterranea: A review. *Bioengineering* **2019**, *6*, 105. [CrossRef] [PubMed]
15. Sharma, P.; Munir, R.I.; Blunt, W.; Dartiailh, C.; Cheng, J.; Charles, T.C.; Levin, D.B. Synthesis and physical properties of polyhydroxyalkanoate polymers with different monomer compositions by recombinant Pseudomonas putida LS46 expressing a novel PHA SYNTHASE (PhaC116) enzyme. *Appl. Sci.* **2017**, *7*, 242. [CrossRef]
16. Gironi, F.; Piemonte, V. Bioplastics and petroleum-based plastics: Strengths and weaknesses. *Energy Sources Part A Recovery Util. Environ. Eff.* **2011**, *33*, 1949–1959. [CrossRef]
17. Ragaert, P.; Buntinx, M.; Maes, C.; Vanheusden, C.; Peeters, R.; Wang, S.; D'Hooge, D.R.; Cardon, L. Polyhydroxyalkanoates for food packaging applications. In *Reference Module in Food Science*; Elsevier: Amsterdam, The Netherlands, 2019.
18. Salvachúa, D.; Rydzak, T.; Auwae, R.; de Capite, A.; Black, B.A.; Bouvier, J.T.; Cleveland, N.S.; Elmore, J.; Furches, A.; Huenemann, J.; et al. Metabolic engineering of Pseudomonas putida for increased polyhydroxyalkanoate production from lignin. *Microb. Biotechnol.* **2020**, *13*, 290–298. [CrossRef]
19. Yu, J. Microbial production of bioplastics from renewable resources. In *Bioprocessing for Value-Added Products from Renewable Resources*; Elsevier: Amsterdam, The Netherlands, 2007; pp. 585–610.
20. Reddy, C.; Ghai, R.; Kalia, V. Polyhydroxyalkanoates: An overview. *Bioresour. Technol.* **2003**, *87*, 137–146. [CrossRef]
21. Singh, A.K.; Sharma, L.; Mallick, N.; Mala, J. Progress and challenges in producing polyhydroxyalkanoate biopolymers from cyanobacteria. *J. Appl. Phycol.* **2016**, *29*, 1213–1232. [CrossRef]
22. Kumar, P.; Singh, M.; Mehariya, S.; Patel, S.K.S.; Lee, J.-K.; Kalia, V.C. Ecobiotechnological approach for exploiting the abilities of bacillus to produce co-polymer of polyhydroxyalkanoate. *Indian J. Microbiol.* **2014**, *54*, 151–157. [CrossRef]
23. Bernard, M. Industrial potential of polyhydroxyalkanoate bioplastic: A brief review. *USURJ Univ. Sask. Undergrad. Res. J.* **2014**, *1*, 1–14. [CrossRef]
24. Ankenbauer, A.; Schäfer, R.A.; Viegas, S.C.; Pobre, V.; Voß, B.; Arraiano, C.M.; Takors, R. Pseudomonas putida KT2440 is naturally endowed to withstand industrial-scale stress conditions. *Microb. Biotechnol.* **2020**, *13*, 1145–1161. [CrossRef] [PubMed]
25. Durner, R.; Witholt, B.; Egli, T. Accumulation of poly[(*R*)-3-hydroxyalkanoates] in *Pseudomonas oleovorans* during growth with octanoate in continuous culture at different dilution rates. *Appl. Environ. Microbiol.* **2000**, *66*, 3408–3414. [CrossRef] [PubMed]
26. Mendonça, T.; Gomez, J.; Buffoni, E.; Rodriguez, R.J.S.; Schripsema, J.; Lopes, M.; Silva, L. Exploring the potential of *Burkholderia sacchari* to produce polyhydroxyalkanoates. *J. Appl. Microbiol.* **2014**, *116*, 815–829. [CrossRef] [PubMed]
27. Pohlmann, A.; Fricke, W.F.; Reinecke, F.; Kusian, B.; Liesegang, H.; Cramm, R.; Eitinger, T.; Ewering, C.; Pötter, M.; Schwartz, E.; et al. Genome sequence of the bioplastic-producing 'Knallgas' bacterium *Ralstonia eutropha* H16. *Nat. Biotechnol.* **2006**, *24*, 1257–1262. [CrossRef] [PubMed]
28. Clark, D.P.; Pazdernik, N.J. *Basics of Biotechnology*, 2nd ed.; Elsevier: Amsterdam, The Netherlands, 2016; pp. 1–31.
29. Wang, Q.; Nomura, C.T. Monitoring differences in gene expression levels and polyhydroxyalkanoate (PHA) production in Pseudomonas putida KT2440 grown on different carbon sources. *J. Biosci. Bioeng.* **2010**, *110*, 653–659. [CrossRef] [PubMed]
30. Brandl, H.; Gross, R.A.; Lenz, R.W.; Fuller, R.C. *Pseudomonas oleovorans* as a source of poly(β-hydroxyalkanoates) for potential applications as biodegradable polyesters. *Appl. Environ. Microbiol.* **1988**, *54*, 1977–1982. [CrossRef]
31. Guamán, L.P.; Barba-Ostria, C.; Zhang, F.; Oliveira-Filho, E.R.; Gomez, J.G.C.; Silva, L.F. Engineering xylose metabolism for production of polyhydroxybutyrate in the non-model bacterium *Burkholderia sacchari*. *Microb. Cell Factories* **2018**, *17*, 74. [CrossRef]
32. Raposo, R.S.; de Almeida, M.C.M.; de Oliveira, M.D.C.M.; da Fonseca, M.M.; Cesário, M.T. A *Burkholderia sacchari* cell factory: Production of poly-3-hydroxybutyrate, xylitol and xylonic acid from xylose-rich sugar mixtures. *New Biotechnol.* **2017**, *34*, 12–22. [CrossRef]
33. Chen, H. Brief introduction to the biotechnology of lignocellulose. In *Biotechnology of Lignocellulose*; Springer Science and Business Media: Berlin/Heidelberg, Germany, 2014; pp. 1–24.
34. Nonato, R.; Mantelatto, P.; Rossell, C. Integrated production of biodegradable plastic, sugar and ethanol. *Appl. Microbiol. Biotechnol.* **2001**, *57*, 1–5.
35. De Sousa Dias, M.; Koller, M.; Puppi, D.; Morelli, A.; Chiellini, F.; Braunegg, G. Fed-batch synthesis of poly(3-hydroxybutyrate) and poly(3-hydroxybutyrate-co-4-hydroxybutyrate) from sucrose and 4-hydroxybutyrate precursors by *Burkholderia sacchari* strain DSM 17165. *Bioengineering* **2017**, *4*, 36. [CrossRef] [PubMed]
36. Windhorst, C.; Gescher, J. Efficient biochemical production of acetoin from carbon dioxide using *Cupriavidus necator* H16. *Biotechnol. Biofuels* **2019**, *12*, 163. [CrossRef] [PubMed]

37. Schwartz, E.; Henne, A.; Cramm, R.; Eitinger, T.; Friedrich, B.; Gottschalk, G. Complete nucleotide sequence of pHG1: A Ralstonia eutropha H16 megaplasmid encoding key enzymes of H2-based lithoautotrophy and anaerobiosis. *J. Mol. Biol.* **2003**, *332*, 369–383. [CrossRef]
38. Krieg, T.; Sydow, A.; Faust, S.; Huth, I.; Holtmann, D. CO_2 to terpenes: Autotrophic and electroautotrophic α-humulene production with *Cupriavidus necator*. *Angew. Chem. Int. Ed.* **2018**, *57*, 1879–1882. [CrossRef] [PubMed]
39. Modelska, M.; Berłowska, J.; Kregiel, D.; Cieciura-Włoch, W.; Antolak, H.; Tomaszewska, J.; Binczarski, M.; Szubiakiewicz, E.; Witońska, I.A. Concept for recycling waste biomass from the sugar industry for chemical and biotechnological purposes. *Molecules* **2017**, *22*, 1544. [CrossRef] [PubMed]
40. Cesário, M.T.; Raposo, R.S.; de Almeida, M.C.M.; van Keulen, F.; Ferreira, B.S.; da Fonseca, M.M.R. Enhanced bioproduction of poly-3-hydroxybutyrate from wheat straw lignocellulosic hydrolysates. *New Biotechnol.* **2014**, *31*, 104–113. [CrossRef]
41. Munawar, K.M.M.; Simarani, K.; Annuar, M.S.M. Bioconversion of mixed free fatty acids to poly-3-hydroxyalkanoates by *Pseudomonas putida* BET001 and modeling of its fermentation in shake flasks. *Electron. J. Biotechnol.* **2016**, *19*, 50–55. [CrossRef]
42. Annuar, M.; Tan, I.; Ibrahim, S.; Ramachandran, K. Production of medium-chain-length poly(3-hydroxyalkanoates) from crude fatty acids mixture by *Pseudomonas putida*. *Food Bioprod. Process.* **2007**, *85*, 104–119. [CrossRef]
43. Kim, S.; Dale, B. Life cycle assessment study of biopolymers (polyhydroxyalkanoates)—Derived from no-tilled corn (11 pp). *Int. J. Life Cycle Assess.* **2004**, *10*, 200–210. [CrossRef]
44. Chanprateep, S. Current trends in biodegradable polyhydroxyalkanoates. *J. Biosci. Bioeng.* **2010**, *110*, 621–632. [CrossRef]
45. Gerngross, T.U. Can biotechnology move us toward a sustainable society? *Nat. Biotechnol.* **1999**, *17*, 541–544. [CrossRef] [PubMed]
46. Gomez, J.G.; Méndez, B.S.; Nikel, P.I.; Pettinari, M.J.; Prieto, M.A.; Silva, L.F. Making green polymers even greener: Towards sustainable production of polyhydroxyalkanoates from agroindustrial by-products. *Adv. Appl. Biotechnol.* **2012**, *3*, 41–62. [CrossRef]
47. Nielsen, C.; Rahman, A.; Rehman, A.U.; Walsh, M.K.; Miller, C.D. Food waste conversion to microbial polyhydroxyalkanoates. *Microb. Biotechnol.* **2017**, *10*, 1338–1352. [CrossRef] [PubMed]
48. Chaudhry, W.N.; Jamil, N.; Ali, I.; Ayaz, M.H.; Hasnain, S. Screening for polyhydroxyalkanoate (PHA)-producing bacterial strains and comparison of PHA production from various inexpensive carbon sources. *Ann. Microbiol.* **2011**, *61*, 623–629. [CrossRef]
49. Kulpreecha, S.; Boonruangthavorn, A.; Meksiriporn, B.; Thongchul, N. Inexpensive fed-batch cultivation for high poly(3-hydroxybutyrate) production by a new isolate of *Bacillus megaterium*. *J. Biosci. Bioeng.* **2009**, *107*, 240–245. [CrossRef]
50. Koller, M.; Sandholzer, D.; Salerno, A.; Braunegg, G.; Narodoslawsky, M. Biopolymer from industrial residues: Life cycle assessment of poly(hydroxyalkanoates) from whey. *Resour. Conserv. Recycl.* **2013**, *73*, 64–71. [CrossRef]
51. Pescuma, M.; de Valdez, G.F.; Mozzi, F. Whey-derived valuable products obtained by microbial fermentation. *Appl. Microbiol. Biotechnol.* **2015**, *99*, 6183–6196. [CrossRef]
52. Povolo, S.; Toffano, P.; Basaglia, M.; Casella, S. Polyhydroxyalkanoates production by engineered *Cupriavidus necator* from waste material containing lactose. *Bioresour. Technol.* **2010**, *101*, 7902–7907. [CrossRef]
53. Alves, M.M.; Pereira, M.A.; Sousa, D.Z.; Cavaleiro, A.; Picavet, M.; Smidt, H.; Stams, A. Waste lipids to energy: How to optimize methane production from long-chain fatty acids (LCFA). *Microb. Biotechnol.* **2009**, *2*, 538–550. [CrossRef]
54. Cerrone, F.; Choudhari, S.K.; Davis, R.; Cysneiros, D.; O'flaherty, V.; Duane, G.; Casey, E.; Guzik, M.W.; Kenny, S.T.; Babu, R.P.; et al. Medium chain length polyhydroxyalkanoate (mcl-PHA) production from volatile fatty acids derived from the anaerobic digestion of grass. *Appl. Microbiol. Biotechnol.* **2014**, *98*, 611–620. [CrossRef]
55. Gui, M.M.; Lee, K.; Bhatia, S. Feasibility of edible oil vs. non-edible oil vs. waste edible oil as biodiesel feedstock. *Energy* **2008**, *33*, 1646–1653. [CrossRef]
56. Ruiz, C.; Kenny, S.T.; Narancic, T.; Babu, R.; Connor, K.O. Conversion of waste cooking oil into medium chain polyhydroxyalkanoates in a high cell density fermentation. *J. Biotechnol.* **2019**, *306*, 9–15. [CrossRef] [PubMed]
57. Chuah, L.F.; Klemeš, J.J.; Yusup, S.; Bokhari, A.; Akbar, M.M. Influence of fatty acids in waste cooking oil for cleaner biodiesel. *Clean Technol. Environ. Policy* **2017**, *19*, 859–868. [CrossRef]
58. Ashby, R.D.; Solaiman, D.K.Y.; Foglia, T.A.; Liu, C.-K. Glucose/lipid mixed substrates as a means of controlling the properties of medium chain length poly(hydroxyalkanoates). *Biomacromolecules* **2001**, *2*, 211–216. [CrossRef] [PubMed]
59. Solaiman, D.K.Y.; Ashby, R.D.; Hotchkiss, A.T.; Foglia, T.A. Biosynthesis of medium-chain-length poly(hydroxyalkanoates) from soy molasses. *Biotechnol. Lett.* **2006**, *28*, 157–162. [CrossRef]
60. Solaiman, D.K.; Ashby, R.D.; Foglia, T.A. Medium-chain-length poly(beta-hydroxyalkanoate) synthesis from triacylglycerols by *Pseudomonas saccharophila*. *Curr. Microbiol.* **1999**, *38*, 151–154. [CrossRef]
61. Solaiman, D.K.Y.; Ashby, R.D.; Foglia, T.A. Production of polyhydroxyalkanoates from intact triacylglycerols by genetically engineered Pseudomonas. *Appl. Microbiol. Biotechnol.* **2001**, *56*, 664–669. [CrossRef]
62. Haba, H.; Lavaud, C.; Harkat, H.; Magid, A.A.; Marcourt, L.; Benkhaled, M. Diterpenoids and triterpenoids from *Euphorbia guyoniana*. *Phytochemistry* **2007**, *68*, 1255–1260. [CrossRef]
63. Song, J.H.; Jeon, C.O.; Choi, M.H.; Yoon, S.C.; Park, W. Polyhydroxyalkanoate (PHA) production using waste vegetable oil by *Pseudomonas* sp. strain DR2. *J. Microbiol. Biotechnol.* **2008**, *18*, 1408–1415.
64. Costa, S.G.; Lépine, F.; Milot, S.; Déziel, E.; Nitschke, M.; Contiero, J. Cassava wastewater as a substrate for the simultaneous production of rhamnolipids and polyhydroxyalkanoates by *Pseudomonas aeruginosa*. *J. Ind. Microbiol. Biotechnol.* **2009**, *36*, 1063–1072. [CrossRef]

65. Fernández, D.; Rodríguez, E.; Bassas, M.; Viñas, M.; Solanas, A.; Llorens, J.; Marqués, A.; Manresa, A. Agro-industrial oily wastes as substrates for PHA production by the new strain *Pseudomonas aeruginosa* NCIB 40045: Effect of culture conditions. *Biochem. Eng. J.* **2005**, *26*, 159–167. [CrossRef]
66. Magdouli, S.; Brar, S.; Blais, J.; Tyagi, R. How to direct the fatty acid biosynthesis towards polyhydroxyalkanoates production? *Biomass Bioenergy* **2015**, *74*, 268–279. [CrossRef]
67. Marsudi, S.; Unno, H.; Hori, K. Palm oil utilization for the simultaneous production of polyhydroxyalkanoates and rhamnolipids by *Pseudomonas aeruginosa*. *Appl. Microbiol. Biotechnol.* **2008**, *78*, 955–961. [CrossRef] [PubMed]
68. Impallomeni, G.; Ballistreri, A.; Carnemolla, G.M.; Guglielmino, S.P.; Nicolò, M.S.; Cambria, M.G. Synthesis and characterization of poly(3-hydroxyalkanoates) from *Brassica carinata* oil with high content of erucic acid and from very long chain fatty acids. *Int. J. Biol. Macromol.* **2011**, *48*, 137–145. [CrossRef]
69. Sun, Y.; Do Young, K.I.M.; Chung, W.C.; Hyung, W.K.; Young, K.Y.; Young, H.R. Characterization of a tacky poly(3-hydroxyalkanoate) produced by *Pseudomonas chlororaphis* HS21 from palm kernel oil. *J. Microbiol.* **2003**, *13*, 64–69.
70. Thakor, N.; Trivedi, U.; Patel, K. Biosynthesis of medium chain length poly(3-hydroxyalkanoates) (mcl-PHAs) by *Comamonas testosteroni* during cultivation on vegetable oils. *Bioresour. Technol.* **2005**, *96*, 1843–1850. [CrossRef]
71. Solaiman, D.K.Y.; Ashby, R.D.; Foglia, T.A. Physiological characterization and genetic engineering of *Pseudomonas corrugata* for medium-chain-length polyhydroxyalkanoates synthesis from triacylglycerols. *Curr. Microbiol.* **2002**, *44*, 189–195. [CrossRef]
72. Ruiz, C.; Kenny, S.T.; Babu, P.R.; Walsh, M.; Narancic, T.; O'Connor, K.E. High cell density conversion of hydrolysed waste cooking oil fatty acids into medium chain length polyhydroxyalkanoate using *Pseudomonas putida* KT2440. *Catalysts* **2019**, *9*, 468. [CrossRef]
73. Kosseva, M.R.; Rusbandi, E. Trends in the biomanufacture of polyhydroxyalkanoates with focus on downstream processing. *Int. J. Biol. Macromol.* **2018**, *107*, 762–778. [CrossRef]
74. Ryckebosch, E.; Drouillon, M.; Vervaeren, H. Techniques for transformation of biogas to biomethane. *Biomass Bioenergy* **2011**, *35*, 1633–1645. [CrossRef]
75. Pérez, V.; Mota, C.R.; Muñoz, R.; Lebrero, R. Polyhydroxyalkanoates (PHA) production from biogas in waste treatment facilities: Assessing the potential impacts on economy, environment and society. *Chemosphere* **2020**, *255*, 126929. [CrossRef] [PubMed]
76. Wellinger, A.; Murphy, J.D.; Baxter, D. *The Biogas Handbook: Science, Production and Applications*; Elsevier: Amsterdam, The Netherlands, 2013.
77. Strong, P.J.; Laycock, B.; Mahamud, S.N.S.; Jensen, P.D.; Lant, P.A.; Tyson, G.; Pratt, S. The opportunity for high-performance biomaterials from methane. *Microorganisms* **2016**, *4*, 11. [CrossRef] [PubMed]
78. Rostkowski, K.H. *Understanding Methanotrophic Polyhydroxybutyrate (PHB) Production across Scale: Life Cycle Assessment, Pure Culture Experimentation, and Pathway/Genome Database Development*; Stanford University: Stanford, CA, USA, 2012.
79. Lee, W.-H.; Azizan, M.N.; Sudesh, K. Effects of culture conditions on the composition of poly(3-hydroxybutyrate-co-4-hydroxybutyrate) synthesized by *Comamonas acidovorans*. *Polym. Degrad. Stab.* **2004**, *84*, 129–134. [CrossRef]
80. Mozejko-Ciesielska, J.; Szacherska, K.; Marciniak, P. *Pseudomonas* species as producers of eco-friendly polyhydroxyalkanoates. *J. Polym. Environ.* **2019**, *27*, 1151–1166. [CrossRef]
81. Chee, J.Y.; Yoga, S.S.; Lau, N.S.; Ling, S.C.; Abed, R.M.; Sudesh, K. Bacterially produced polyhydroxyalkanoate (PHA): Converting renewable resources into bioplastic. In *Current Research, Technology and Education Topics in Applied Microbiology and Microbial Biotechnology*; Formatex Research Center: Badajoz, Spain, 2010; Volume 2, pp. 1395–1404.
82. Lee, S.Y. Plastic bacteria? Progress and prospects for polyhydroxyalkanoate production in bacteria. *Trends Biotechnol.* **1996**, *14*, 431–438. [CrossRef]
83. Koller, M. A review on established and emerging fermentation schemes for microbial production of polyhydroxyalkanoate (PHA) biopolyesters. *Fermentation* **2018**, *4*, 30. [CrossRef]
84. Srivastava, A.; Gupta, S. Fed-batch fermentation—Design strategies. In *Comprehensive Biotechnology*; Academic Press: Cambridge, MA, USA, 2011; pp. 515–526.
85. Ienczak, J.L.; Schmidell, W.; de Aragão, G.M.F. High-cell-density culture strategies for polyhydroxyalkanoate production: A review. *J. Ind. Microbiol. Biotechnol.* **2013**, *40*, 275–286. [CrossRef]
86. Rocha, R.C.S.; da Silva, L.F.; Taciro, M.K.; Pradella, J.G.C. Production of poly(3-hydroxybutyrate-co-3-hydroxyvalerate) P(3HB-co-3HV) with a broad range of 3HV content at high yields by *Burkholderia sacchari* IPT 189. *World J. Microbiol. Biotechnol.* **2008**, *24*, 427–431. [CrossRef]
87. Sun, Z.; Ramsay, J.; Guay, M.; Ramsay, B. Enhanced yield of medium-chain-length polyhydroxyalkanoates from nonanoic acid by co-feeding glucose in carbon-limited, fed-batch culture. *J. Biotechnol.* **2009**, *143*, 262–267. [CrossRef]
88. Zinn, M. Biosynthesis of medium-chain-length poly[(*R*)-3-hydroxyalkanoates]. In *Plastics from Bacteria*; Springer Science and Business Media: Berlin/Heidelberg, Germany, 2010; pp. 213–236.
89. Zinn, M.; Egli, T.; Herwig, C.; Narang, A. Editorial: Recent advances in continuous cultivation. *Front. Bioeng. Biotechnol.* **2021**, *9*, 641249. [CrossRef]
90. Lillo, J.G.; Rodriguez-Valera, F. Effects of culture conditions on poly(beta-hydroxybutyric acid) production by *Haloferax mediterranei*. *Appl. Environ. Microbiol.* **1990**, *56*, 2517–2521. [CrossRef] [PubMed]
91. Koller, M.; Braunegg, G. Potential and prospects of continuous polyhydroxyalkanoate (PHA) production. *Bioengineering* **2015**, *2*, 94–121. [CrossRef] [PubMed]

92. Durner, R.; Zinn, M.; Witholt, B.; Egli, T. Accumulation of poly[(*R*)-3-hydroxyalkanoates] in *Pseudomonas oleovorans* during growth in batch and chemostat culture with different carbon sources. *Biotechnol. Bioeng.* **2001**, *72*, 278–288. [CrossRef]
93. Choi, J.-I.; Lee, S.Y. Efficient and economical recovery of poly(3-hydroxybutyrate) from recombinant *Escherichia coli* by simple digestion with chemicals. *Biotechnol. Bioeng.* **1999**, *62*, 546–553. [CrossRef]
94. Chen, G.-Q. Industrial production of PHA. In *Plastics from Bacteria: Natural Functions and Applications*; Springer Science and Business Media: Berlin/Heidelberg, Germany, 2009; pp. 121–132. [CrossRef]
95. Chen, G.-Q. Production of Poly-D(-)-3-Hydroxybutyrate and Poly-D(-)-3-Hydroxyvalerate by strains of Alcaligenes latus. *Antonie Leeuwenhoek* **1991**, *60*, 61–66. [CrossRef] [PubMed]
96. Chen, G.; Zhang, G.; Park, S.; Lee, S. Industrial scale production of poly(3-hydroxybutyrate-co-3-hydroxyhexanoate). *Appl. Microbiol. Biotechnol.* **2001**, *57*, 50–55. [PubMed]
97. Quillaguamán, J.; Guzmán, H.; Van-Thuoc, D.; Hatti-Kaul, R. Synthesis and production of polyhydroxyalkanoates by halophiles: Current potential and future prospects. *Appl. Microbiol. Biotechnol.* **2010**, *85*, 1687–1696. [CrossRef]
98. Mitra, R.; Xu, T.; Xiang, H.; Han, J. Current developments on polyhydroxyalkanoates synthesis by using halophiles as a promising cell factory. *Microb. Cell Factories* **2020**, *19*, 86. [CrossRef]
99. Liu, C.; Baffoe, D.K.; Zhan, Y.; Zhang, M.; Li, Y.; Zhang, G. Halophile, an essential platform for bioproduction. *J. Microbiol. Methods* **2019**, *166*, 105704. [CrossRef]
100. Liu, L.-Y.; Xie, G.-J.; Xing, D.-F.; Liu, B.-F.; Ding, J.; Ren, N.-Q. Biological conversion of methane to polyhydroxyalkanoates: Current advances, challenges, and perspectives. *Environ. Sci. Ecotechnol.* **2020**, *2*, 100029. [CrossRef]
101. Lee, J.; Yasin, M.; Park, S.; Chang, I.S.; Ha, K.S.; Lee, E.Y.; Lee, J.; Kim, C. Gas-liquid mass transfer coefficient of methane in bubble column reactor. *Korean J. Chem. Eng.* **2015**, *32*, 1060–1063. [CrossRef]
102. Rostkowski, K.H.; Criddle, C.S.; Lepech, M.D. Cradle-to-gate life cycle assessment for a cradle-to-cradle cycle: Biogas-to-bioplastic (and back). *Environ. Sci. Technol.* **2012**, *46*, 9822–9829. [CrossRef] [PubMed]
103. Levett, I.; Birkett, G.; Davies, N.; Bell, A.; Langford, A.; Laycock, B.; Lant, P.; Pratt, S. Techno-economic assessment of poly-3-hydroxybutyrate (PHB) production from methane—The case for thermophilic bioprocessing. *J. Environ. Chem. Eng.* **2016**, *4*, 3724–3733. [CrossRef]
104. Listewnik, H.-F.; Wendlandt, K.-D.; Jechorek, M.; Mirschel, G. Process design for the microbial synthesis of poly-β-hydroxybutyrate (PHB) from natural gas. *Eng. Life Sci.* **2007**, *7*, 278–282. [CrossRef]
105. Jiang, G.; Hill, D.J.; Kowalczuk, M.; Johnston, B.; Adamus, G.; Irorere, V.; Radecka, I. Carbon sources for polyhydroxyalkanoates and an integrated biorefinery. *Int. J. Mol. Sci.* **2016**, *17*, 1157. [CrossRef] [PubMed]
106. Kim, B.S. Production of poly(3-hydroxybutyrate) from inexpensive substrates. *Enzym. Microb. Technol.* **2000**, *27*, 774–777. [CrossRef]
107. Jiang, G.; Johnston, B.; Townrow, D.E.; Radecka, I.; Koller, M.; Chaber, P.; Adamus, G.; Kowalczuk, M. Biomass extraction using non-chlorinated solvents for biocompatibility improvement of polyhydroxyalkanoates. *Polymers* **2018**, *10*, 731. [CrossRef]
108. Ramsay, J.A.; Berger, E.; Voyer, R.; Chavarie, C. Extraction of poly-3-hydroxybutyrate using chlorinated solvents. *Biotechnol. Tech.* **1994**, *8*, 589–594. [CrossRef]
109. Rebocho, A.T.; Pereira, J.R.; Neves, L.A.; Alves, V.D.; Sevrin, C.; Grandfils, C.; Freitas, F.; Reis, M.A. Preparation and characterization of films based on a natural P(3HB)/mcl-PHA blend obtained through the co-culture of cupriavidus necator and pseudomonas citronellolis in apple pulp waste. *Bioengineering* **2020**, *7*, 34. [CrossRef]
110. Kunasundari, B.; Sudesh, K. Isolation and recovery of microbial polyhydroxyalkanoates. *Express Polym. Lett.* **2011**, *5*, 620–634. [CrossRef]
111. Jacquel, N.; Lo, C.-W.; Wei, Y.-H.; Wu, H.-S.; Wang, S.S. Isolation and purification of bacterial poly(3-hydroxyalkanoates). *Biochem. Eng. J.* **2008**, *39*, 15–27. [CrossRef]
112. De Koning, G.J.M.; Kellerhals, M.; van Meurs, C.; Witholt, B. A process for the recovery of poly(hydroxyalkanoates) from Pseudomonads part 2: Process development and economic evaluation. *Bioprocess Biosyst. Eng.* **1997**, *17*, 15–21. [CrossRef]
113. Marudkla, J.; Patjawit, A.; Chuensangjun, C.; Sirisansaneeyakul, S. Optimization of poly(3-hydroxybutyrate) extraction from *Cupriavidus necator* DSM 545 using sodium dodecyl sulfate and sodium hypochlorite. *Agric. Nat. Resour.* **2018**, *52*, 266–273. [CrossRef]
114. Berger, E.; Ramsay, B.A.; Chavarie, C.; Braunegg, G. PHB recovery by hypochlorite digestion of non-PHB biomass. *Biotechnol. Tech.* **1989**, *3*, 227–232. [CrossRef]
115. Holmes, P.A.; Lim, G.B. Separation Process. U.S. Patent 4,910,145, 20 March 1990.
116. Mondal, S.; Subramaniam, C. Xenobiotic contamination of water by plastics and pesticides revealed through real-time, ultrasensitive, and reliable surface-enhanced raman scattering. *ACS Sustain. Chem. Eng.* **2020**, *8*, 7639–7648. [CrossRef]
117. Bugnicourt, E.; Cinelli, P.; Lazzeri, A.; Álvarez, V. Polyhydroxyalkanoate (PHA): Review of synthesis, characteristics, processing and potential applications in packaging. *Express Polym. Lett.* **2014**, *8*, 791–808. [CrossRef]
118. Pötter, M.; Steinbüchel, A. Biogenesis and structure of polyhydroxyalkanoate granules. In *Inclusions in Prokaryotes*; Springer: Berlin/Heidelberg, Germany, 2006; pp. 109–136.
119. Muneer, F.; Rasul, I.; Azeem, F.; Siddique, M.H.; Zubair, M.; Nadeem, H. Microbial polyhydroxyalkanoates (PHAs): Efficient replacement of synthetic polymers. *J. Polym. Environ.* **2020**, *28*, 2301–2323. [CrossRef]
120. Li, Z.; Yang, J.; Loh, X.J. Polyhydroxyalkanoates: Opening doors for a sustainable future. *NPG Asia Mater.* **2016**, *8*, e265. [CrossRef]

121. Rigouin, C.; Lajus, S.; Ocando, C.; Borsenberger, V.; Nicaud, J.M.; Marty, A.; Avérous, L.; Bordes, F. Production and characterization of two medium-chain-length polydroxyalkanoates by engineered strains of *Yarrowia lipolytica*. *Microb. Cell Factories* **2019**, *18*, 99. [CrossRef]
122. Turco, R.; Santagata, G.; Corrado, I.; Pezzella, C.; di Serio, M. In vivo and post-synthesis strategies to enhance the properties of PHB-based materials: A review. *Front. Bioeng. Biotechnol.* **2021**, *8*, 1454. [CrossRef]
123. Modi, S.J. Assessing the Feasibility of Poly-(3-Hydroxybutyrate-Co-3-Hydroxyvalerate) (PHBV) and Poly-(Lactic Acid) for Potential Food Packaging Applications. Ph.D. Dissertation, The Ohio State University, Columbus, OH, USA, 2010.
124. Wei, L.; Stark, N.M.; Mc Donald, A.G. Interfacial improvements in biocomposites based on poly(3-hydroxybutyrate) and poly(3-hydroxybutyrate-co-3-hydroxyvalerate) bioplastics reinforced and grafted with α-cellulose fibers. *Green Chem.* **2015**, *17*, 4800–4814. [CrossRef]
125. Pradhan, S.; Dikshit, P.K. Production, ultrasonic extraction, and characterization of poly(3-hydroxybutyrate) (PHB) using *Bacillus megaterium* and *Cupriavidus necator*. *Polym. Adv. Technol.* **2018**, *29*, 2392–2400. [CrossRef]
126. Sharma, R.; Ray, A.R. Polyhydroxybutyrate, its copolymers and blends. *J. Macromol. Sci. Part C* **1995**, *35*, 327–359. [CrossRef]
127. Srubar, W.; Wright, Z.; Tsui, A.; Michel, A.; Billington, S.; Frank, C. Characterizing the effects of ambient aging on the mechanical and physical properties of two commercially available bacterial thermoplastics. *Polym. Degrad. Stab.* **2012**, *97*, 1922–1929. [CrossRef]
128. Mc Adam, B.; Fournet, M.B.; Mc Donald, P.; Mojicevic, M. Production of polyhydroxybutyrate (PHB) and factors impacting its chemical and mechanical characteristics. *Polymers* **2020**, *12*, 2908. [CrossRef]
129. El-Hadi, A.M.; Schnabel, R.; Straube, E.; Müller, G.; Henning, S. Correlation between degree of crystallinity, morphology, glass temperature, mechanical properties and biodegradation of poly(3-hydroxyalkanoate) PHAs and their blends. *Polym. Test.* **2002**, *21*, 665–674. [CrossRef]
130. Mohanty, A.K.; Misra, M.; Drzal, L.T. Sustainable bio-composites from renewable resources: Opportunities and challenges in the green materials world. *J. Polym. Environ.* **2002**, *10*, 19–26. [CrossRef]
131. Rosengart, A.; Cesário, M.; de Almeida, C.D.; Raposo, R.S.; Espert, A.; de Apodaca, E.D.; da Fonseca, M.M. Efficient P(3HB) extraction from *Burkholderia sacchari* cells using non-chlorinated solvents. *Biochem. Eng. J.* **2015**, *103*, 39–46. [CrossRef]
132. Zhu, C.; Nomura, C.T.; Perrotta, J.A.; Stipanovic, A.J.; Nakas, J.P. The effect of nucleating agents on physical properties of poly-3-hydroxybutyrate (PHB) and poly-3-hydroxybutyrate-co-3-hydroxyvalerate (PHB-co-HV) produced by *Burkholderia cepacia* ATCC 17759. *Polym. Test.* **2012**, *31*, 579–585. [CrossRef]
133. Seoane, I.T.; Manfredi, L.B.; Cyras, V.P. Effect of two different plasticizers on the properties of poly(3-hydroxybutyrate) binary and ternary blends. *J. Appl. Polym. Sci.* **2018**, *135*, 46016. [CrossRef]
134. Barouti, G.; Guillaume, S.M. Polyhydroxybutyrate (PHB)-based triblock copolymers: Synthesis of hydrophobic PHB/poly(benzyl β-malolactonate) and amphiphilic PHB/poly(malic acid) analogues by ring-opening polymerization. *Polym. Chem.* **2016**, *7*, 4603–4608. [CrossRef]
135. Huang, Y.; Mei, L.; Chen, X.; Wang, Q. Recent developments in food packaging based on nanomaterials. *Nanomaterials* **2018**, *8*, 830. [CrossRef] [PubMed]
136. Savenkova, L.; Gercberga, Z.; Nikolaeva, V.; Dzene, A.; Bibers, I.; Kalnin, M. Mechanical properties and biodegradation characteristics of PHB-based films. *Process Biochem.* **2000**, *35*, 573–579. [CrossRef]
137. Puente, J.A.S.; Esposito, A.; Chivrac, F.; Dargent, E. Effect of boron nitride as a nucleating agent on the crystallization of bacterial poly(3-hydroxybutyrate). *J. Appl. Polym. Sci.* **2013**, *128*, 2586–2594. [CrossRef]
138. Androsch, R. Surface structure of folded-chain crystals of poly(R-3-hydroxybutyrate) of different chain length. *Polymer* **2008**, *49*, 4673–4679. [CrossRef]
139. Wunderlich, B. Reversible crystallization and the rigid–amorphous phase in semicrystalline macromolecules. *Prog. Polym. Sci.* **2002**, *28*, 383–450. [CrossRef]
140. Sauvageau, D.; Cooper, D.G.; Nicell, J.A. Relative rates and mechanisms of biodegradation of diester plasticizers mediated by *Rhodococcus rhodochrous*. *Can. J. Chem. Eng.* **2009**, *87*, 499–506. [CrossRef]
141. Rai, R.; Keshavarz, T.; Roether, J.A.; Boccaccini, A.R.; Roy, I. Medium chain length polyhydroxyalkanoates, promising new biomedical materials for the future. *Mater. Sci. Eng. R Rep.* **2011**, *72*, 29–47. [CrossRef]
142. Grigore, M.E.; Grigorescu, R.M.; Iancu, L.; Ion, R.-M.; Zaharia, C.; Andrei, E.R. Methods of synthesis, properties and biomedical applications of polyhydroxyalkanoates: A review. *J. Biomater. Sci. Polym. Ed.* **2019**, *30*, 695–712. [CrossRef]
143. Chen, G.G.Q. *Plastics from Bacteria: Natural Functions and Applications*; Springer: Berlin/Heidelberg, Germany, 2009.
144. Choi, J.S.; Park, W.H. Effect of biodegradable plasticizers on thermal and mechanical properties of poly(3-hydroxybutyrate). *Polym. Test.* **2004**, *23*, 455–460. [CrossRef]
145. Mangeon, C.; Michely, L.; de Anda, A.R.; Thevenieau, F.; Renard, E.; Langlois, V. Natural terpenes used as plasticizers for poly(3-hydroxybutyrate). *ACS Sustain. Chem. Eng.* **2018**, *6*, 16160–16168. [CrossRef]
146. Lee, E.Y.; Jendrossek, D.; Schirmer, A.; Choi, C.Y.; Steinbüchel, A. Biosynthesis of copolyesters consisting of 3-hydroxybutyric acid and medium-chain-length 3-hydroxyalkanoic acids from 1, 3-butanediol or from 3-hydroxybutyrate by *Pseudomonas* sp. A33. *Appl. Microbiol. Biotechnol.* **1995**, *42*, 901–909. [CrossRef]
147. Sudesh, K.; Abe, H.; Doi, Y. Synthesis, structure and properties of polyhydroxyalkanoates: Biological polyesters. *Prog. Polym. Sci.* **2000**, *25*, 1503–1555. [CrossRef]

148. Hartmann, R.; Hany, R.; Pletscher, E.; Ritter, A.; Witholt, B.; Zinn, M. Tailor-made olefinic medium-chain-length poly[(*R*)-3-hydroxyalkanoates] by *Pseudomonas putida* GPo1: Batch versus chemostat production. *Biotechnol. Bioeng.* **2006**, *93*, 737–746. [CrossRef] [PubMed]
149. Lamberti, F.M.; Román-Ramírez, L.A.; Wood, J. Recycling of bioplastics: Routes and benefits. *J. Polym. Environ.* **2020**, *28*, 2551–2571. [CrossRef]
150. Mirjalili, F.; Chuah, L.; Salahi, E. Mechanical and morphological properties of polypropylene/nano α-Al_2O_3 composites. *Sci. World J.* **2014**, *2014*, 718765. [CrossRef]
151. Zhao, K.; Deng, Y.; Chen, J.C.; Chen, G.-Q. Polyhydroxyalkanoate (PHA) scaffolds with good mechanical properties and biocompatibility. *Biomaterials* **2003**, *24*, 1041–1045. [CrossRef]
152. Corre, Y.-M.; Bruzaud, S.; Audic, J.-L.; Grohens, Y. Morphology and functional properties of commercial polyhydroxyalkanoates: A comprehensive and comparative study. *Polym. Test.* **2012**, *31*, 226–235. [CrossRef]
153. De Koning, G.; Lemstra, P.; Hill, D.; Carswell, T.; O'Donnell, J. Ageing phenomena in bacterial poly[(*R*)-3-hydroxybutyrate]: 1. A study on the mobility in poly[(*R*)-3-hydroxybutyrate] powders by monitoring the radical decay with temperature after γ-radiolysis at 77 K. *Polymer* **1992**, *33*, 3295–3297. [CrossRef]
154. Sawyer, L.C.; Grubb, D.T.; Meyers, G.F. (Eds.) *Polymer Microscopy*, 3rd ed.; Springer: New York, NY, USA, 2008; 540p, ISBN 978-0-387-72627-4.
155. Struik, L.C.E. *Physical Aging in Amorphous Polymers and Other Materials*; Elsevier: Amsterdam, The Netherlands, 1977.
156. Biddlestone, F.; Harris, A.; Hay, J.N.; Hammond, T. The physical ageing of amorphous poly(hydroxybutyrate). *Polym. Int.* **1996**, *39*, 221–229. [CrossRef]
157. Hankermeyer, C.R.; Tjeerdema, R.S. Polyhydroxybutyrate: Plastic made and degraded by microorganisms. In *Reviews of Environmental Contamination and Toxicology*; Ware, G.W., Ed.; Springer: New York, NY, USA, 1999; pp. 1–24.
158. Zhong, Y.; Godwin, P.; Jin, Y.; Xiao, H. Biodegradable polymers and green-based antimicrobial packaging materials: A mini-review. *Adv. Ind. Eng. Polym. Res.* **2020**, *3*, 27–35. [CrossRef]
159. Thellen, C.; Coyne, M.; Froio, D.; Auerbach, M.; Wirsen, C.; Ratto, J.A. A processing, characterization and marine biodegradation study of melt-extruded polyhydroxyalkanoate (PHA) films. *J. Polym. Environ.* **2008**, *16*, 1–11. [CrossRef]
160. Hazer, B.; Steinbüchel, A. Increased diversification of polyhydroxyalkanoates by modification reactions for industrial and medical applications. *Appl. Microbiol. Biotechnol.* **2007**, *74*, 1–12. [CrossRef] [PubMed]
161. Silva, J.B.; Pereira, J.R.; Marreiros, B.C.; Reis, M.A.; Freitas, F. Microbial production of medium-chain length polyhydroxyalkanoates. *Process Biochem.* **2021**, *102*, 393–407. [CrossRef]
162. Abe, H.; Ishii, N.; Sato, S.; Tsuge, T. Thermal properties and crystallization behaviors of medium-chain-length poly(3-hydroxyalkanoate)s. *Polymer* **2012**, *53*, 3026–3034. [CrossRef]
163. Sánchez, R.J.; Schripsema, J.; da Silva, L.F.; Taciro, M.K.; Pradella, J.G.; Gomez, J.C. Medium-chain-length polyhydroxyalkanoic acids (PHAmcl) produced by *Pseudomonas putida* IPT 046 from renewable sources. *Eur. Polym. J.* **2003**, *39*, 1385–1394. [CrossRef]
164. Wang, Y.; Chung, A.; Chen, G.-Q. Synthesis of medium-chain-length polyhydroxyalkanoate homopolymers, random copolymers, and block copolymers by an engineered strain of *Pseudomonas entomophila*. *Adv. Health Mater.* **2017**, *6*, 1601017. [CrossRef]
165. Chen, S.; Liu, Q.; Wang, H.; Zhu, B.; Yu, F.; Chen, G.-Q.; Inoue, Y. Polymorphic crystallization of fractionated microbial medium-chain-length polyhydroxyalkanoates. *Polymer* **2009**, *50*, 4378–4388. [CrossRef]
166. Mizuno, S.; Hiroe, A.; Fukui, T.; Abe, H.; Tsuge, T. Fractionation and thermal characteristics of biosynthesized polyhydoxyalkanoates bearing aromatic groups as side chains. *Polym. J.* **2017**, *49*, 557–565. [CrossRef]
167. Gross, R.A.; de Mello, C.; Lenz, R.W.; Brandl, H.; Fuller, R.C. The biosynthesis and characterization of poly(β-hydroxyalkanoates) produced by *Pseudomonas oleovorans*. *Macromolecules* **1989**, *22*, 1106–1115. [CrossRef]
168. Gagnon, K.D.; Lenz, R.W.; Farris, R.J.; Fuller, R.C. Crystallization behavior and its influence on the mechanical properties of a thermoplastic elastomer produced by *Pseudomonas oleovorans*. *Macromolecules* **1992**, *25*, 3723–3728. [CrossRef]
169. Larrañaga, A.; Fernández, J.; Vega, A.; Etxeberria, A.; Ronchel, C.; Adrio, J.; Sarasua, J. Crystallization and its effect on the mechanical properties of a medium chain length polyhydroxyalkanoate. *J. Mech. Behav. Biomed. Mater.* **2014**, *39*, 87–94. [CrossRef] [PubMed]
170. Marschessault, R.; Monasterios, C.; Morin, F.; Sundararajan, P. Chiral poly(β-hydroxyalkanoates): An adaptable helix influenced by the alkane side-chain. *Int. J. Biol. Macromol.* **1990**, *12*, 158–165. [CrossRef]
171. Posen, I.D.; Jaramillo, P.; Landis, A.E.; Griffin, W.M. Greenhouse gas mitigation for U.S. plastics production: Energy first, feedstocks later. *Environ. Res. Lett.* **2017**, *12*, 034024. [CrossRef]
172. Plackett, D.; Siró, I. Polyhydroxyalkanoates (PHAs) for food packaging. In *Multifunctional and Nanoreinforced Polymers for Food Packaging*; Elsevier: Amsterdam, The Netherlands, 2011; pp. 498–526.
173. Asrar, J.; Valentin, H.E.; Berger, P.A.; Tran, M.; Padgette, S.R.; Garbow, J.R. Biosynthesis and properties of poly(3-hydroxybutyrate-co-3-hydroxyhexanoate) polymers. *Biomacromolecules* **2002**, *3*, 1006–1012. [CrossRef] [PubMed]
174. Kai, D.; Loh, X.J. Polyhydroxyalkanoates: Chemical modifications toward biomedical applications. *ACS Sustain. Chem. Eng.* **2014**, *2*, 106–119. [CrossRef]
175. Larsson, M.; Markbo, O.; Jannasch, P. Melt processability and thermomechanical properties of blends based on polyhydroxyalkanoates and poly(butylene adipate-co-terephthalate). *RSC Adv.* **2016**, *6*, 44354–44363. [CrossRef]

176. Nerkar, M.; Ramsay, J.A.; Ramsay, B.A.; Kontopoulou, M. Melt compounded blends of short and medium chain-length poly-3-hydroxyalkanoates. *J. Polym. Environ.* **2014**, *22*, 236–243. [CrossRef]
177. Xiang, H.; Chen, W.; Chen, Z.; Sun, B.; Zhu, M. Significant accelerated crystallization of long chain branched poly(3-hydroxybutyrate-co-3-hydroxyvalerate) with high nucleation temperature under fast cooling rate. *Compos. Sci. Technol.* **2017**, *142*, 207–213. [CrossRef]
178. Gopi, S.; Kontopoulou, M.; Ramsay, B.A.; Ramsay, J.A. Manipulating the structure of medium-chain-length polyhydroxyalkanoate (MCL-PHA) to enhance thermal properties and crystallization kinetics. *Int. J. Biol. Macromol.* **2018**, *119*, 1248–1255. [CrossRef]
179. Bhatia, S.K.; Wadhwa, P.; Hong, J.W.; Hong, Y.G.; Jeon, J.-M.; Lee, E.S.; Yang, Y.-H. Lipase mediated functionalization of poly(3-hydroxybutyrate-co-3-hydroxyvalerate) with ascorbic acid into an antioxidant active biomaterial. *Int. J. Biol. Macromol.* **2019**, *123*, 117–123. [CrossRef] [PubMed]
180. Timbart, L.; Renard, E.; Tessier, M.; Langlois, V. Monohydroxylated poly(3-hydroxyoctanoate) oligomers and its functionalized derivatives used as macroinitiators in the synthesis of degradable diblock copolyesters. *Biomacromolecules* **2007**, *8*, 1255–1265. [CrossRef] [PubMed]
181. Chen, Z.; Cheng, S.; Xu, K. Block poly(ester-urethane)s based on poly(3-hydroxybutyrate-co-4-hydroxybutyrate) and poly(3-hydroxyhexanoate-co-3-hydroxyoctanoate). *Biomaterials* **2009**, *30*, 2219–2230. [CrossRef] [PubMed]
182. Saad, G.R.; Lee, Y.J.; Seliger, H. Synthesis and characterization of biodegradable poly(ester-urethanes) based on bacterial poly(R-3-hydroxybutyrate). *J. Appl. Polym. Sci.* **2002**, *83*, 703–718. [CrossRef]
183. Ravenelle, F.; Marchessault, R.H. Self-assembly of poly([*R*]-3-hydroxybutyric acid)-block-poly(ethylene glycol) diblock copolymers. *Biomacromolecules* **2003**, *4*, 856–858. [CrossRef]
184. Mrkić, S.; Galić, K.; Ivanković, M.; Hamin, S.; Ciković, N. Gas transport and thermal characterization of mono- and di-polyethylene films used for food packaging. *J. Appl. Polym. Sci.* **2006**, *99*, 1590–1599. [CrossRef]
185. Luo, F.; Fortenberry, A.; Ren, J.; Qiang, Z. Recent Progress in enhancing poly(lactic acid) stereocomplex formation for material property improvement. *Front. Chem.* **2020**, *8*, 688. [CrossRef]
186. Gupta, A.; Mulchandani, N.; Shah, M.; Kumar, S.; Katiyar, V. Functionalized chitosan mediated stereocomplexation of poly(lactic acid): Influence on crystallization, oxygen permeability, wettability and biocompatibility behavior. *Polymer* **2018**, *142*, 196–208. [CrossRef]
187. Arrieta, M.P.; Samper, M.D.; Lopez, J.; Jiménez, A. Combined effect of poly(hydroxybutyrate) and plasticizers on polylactic acid properties for film intended for food packaging. *J. Polym. Environ.* **2014**, *22*, 460–470. [CrossRef]
188. Arrieta, M.P.; Samper, M.D.; Aldas, M.; López, J. On the use of PLA-PHB blends for sustainable food packaging applications. *Materials* **2017**, *10*, 1008. [CrossRef]
189. Loureiro, N.C.; Ghosh, S.; Viana, J.C.; Esteves, J.L. Thermal characterization of polyhydroxyalkanoates and poly(lactic acid) blends obtained by injection molding. *Polym. Technol. Eng.* **2015**, *54*, 350–356. [CrossRef]
190. Abdelwahab, M.; Flynn, A.; Chiou, B.-S.; Imam, S.; Orts, W.; Chiellini, E. Thermal, mechanical and morphological characterization of plasticized PLA–PHB blends. *Polym. Degrad. Stab.* **2012**, *97*, 1822–1828. [CrossRef]
191. Loureiro, N.; Esteves, J.L.; Viana, J.C.; Ghosh, S. Mechanical characterization of polyhydroxyalkanoate and poly(lactic acid) blends. *J. Thermoplast. Compos. Mater.* **2015**, *28*, 195–213. [CrossRef]
192. Armentano, I.; Fortunati, E.; Burgos, N.; Dominici, F.; Luzi, F.; Fiori, S.; Jiménez, A.; Yoon, K.; Ahn, J.; Kang, S.; et al. Bio-based PLA_PHB plasticized blend films: Processing and structural characterization. *LWT* **2015**, *64*, 980–988. [CrossRef]
193. Arrieta, M.; Fortunati, E.; Dominici, F.; Rayón, E.; López, J.; Kenny, J. PLA-PHB/cellulose based films: Mechanical, barrier and disintegration properties. *Polym. Degrad. Stab.* **2014**, *107*, 139–149. [CrossRef]
194. Arrieta, M.P.; Castro-Lopez, M.D.M.; Rayón, E.; Barral-Losada, L.F.; López-Vilariño, J.M.; López, J.; González-Rodríguez, M.V. Plasticized poly(lactic acid)–poly(hydroxybutyrate) (PLA–PHB) blends incorporated with catechin intended for active food-packaging applications. *J. Agric. Food Chem.* **2014**, *62*, 10170–10180. [CrossRef]
195. Noda, I.; Green, P.R.; Satkowski, M.M.; Schechtman, L.A. Preparation and properties of a novel class of polyhydroxyalkanoate copolymers. *Biomacromolecules* **2005**, *6*, 580–586. [CrossRef]
196. Schreck, K.M.; Hillmyer, M.A. Block copolymers and melt blends of polylactide with Nodax, TM microbial polyesters: Preparation and mechanical properties. *J. Biotechnol.* **2007**, *132*, 287–295. [CrossRef]
197. Noda, I.; Satkowski, M.M.; Dowrey, A.E.; Marcott, C. Polymer alloys of nodax copolymers and poly(lactic acid). *Macromol. Biosci.* **2004**, *4*, 269–275. [CrossRef]
198. Botta, L.; Mistretta, M.C.; Palermo, S.; Fragala, M.; Pappalardo, F. Characterization and processability of blends of polylactide acid with a new biodegradable medium-chain-length polyhydroxyalkanoate. *J. Polym. Environ.* **2015**, *23*, 478–486. [CrossRef]
199. Gassner, F.; Owen, A. Physical properties of poly(β-hydroxybutyrate)-poly(ε-caprolactone) blends. *Polymer* **1994**, *35*, 2233–2236. [CrossRef]
200. Antunes, M.C.M.; Felisberti, M.I. Blends of poly(hydroxybutyrate) and poly(ε-caprolactone) obtained from melting mixture. *Polímeros* **2005**, *15*, 134–138. [CrossRef]
201. Vogel, C.; Wessel, E.; Siesler, H.W. FT-IR imaging spectroscopy of phase separation in blends of poly(3-hydroxybutyrate) with poly(l-lactic acid) and poly(ε-caprolactone). *Biomacromolecules* **2008**, *9*, 523–527. [CrossRef] [PubMed]

202. Feng, L.; Watanabe, T.; Wang, Y.; Kichise, T.; Fukuchi, T.; Chen, G.-Q.; Doi, Y.; Inoue, Y. Studies on comonomer compositional distribution of bacterial poly(3-hydroxybutyrate-co-3-hydroxyhexanoate)s and thermal characteristics of their factions. *Biomacromolecules* **2002**, *3*, 1071–1077. [CrossRef] [PubMed]
203. Katsumata, K.; Saito, T.; Yu, F.; Nakamura, N.; Inoue, Y. The toughening effect of a small amount of poly(ε-caprolactone) on the mechanical properties of the poly(3-hydroxybutyrate-co-3-hydroxyhexanoate)/PCL blend. *Polym. J.* **2011**, *43*, 484–492. [CrossRef]
204. Nishida, M.; Tanaka, T.; Hayakawa, Y.; Ogura, T.; Ito, Y.; Nishida, M. Multi-scale instrumental analyses of plasticized polyhydroxyalkanoates (PHA) blended with polycaprolactone (PCL) and the effects of crosslinkers and graft copolymers. *RSC Adv.* **2019**, *9*, 1551–1561. [CrossRef]
205. Garcia-Garcia, D.; Ferri, J.M.; Boronat, T.; Lopez-Martinez, J.; Balart, R. Processing and characterization of binary poly(hydroxybutyrate) (PHB) and poly(caprolactone) (PCL) blends with improved impact properties. *Polym. Bull.* **2016**, *73*, 3333–3350. [CrossRef]
206. Przybysz, M.; Marć, M.; Klein, M.; Saeb, M.R.; Formela, K. Structural, mechanical and thermal behavior assessments of PCL/PHB blends reactively compatibilized with organic peroxides. *Polym. Test.* **2018**, *67*, 513–521. [CrossRef]
207. Javadi, A.; Kramschuster, A.J.; Pilla, S.; Lee, J.; Gong, S.; Turng, L.-S. Processing and characterization of microcellular PHBV/PBAT blends. *Polym. Eng. Sci.* **2010**, *50*, 1440–1448. [CrossRef]
208. Pal, A.K.; Wu, F.; Misra, M.; Mohanty, A.K. Reactive extrusion of sustainable PHBV/PBAT-based nanocomposite films with organically modified nanoclay for packaging applications: Compression moulding vs. cast film extrusion. *Compos. Part B Eng.* **2020**, *198*, 108141. [CrossRef]
209. Javadi, A.; Srithep, Y.; Lee, J.; Pilla, S.; Clemons, C.; Gong, S.; Turng, L.-S. Processing and characterization of solid and microcellular PHBV/PBAT blend and its RWF/nanoclay composites. *Compos. Part A Appl. Sci. Manuf.* **2010**, *41*, 982–990. [CrossRef]
210. Nagarajan, V.; Misra, M.; Mohanty, A.K. New engineered biocomposites from poly(3-hydroxybutyrate-co-3-hydroxyvalerate) (PHBV)/poly(butylene adipate-co-terephthalate) (PBAT) blends and switchgrass: Fabrication and performance evaluation. *Ind. Crop. Prod.* **2013**, *42*, 461–468. [CrossRef]
211. Ibrahim, N.A.; Chieng, B.W.; Yunus, W.M.Z.W. Morphology, thermal and mechanical properties of biodegradable poly(butylene succinate)/poly(butylene adipate-co-terephthalate)/clay nanocomposites. *Polym. Plast. Technol. Eng.* **2010**, *49*, 1571–1580. [CrossRef]
212. Kennouche, S.; le Moigne, N.; Kaci, M.; Quantin, J.C.; Caro-Bretelle, A.S.; Delaite, C.; Lopez-Cuesta, J.M. Morphological characterization and thermal properties of compatibilized poly(3-hydroxybutyrate-co-3-hydroxyvalerate) (PHBV)/poly(butylene succinate) (PBS)/halloysite ternary nanocomposites. *Eur. Polym. J.* **2016**, *75*, 142–162. [CrossRef]
213. Bhatia, A.; Gupta, R.; Bhattacharya, S.; Choi, H. Compatibility of biodegradable poly(lactic acid) (PLA) and poly(butylene succinate) (PBS) blends for packaging application. *Korea-Aust. Rheol. J.* **2007**, *19*, 125–131.
214. Qiu, Z.; Ikehara, T.; Nishi, T. Miscibility and crystallization behaviour of biodegradable blends of two aliphatic polyesters. poly(3-hydroxybutyrate-co-hydroxyvalerate) and poly(butylene succinate) blends. *Polymer* **2003**, *44*, 7519–7527. [CrossRef]
215. Ma, P.P.; Hristova-Bogaerds, D.G.; Lemstra, P.J.; Zhang, Y.; Wang, S. Toughening of PHBV/PBS and PHB/PBS Blends via in situ compatibilization using dicumyl peroxide as a free-radical grafting initiator. *Macromol. Mater. Eng.* **2011**, *297*, 402–410. [CrossRef]
216. Righetti, M.C.; Cinelli, P.; Aliotta, L.; Bianchi, E.; Tricoli, F.; Seggiani, M.; Lazzeri, A. Immiscible PHB/PB S and PHB/PBSA blends: Morphology, phase composition and modelling of elastic modulus. *Polym. Int.* **2021**, *71*, 47–56. [CrossRef]
217. Meereboer, K.W.; Pal, A.K.; Misra, M.; Mohanty, A.K. Sustainable PHBV/cellulose acetate blends: Effect of a chain extender and a plasticizer. *ACS Omega* **2020**, *5*, 14221–14231. [CrossRef]
218. Tomasi, G.; Scandola, M. Blends of bacterial poly(3-hydroxybutyrate) with cellulose acetate butyrate in activated sludge. *J. Macromol. Sci. Part A* **1995**, *32*, 671–681. [CrossRef]
219. El-Shafee, E.; Saad, G.R.; Fahmy, S.M. Miscibility, crystallization and phase structure of poly(3-hydroxybutyrate)/cellulose acetate butyrate blends. *Eur. Polym. J.* **2001**, *37*, 2091–2104. [CrossRef]
220. Chiulan, I.; Panaitescu, D.M.; Frone, A.N.; Teodorescu, M.; Nicolae, C.A.; Căşărică, A.; Tofan, V.; Sălăgeanu, A. Biocompatible polyhydroxyalkanoates/bacterial cellulose composites: Preparation, characterization, andin vitroevaluation. *J. Biomed. Mater. Res. Part A* **2016**, *104*, 2576–2584. [CrossRef] [PubMed]
221. Bhatt, R.; Shah, D.; Patel, K.; Trivedi, U. PHA–rubber blends: Synthesis, characterization and biodegradation. *Bioresour. Technol.* **2008**, *99*, 4615–4620. [CrossRef] [PubMed]
222. Melendez-Rodriguez, B.; Torres-Giner, S.; Angulo, I.; Pardo-Figuerez, M.; Hilliou, L.; Escuin, J.; Cabedo, L.; Nevo, Y.; Prieto, C.; Lagaron, J. High-oxygen-barrier multilayer films based on polyhydroxyalkanoates and cellulose nanocrystals. *Nanomaterials* **2021**, *11*, 1443. [CrossRef] [PubMed]
223. Bhardwaj, U.; Dhar, P.; Kumar, A.; Katiyar, V. Polyhydroxyalkanoates (PHA)-cellulose based nanobiocomposites for food packaging applications. In *Food Additives and Packaging*; Komolprasert, V., Turowski, P., Eds.; ACS Publications: Washington, DC, USA, 2014; Volume 1162, pp. 275–314. [CrossRef]
224. Zhang, M.; Thomas, N.L. Preparation and properties of polyhydroxybutyrate blended with different types of starch. *J. Appl. Polym. Sci.* **2009**, *116*, 688–694. [CrossRef]
225. Lai, S.-M.; Sun, W.-W.; Don, T.-M. Preparation and characterization of biodegradable polymer blends from poly(3-hydroxybutyrate)/poly(vinyl acetate)-modified corn starch. *Polym. Eng. Sci.* **2015**, *55*, 1321–1329. [CrossRef]

226. Sun, S.; Liu, P.; Ji, N.; Hou, H.; Dong, H. Effects of various cross-linking agents on the physicochemical properties of starch/PHA composite films produced by extrusion blowing. *Food Hydrocoll.* **2018**, *77*, 964–975. [CrossRef]
227. Wong, S.; Shanks, R.; Hodzic, A. Properties of poly(3-hydroxybutyric acid) composites with flax fibres modified by plasticiser absorption. *Macromol. Mater. Eng.* **2002**, *287*, 647–655. [CrossRef]
228. Barkoula, N.; Garkhail, S.; Peijs, T. Biodegradable composites based on flax/polyhydroxybutyrate and its copolymer with hydroxyvalerate. *Ind. Crop. Prod.* **2010**, *31*, 34–42. [CrossRef]
229. Khan, M.A.; Ali, K.M.I.; Hinrichsen, G.; Kopp, C.; Kropke, S. Study on physical and mechanical properties of biopol-jute composite. *Polym. Technol. Eng.* **1999**, *38*, 99–112. [CrossRef]
230. Avella, M.; la Rota, G.; Martuscelli, E.; Raimo, M.; Sadocco, P.; Elegir, G.; Riva, R. Poly(3-hydroxybutyrate-co-3-hydroxyvalerate) and wheat straw fibre composites: Thermal, mechanical properties and biodegradation behaviour. *J. Mater. Sci.* **2000**, *35*, 829–836. [CrossRef]
231. Singh, S.; Mohanty, A.K.; Sugie, T.; Takai, Y.; Hamada, H. Renewable resource based biocomposites from natural fiber and polyhydroxybutyrate-co-valerate (PHBV) bioplastic. *Compos. Part A Appl. Sci. Manuf.* **2008**, *39*, 875–886. [CrossRef]
232. Persico, P.; Acierno, D.; Carfagna, C.; Cimino, F. Mechanical and thermal behaviour of ecofriendly composites reinforced by kenaf and caro fibers. *Int. J. Polym. Sci.* **2011**, *2011*, 841812. [CrossRef]
233. Joyyi, L.; Thirmizir, M.Z.A.; Salim, M.S.; Han, L.; Murugan, P.; Kasuya, K.-I.; Maurer, F.H.; Arifin, M.I.Z.; Sudesh, K. Composite properties and biodegradation of biologically recovered P(3HB-co-3HHx) reinforced with short kenaf fibers. *Polym. Degrad. Stab.* **2017**, *137*, 100–108. [CrossRef]
234. Shibata, M.; Takachiyo, K.-I.; Ozawa, K.; Yosomiya, R.; Takeishi, H. Biodegradable polyester composites reinforced with short abaca fiber. *J. Appl. Polym. Sci.* **2002**, *85*, 129–138. [CrossRef]
235. Moliner, C.; Badia, J.D.; Bosio, B.; Arato, E.; Kittikorn, T.; Strömberg, E.; Juanes, R.T.; Ek, M.; Karlsson, S.; Ribes-Greus, A. Thermal and thermo-oxidative stability and kinetics of decomposition of PHBV/sisal composites. *Chem. Eng. Commun.* **2017**, *205*, 226–237. [CrossRef]
236. Hermida Élida, B.; Mega, V.I. Transcrystallization kinetics at the poly(3-hydroxybutyrate-co-3-hydroxyvalerate)/hemp fibre interface. *Compos. Part A Appl. Sci. Manuf.* **2007**, *38*, 1387–1394. [CrossRef]
237. Ramsay, B.A.; Langlade, V.; Carreau, P.J. Biodegradability and mechanical properties of poly-(beta-hydroxybutyrate-co-beta-hydroxyvalerate)-starch blends. *Appl. Environ. Microbiol.* **1993**, *59*, 1242–1246. [CrossRef]
238. Gatenholm, P.; Kubát, J.; Mathiasson, A. Biodegradable natural composites. I. Processing and properties. *J. Appl. Polym. Sci.* **1992**, *45*, 1667–1677. [CrossRef]
239. Seggiani, M.; Cinelli, P.; Balestri, E.; Mallegni, N.; Stefanelli, E.; Rossi, A.; Lardicci, C.; Lazzeri, A. Novel sustainable composites based on poly(hydroxybutyrate-co-hydroxyvalerate) and seagrass beach-CAST fibers: Performance and degradability in marine environments. *Materials* **2018**, *11*, 772. [CrossRef]
240. Maiti, P.; Batt, C.A.; Giannelis, E.P. New biodegradable polyhydroxybutyrate/layered silicate nanocomposites. *Biomacromolecules* **2007**, *8*, 3393–3400. [CrossRef] [PubMed]
241. Choi, W.M.; Kim, T.W.; Park, O.O.; Chang, Y.K.; Lee, J.W. Preparation and characterization of poly(hydroxybutyrate-co-hydroxyvalerate)-organoclay nanocomposites. *J. Appl. Polym. Sci.* **2003**, *90*, 525–529. [CrossRef]
242. Bordes, P.; Pollet, E.; Bourbigot, S.; Avérous, L. Structure and properties of PHA/clay nano-biocomposites prepared by melt intercalation. *Macromol. Chem. Phys.* **2008**, *209*, 1473–1484. [CrossRef]
243. Xu, P.; Yang, W.; Niu, D.; Yu, M.; Du, M.; Dong, W.; Chen, M.; Jan Lemstra, P.; Ma, P. Multifunctional and robust polyhydroxyalkanoate nanocomposites with superior gas barrier, heat resistant and inherent antibacterial performances. *Chem. Eng. J.* **2020**, *382*, 122864. [CrossRef]
244. Tamiya, T.; Hsu, Y.-I.; Asoh, T.-A.; Uyama, H. Improvement of interfacial adhesion between poly(3-hydroxybutyrate-co-3-hydroxyhexanoate) and silica particles. *Ind. Eng. Chem. Res.* **2020**, *59*, 13595–13602. [CrossRef]
245. Jaques, N.; Silva, I.D.D.S.; Neto, M.D.C.; Diniz, R.K.M.; Wellen, R.M.R.; Canedo, E.L. Comparative study of the effect of TiO_2 and ZnO on the crystallization of PHB. *Matéria Rio Jan.* **2017**, *22*. [CrossRef]
246. Kirboga, S.; Öner, M. The properties of PHBV/$CaCO_3$ composites prepared by melt processing. In Proceedings of the 6th International Conference on New Trends in Chemistry, Online, 17–18 October 2020.
247. Visakh, P.M. Chapter 1. Polyhydroxyalkanoates (PHAs), their blends, composites and nanocomposites: State of the art, new challenges and opportunities. In *Polyhydroxyalkanoate (PHA) Based Blends, Composites and Nanocomposites*; Royal Society of Chemistry: London, UK, 2014; pp. 1–17.
248. Sharma, V.; Sehgal, R.; Gupta, R. Polyhydroxyalkanoate (PHA): Properties and modifications. *Polymer* **2021**, *212*, 123161. [CrossRef]
249. Xu, Z.; Pan, C.; Li, X.; Hao, N.; Zhang, T.; Gaffrey, M.J.; Pu, Y.; Cort, J.R.; Ragauskas, A.J.; Qian, W.-J.; et al. Enhancement of polyhydroxyalkanoate production by co-feeding lignin derivatives with glycerol in *Pseudomonas putida* KT2440. *Biotechnol. Biofuels* **2021**, *14*, 11. [CrossRef]
250. Kim, Y.; Kim, H.W.; Chung, M.G.; Rhee, Y.H. Biosynthesis, modification, and biodegradation of bacterial medium-chain-length polyhydroxyalkanoates. *J. Microbiol.* **2007**, *45*, 87–97.
251. Chen, G.-Q.; Jiang, X.-R.; Guo, Y. Synthetic biology of microbes synthesizing polyhydroxyalkanoates (PHA). *Synth. Syst. Biotechnol.* **2016**, *1*, 236–242. [CrossRef]

252. Meng, D.-C.; Shen, R.; Yao, H.; Chen, J.-C.; Wu, Q.; Chen, G.-Q. Engineering the diversity of polyesters. *Curr. Opin. Biotechnol.* **2014**, *29*, 24–33. [CrossRef] [PubMed]
253. Zan, L.; Tian, L.; Liu, Z.; Peng, Z. A new polystyrene–TiO_2 nanocomposite film and its photocatalytic degradation. *Appl. Catal. A Gen.* **2004**, *264*, 237–242. [CrossRef]
254. Lagaron, J.M.; Lopez-Rubio, A. Nanotechnology for bioplastics: Opportunities, challenges and strategies. *Trends Food Sci. Technol.* **2011**, *22*, 611–617. [CrossRef]
255. Ray, S.; Kalia, V.C. Biomedical applications of polyhydroxyalkanoates. *Indian J. Microbiol.* **2017**, *57*, 261–269. [CrossRef] [PubMed]
256. Karkhanis, S.S.; Stark, N.M.; Sabo, R.C.; Matuana, L.M. Water vapor and oxygen barrier properties of extrusion-blown poly(lactic acid)/cellulose nanocrystals nanocomposite films. *Compos. Part A Appl. Sci. Manuf.* **2018**, *114*, 204–211. [CrossRef]
257. Malmir, S.; Montero, B.; Rico, M.; Barral, L.; Bouza, R. Morphology, thermal and barrier properties of biodegradable films of poly(3-hydroxybutyrate-co-3-hydroxyvalerate) containing cellulose nanocrystals. *Compos. Part A Appl. Sci. Manuf.* **2017**, *93*, 41–48. [CrossRef]
258. Norrrahim, F.; Ariffin, H.; Hassan, M.; Ibrahim, N.; Nishida, H. Performance evaluation and chemical recyclability of a polyethyene/poly-(3-hydroxybutyrate-co-3-hydroxyvalerate) blend for sustainable packaging. *RSC Adv.* **2013**, *3*, 24378. [CrossRef]
259. Nazrin, A.; Sapuan, S.M.; Zuhri, M.Y.M.; Ilyas, R.; Syafiq, R.; Sherwani, S.F.K. Nanocellulose reinforced thermoplastic starch (TPS), polylactic acid (PLA), and polybutylene succinate (PBS) for food packaging applications. *Front. Chem.* **2020**, *8*, 213. [CrossRef]
260. Rodríguez, F.J.; Galotto, M.J.; Guarda, A.; Bruna, J.E. Modification of cellulose acetate films using nanofillers based on organoclays. *J. Food Eng.* **2012**, *110*, 262–268. [CrossRef]
261. Keskin, G.; Kızıl, G.; Bechelany, M.; Pochat-Bohatier, C.; Öner, M. Potential of polyhydroxyalkanoate (PHA) polymers family as substitutes of petroleum based polymers for packaging applications and solutions brought by their composites to form barrier materials. *Pure Appl. Chem.* **2017**, *89*, 1841–1848. [CrossRef]
262. Asrar, J.; Gruys, K.J. Biodegradable polymer (Biopol®). *Biopolym. Online* **2005**, *4*. [CrossRef]
263. Fabra, M.J.; Lopez-Rubio, A.; Lagaron, J.M. Nanostructured interlayers of zein to improve the barrier properties of high barrier polyhydroxyalkanoates and other polyesters. *J. Food Eng.* **2014**, *127*, 1–9. [CrossRef]
264. Yang, X.; Zhao, K.; Chen, G.-Q. Effect of surface treatment on the biocompatibility of microbial polyhydroxyalkanoates. *Biomaterials* **2002**, *23*, 1391–1397. [CrossRef]
265. Deng, Y.; Zhao, K.; Zhang, X.-F.; Hu, P.; Chen, G.-Q. Study on the three-dimensional proliferation of rabbit articular cartilage-derived chondrocytes on polyhydroxyalkanoate scaffolds. *Biomaterials* **2002**, *23*, 4049–4056. [CrossRef]
266. Kassab, A.C.; Xu, K.; Denkbas, E.; Dou, Y.; Zhao, S.; Piskin, E. Rifampicin carrying polyhydroxybutyrate microspheres as a potential chemoembolization agent. *J. Biomater. Sci. Polym. Ed.* **1997**, *8*, 947–961. [CrossRef] [PubMed]
267. Williams, S.F.; Martin, D.; Horowitz, D.M.; Peoples, O.P. PHA applications: Addressing the price performance issue: I. Tissue engineering. *Int. J. Biol. Macromol.* **1999**, *25*, 111–121. [CrossRef]
268. Rhodes, C.J. Plastic pollution and potential solutions. *Sci. Prog.* **2018**, *101*, 207–260. [CrossRef]
269. Geyer, R.; Jambeck, J.R.; Law, K.L. Production, use, and fate of all plastics ever made. *Sci. Adv.* **2017**, *3*, e1700782. [CrossRef]
270. Sudesh, K.; Iwata, T. Sustainability of biobased and biodegradable plastics. *CLEAN—Soil Air Water* **2008**, *36*, 433–442. [CrossRef]
271. Yang, S.-T.; Liu, X.; Zhang, Y. Metabolic engineering—Applications, methods, and challenges. In *Bioprocessing for Value-Added Products from Renewable Resources*; Elsevier: Amsterdam, The Netherlands, 2007; pp. 73–118.
272. Minteer, S. Biochemical production of other bioalcohols: Biomethanol, biopropanol, bioglycerol, and bioethylene glycol. In *Handbook of Biofuels Production*; Elsevier: Amsterdam, The Netherlands, 2011; pp. 258–265.
273. Pang, J.; Zheng, M.; Sun, R.; Wang, A.; Wang, X.; Zhang, T. Synthesis of ethylene glycol and terephthalic acid from biomass for producing PET. *Green Chem.* **2016**, *18*, 342–359. [CrossRef]
274. Upare, P.P.; Hwang, D.W.; Hwang, Y.K.; Lee, U.-H.; Hong, D.-Y.; Chang, J.-S. An integrated process for the production of 2,5-dimethylfuran from fructose. *Green Chem.* **2015**, *17*, 3310–3313. [CrossRef]
275. Talebian-Kiakalaieh, A.; Amin, N.A.S.; Hezaveh, H. Glycerol for renewable acrolein production by catalytic dehydration. *Renew. Sustain. Energy Rev.* **2014**, *40*, 28–59. [CrossRef]
276. Morais, A.R.; Dworakowska, S.; Reis, A.; Gouveia, L.; Matos, C.T.; Bogdał, D.; Bogel-Łukasik, R. Chemical and biological-based isoprene production: Green metrics. *Catal. Today* **2015**, *239*, 38–43. [CrossRef]
277. Andreeßen, C.; Steinbüchel, A. Recent developments in non-biodegradable biopolymers: Precursors, production processes, and future perspectives. *Appl. Microbiol. Biotechnol.* **2019**, *103*, 143–157. [CrossRef] [PubMed]
278. Narancic, T.; Verstichel, S.; Chaganti, S.R.; Morales-Gamez, L.; Kenny, S.T.; de Wilde, B.; Padamati, R.B.; O'Connor, K.E. Biodegradable plastic blends create new possibilities for end-of-life management of plastics but they are not a panacea for plastic pollution. *Environ. Sci. Technol.* **2018**, *52*, 10441–10452. [CrossRef] [PubMed]
279. Storz, H.; Vorlop, K.-D. Bio-based plastics: Status, challenges and trends. *Landbauforsch. Volkenrode* **2013**, *63*, 321–332.
280. Meereboer, K.W.; Misra, M.; Mohanty, A.K. Review of recent advances in the biodegradability of polyhydroxyalkanoate (PHA) bioplastics and their composites. *Green Chem.* **2020**, *22*, 5519–5558. [CrossRef]
281. Boyandin, A.N.; Prudnikova, S.V.; Filipenko, M.L.; Khrapov, E.A.; Vasil'ev, A.D.; Volova, T.G. Biodegradation of polyhydroxyalkanoates by soil microbial communities of different structures and detection of PHA degrading microorganisms. *Appl. Biochem. Microbiol.* **2011**, *48*, 28–36. [CrossRef]

282. Kim, D.Y.; Rhee, Y.H. Biodegradation of microbial and synthetic polyesters by fungi. *Appl. Microbiol. Biotechnol.* **2003**, *61*, 300–308. [CrossRef]
283. Kobayashi, T.; Uchino, K.; Abe, T.; Yamazaki, Y.; Saito, T. Novel intracellular 3-hydroxybutyrate-oligomer hydrolase in *Wautersia eutropha* H16. *J. Bacteriol.* **2005**, *187*, 5129–5135. [CrossRef]
284. Mukai, K.; Yamada, K.; Doi, Y. Enzymatic degradation of poly(hydroxyalkanoates) by a marine bacterium. *Polym. Degrad. Stab* **1993**, *41*, 85–91. [CrossRef]
285. Numata, K.; Abe, H.; Doi, Y. Enzymatic processes for biodegradation of poly(hydroxyalkanoate)s crystals. *Can. J. Chem.* **2008**, *86*, 471–483. [CrossRef]
286. Vroman, I.; Tighzert, L. Biodegradable polymers. *Materials* **2009**, *2*, 307–344. [CrossRef]
287. Numata, K.; Kikkawa, Y.; Tsuge, T.; Iwata, T.; Doi, Y.; Abe, H. Enzymatic degradation processes of poly[(*R*)-3-hydroxybutyric acid] and poly[(*R*)-3-hydroxybutyric acid-co-(*R*)-3-hydroxyvaleric acid] single crystals revealed by atomic force microscopy Effects of molecular weight and second-monomer composition on erosion rates. *Biomacromolecules* **2005**, *6*, 2008–2016. [CrossRef]
288. Huang, Y.; Zhang, C.; Pan, Y.; Zhou, Y.; Jiang, L.; Dan, Y. Effect of NR on the hydrolytic degradation of PLA. *Polym. Degrad. Stab.* **2013**, *98*, 943–950. [CrossRef]
289. Ambrières, W. Plastics recycling worldwide: Current overview and desirable changes. *J. Field Actions* **2019**, *19*, 12–21.
290. Ragaert, K.; Delva, L.; van Geem, K. Mechanical and chemical recycling of solid plastic waste. *Waste Manag.* **2017**, *69*, 24–58 [CrossRef]
291. Hatti-Kaul, R.; Nilsson, L.J.; Zhang, B.; Rehnberg, N.; Lundmark, S. Designing biobased recyclable polymers for plastics. *Trends Biotechnol.* **2020**, *38*, 50–67. [CrossRef]
292. Rahimi, A.; García, J.M. Chemical recycling of waste plastics for new materials production. *Nat. Rev. Chem.* **2017**, *1*, 0046. [CrossRef]
293. Arikan, E.B.; Ozsoy, H.D. A review: Investigation of bioplastics. *J. Civ. Eng. Archit.* **2015**, *9*, 188–192.
294. Prieto, A. To be, or not to be biodegradable … That is the question for the bio-based plastics. *Microb. Biotechnol.* **2016**, *9*, 652–657. [CrossRef] [PubMed]
295. Niaounakis, M. Recycling of biopolymers—The patent perspective. *Eur. Polym. J.* **2019**, *114*, 464–475. [CrossRef]
296. Rivas, L.F.; Casarin, S.A.; Nepomuceno, N.C.; Alencar, M.I.; Agnelli, J.A.M.; de Medeiros, E.S.; Neto, A.D.O.W.; Oliveira, M.; de Medeiros, A.M.; Santos, A.S.F. Reprocessability of PHB in extrusion: ATR-FTIR, tensile tests and thermal studies. *Polímeros* **2017**, *27*, 122–128. [CrossRef]
297. Yang, X.; Clénet, J.; Xu, H.; Odelius, K.; Hakkarainen, M. Two step extrusion process: From thermal recycling of PHB to plasticized PLA by reactive extrusion grafting of PHB degradation products onto PLA chains. *Macromolecules* **2015**, *48*, 2509–2518. [CrossRef]
298. Zembouai, I.; Bruzaud, S.; Kaci, M.; Benhamida, A.; Corre, Y.M.; Grohens, Y. Mechanical recycling of poly(3-hydroxybutyrate-co-3-hydroxyvalerate)/polylactide based blends. *J. Polym. Environ.* **2014**, *22*, 449–459. [CrossRef]
299. Soroudi, A.; Jakubowicz, I. Recycling of bioplastics, their blends and biocomposites: A review. *Eur. Polym. J.* **2013**, *49*, 2839–2858. [CrossRef]
300. Vu, D.; Åkesson, D.; Taherzadeh, M.J.; Ferreira, J.A. Recycling strategies for polyhydroxyalkanoate-based waste materials: An overview. *Bioresour. Technol.* **2020**, *298*, 122393. [CrossRef]
301. Zaverl, M.; Seydibeyoğlu, M.Ö.; Misra, M.; Mohanty, A. Studies on recyclability of polyhydroxybutyrate-co-valerate bioplastic: Multiple melt processing and performance evaluations. *J. Appl. Polym. Sci.* **2012**, *125*, E324–E331. [CrossRef]
302. Aoyagi, Y.; Yamashita, K.; Doi, Y. Thermal degradation of poly[(*R*)-3-hydroxybutyrate], poly[ε-caprolactone], and poly[(*S*)-lactide]. *Polym. Degrad. Stab.* **2002**, *76*, 53–59. [CrossRef]
303. Ariffin, H.; Nishida, H.; Shirai, Y.; Hassan, M.A. Highly selective transformation of poly[(*R*)-3-hydroxybutyric acid] into trans-crotonic acid by catalytic thermal degradation. *Polym. Degrad. Stab.* **2010**, *95*, 1375–1381. [CrossRef]
304. Sato, S.; Ishii, N.; Hamada, Y.; Abe, H.; Tsuge, T. Utilization of 2-alkenoic acids for biosynthesis of medium-chain-length polyhydroxyalkanoates in metabolically engineered *Escherichia coli* to construct a novel chemical recycling system. *Polym. Degrad. Stab.* **2012**, *97*, 329–336. [CrossRef]
305. Yang, X.; Odelius, K.; Hakkarainen, M. Microwave-assisted reaction in green solvents recycles PHB to functional chemicals. *ACS Sustain. Chem. Eng.* **2014**, *2*, 2198–2203. [CrossRef]
306. Zaheer, M.R.; Kuddus, M. PHB (poly-β-hydroxybutyrate) and its enzymatic degradation. *Polym. Adv. Technol.* **2018**, *29*, 30–40. [CrossRef]
307. Wang, S.; Lydon, K.A.; White, E.M.; Grubbs, J.B.; Lipp, E.K.; Locklin, J.; Jambeck, J.R. Biodegradation of poly(3-hydroxybutyrate-co-3-hydroxyhexanoate) plastic under anaerobic sludge and aerobic seawater conditions: Gas evolution and microbial diversity. *Environ. Sci. Technol.* **2018**, *52*, 5700–5709. [CrossRef]
308. Shogren, R.; Doane, W.; Garlotta, D.; Lawton, J.; Willett, J. Biodegradation of starch/polylactic acid/poly(hydroxyester-ether) composite bars in soil. *Polym. Degrad. Stab.* **2003**, *79*, 405–411. [CrossRef]
309. Mandic, M.; Spasic, J.; Ponjavic, M.; Nikolic, M.S.; Cosovic, V.R.; O'Connor, K.E.; Nikodinovic-Runic, J.; Djokic, L.; Jeremic, S. Biodegradation of poly(ε-caprolactone) (PCL) and medium chain length polyhydroxyalkanoate (mcl-PHA) using whole cells and cell free protein preparations of Pseudomonas and Streptomyces strains grown on waste cooking oil. *Polym. Degrad. Stab.* **2019**, *162*, 160–168. [CrossRef]

310. Sikorska, W.; Musiol, M.; Nowak, B.; Pajak, J.; Labuzek, S.; Kowalczuk, M.; Adamus, G. Degradability of polylactide and its blend with poly[(R,S)-3-hydroxybutyrate] in industrial composting and compost extract. *Int. Biodeterior. Biodegrad.* **2015**, *101*, 32–41. [CrossRef]
311. Weng, Y.-X.; Wang, X.-L.; Wang, Y.-Z. Biodegradation behavior of PHAs with different chemical structures under controlled composting conditions. *Polym. Test.* **2011**, *30*, 372–380. [CrossRef]

MDPI
St. Alban-Anlage 66
4052 Basel
Switzerland
www.mdpi.com

Catalysts Editorial Office
E-mail: catalysts@mdpi.com
www.mdpi.com/journal/catalysts

www.ingramcontent.com/pod-product-compliance
Lightning Source LLC
Chambersburg PA
CBHW040823250926
51850CB00007BA/1250
* 9 7 8 3 7 2 5 8 0 6 4 4 7 *